Overshoot

Overshoot

*How the World Surrendered
to Climate Breakdown*

Andreas Malm
and
Wim Carton

VERSO
London · New York

First published by Verso 2024
© Andreas Malm and Wim Carton 2024

1 3 5 7 9 10 8 6 4 2

Verso
UK: 6 Meard Street, London W1F 0EG
US: 388 Atlantic Avenue, Brooklyn, NY 11217
versobooks.com

Verso is the imprint of New Left Books

ISBN-13: 978-1-80429-398-0
ISBN-13: 978-1-80429-399-7 (UK EBK)
ISBN-13: 978-1-80429-400-0 (US EBK)

British Library Cataloguing in Publication Data
A catalogue record for this book is available from the British Library

Library of Congress Cataloging-in-Publication Data
A catalog record for this book is available from the Library of Congress

Typeset in Sabon by MJ & N Gavan, Truro, Cornwall
Printed and bound by CPI Group (UK) Ltd, Croydon, CR0 4YY

For Ebbe, Nadim and Latifa

Contents

Preface

What do we do when catastrophic climate chaos is a fact? Are we already there? Is it now too late to avoid warming of one and a half degrees and maybe two as well? How did we even reach the point where these questions seem relevant to ask? What follows is a history of what we shall call the overshoot conjuncture, or the period when officially declared limits to global warming are exceeded – or in the process of being so – and the dominant classes responsible for the excess throw up their hands in resignation and accept that intolerable heat is coming. This acceptance can be tacit or explicit. It is often couched in the idea of a promised return to safer levels: we can let the warming pass 1.5 °C or 2 °C and then, at a later date, reverse it and turn the temperatures down to where they should be. Too much heat is acceptable, because it can be undone *post factum* with technologies for cooling the Earth. Overshoot is here not a fate passively acquiesced to. It is an actively championed programme for how to deal with the rush into catastrophe: let it continue for the time being, and then we shall sort things out towards the end of this century.

Programmatic overshoot became, as we shall see, hegemonic in mainstream science and policy in the years surrounding the Paris Agreement; but this did not happen because the idea was so strikingly brilliant. Rather it represented an alignment with the power of business as usual. The idea corresponded to real material forces pushing the Earth towards 1.5 °C and beyond, and here a degree of naïve puzzlement must be registered. Why couldn't it just stop? What was it that drove the world into the heat, even as the consequences were plain to see? How was it that – despite all the reports, summits, pledges, agreements and, above all, observations and experiences of disasters striking harder and harder – the curves were *still* pointing in the wrong direction, the emissions still growing, the expansion of fossil fuel infrastructure proceeding apace as if nothing was happening? What spell had been cast on this world that just would not be broken? This is the question of why the world surrendered to climate breakdown, of why the warming was not contained at a level that might have been tolerable. But it is not an exercise in brooding historiography. This is a history of the present and near future: an attempt to gauge the power of the forces that destroy the conditions of life on Earth and that must be contended with in the coming years, if any such conditions are to be preserved. The heat is rapidly becoming too much to bear, and precisely for that reason, it is too late to give up this struggle. There is, henceforth, no path to a liveable planet that does not pass through the complete destruction of business as usual. What would that look like?

Overshoot is thus a term with several valences. Before it entered the climate lexicon, it denoted the overuse of resources on Earth in general, but we shall set that broader phenomenon aside and focus on overshoot in a rapidly warming world.[1] Here it can mean simply a rise in temperatures above a declared limit; a programme for going ahead with it and then annulling the rise; a conjuncture when these things are occurring, in the physical world as well as the realm of ideas – we shall slide between them. The years between 2018 and 2022 marked a kind of beginning. It was in 2018 that the Intergovernmental Panel on Climate Change (IPCC) released its Special Report on Global Warming of 1.5°C, laying out the dangers of crossing this boundary and inspiring the 'international community' to confirm its commitment to it. Warming in excess of 1.5°C, the world learned in this year, would be too unsafe to live with. Before the ink had dried on the Special Report, however, this very boundary came into view; and what then transpired on the ground was a revving of the engine. The half-decade between the Special Report and the events of 2022 formed the start of the overshoot conjuncture – a limit officially understood and proclaimed, only to immediately become an object of transgression. These two elements had not been conjoined before. But they may well be so again, as the crisis deepens: this conjuncture inclines towards renewal and deterioration. If 1.5°C was in the spotlight during those years, it will be 1.7°C next, then 2°C, and so on, the logic of overshoot potentially reappearing at every identified limit, more ahead of us than behind.

Writing history in this moment is to try to catch it as it flies, or to follow long-distance projectiles gliding through the sky; it takes years for one investment in fossil fuel infrastructure to strike as a disaster somewhere on Earth; each disaster in the present, conversely, is the lagging result of emissions in the past. Between the writing and publication of this book, more projectiles will have been fired off. Others will have landed. Writing the history of this process is then essentially impossible, since it has no closed chapters, in the manner of, say, the Abbasid Caliphate or the French Revolution, whatever their causal afterlives: overshoot has only just begun. And yet a historical approach to it is called for, precisely because of the manner in which it unfolds over time. Sources of emissions produce effects along an atmospheric arc between past and future, and the focus here is on the former: on the problem of mitigation, as traditionally understood – the sources and what it would mean to close them down. Can the transgression of the temperature limits from the Paris Agreement still be prevented? Could it be done without much pain? Is it merely a segment of oil and gas companies that stand between the rest of us and a stabilised climate? Or is it technical necessity that still ties us all to fossil fuels, at this late hour? These are some points of entry into the problem, which is one of political economy as much as history. The barrage is kept up by forces at work in the depths of capital accumulation.

But if the field of mitigation constitutes overshoot – every increase in temperatures the product of an enemy that has still never ceased to be victorious – the conjuncture is also defined by other fields having their moment.[2] The dominant classes have to come up with secondary, backup measures for managing the consequences of excess heat. Three stand out. Adaptation is the pursuit of adjusting life to a climate of disasters and making it less disastrous. Carbon dioxide removal is designed to draw the gas down from the atmosphere after the fact. Geoengineering is the art of blocking incoming sunlight to reduce the influx of heat to the Earth. All three centre on technologies; all come to the fore because fossil capital has been so successful in staving off challenges; all depart from the refusal of states to undertake any meaningful mitigation: the temperature rise overtops the barriers and gets channelled into new political projects. All three are also replete with repercussions, ranging from the annoying to the apocalyptic. A second instalment of this study, *The Long Heat: Climate Politics When It's Too Late*, will deal with adaptation,

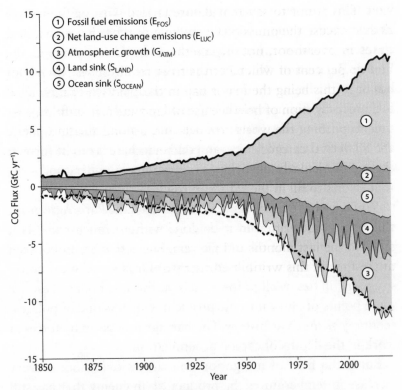

Figure 1 Carbon Emissions (+ve) and their Partitioning (−ve)

Source: Pierre Friedlingstein, Michael O'Sullivan, Matthew W. Jones et al., 'Global Carbon Budget 2022', *Earth System Science Data* (2022) 14: 4827.[4]

removal and geoengineering in some detail. It will pay special attention to the psychic dimensions of the climate crisis, notably the tremendous capacity of people in capitalist society to deny and, when this no longer works, repress it. The emphasis will be on the multiplying fronts of climate politics, the emerging struggles over adaptation and geoengineering, but carbon dioxide removal in particular: there won't be less to fight for when it's too late. There will be more. In the present volume, the emphasis is on the original front, which will always remain central.

Within this front, our focus is exclusively on fossil fuels. We do not address deforestation, the second main source of CO_2 emissions, but stay with the first, for reasons seen in the graph above: its preponderance is conspicuous and rises over time. The overburdening of the atmospheric sink is mainly a function of emissions shooting up and up in the department of fuel combustion. This is not to say that deforestation is insignificant. Nor does it excuse the omission of the role of the other greenhouse gases in overshoot, notably methane emissions (between 25 and 40 per cent of which derives from fossil fuels).[3] We readily admit to this being the major gap in the following pages: there is no investigation of how the use of land and livestock contributes to pushing the world towards and beyond 1.5°C. Some of the analytical categories developed here might be applicable to them – particularly the defence of property – and it is our hope that efforts to fill in the gaps will follow.

While we have been working on these issues for several years, this book was written in mid-2023. At its end, we update the story to the last months of that year. Business as usual remains, at the time of this writing, distinctly unfinished.

PART I

The Limit Is Not a Limit

And when the third decade of the millennium dawned, the relationship remained firmly in place: the warmer the globe became, the more fossil fuels were poured on the fire. The higher the temperatures, the larger the emissions. The closer the Earth came to being engulfed in flames – literally and figuratively – the harder companies worked to get oil and gas and coal out of the ground and ferry them off to combustion. And so a notion began to take hold: that humanity had to reconcile itself to this state of affairs, accept what must not happen as a *fait accompli*, give up on the idea that emissions can be slashed at the speed required and instead try something else. The time had come, it was said, to postpone the lost cause. If the house was on fire, putting that fire out would have to wait. The age of overshoot was upon us.

1

Chronicle of Three Years Out of Control

The decade opened, however, on a different note. In 2020, the first year of the Covid-19 pandemic, something highly unusual took place: global CO_2 emissions *fell*. The lockdowns that closed the highways of the world economy cut their total by some 5 or 6 per cent.[1] Climate policy had nothing to do with this occurrence; it was squarely the by-product of measures for slowing down the virus. But coincidentally, the pandemic broke out just as the wave of climate mobilisations on streets from Berlin to Bogotá and Luanda to London crested – in 2019, this had been 'the fastest-growing social movement in history' – and so proposals were floated for using the pandemic to start the transition by then long overdue.[2] These came to nothing.

Like an arsonist briefly interrupted by the sound of voices in the building, fossil capital resumed its work. Already in December 2020, as the lockdowns receded, it was back in full swing, and in the following year it surpassed itself: according to the International Energy Agency, it not only rebounded but went beyond the levels preceding the pandemic, as total CO_2 emissions rose by 6 per cent, or two gigatonnes.[3] What is a gigatonne? It is a unit of gigantic mass, one of which equals the weight of over 100,000,000 African elephants.[4] With two more gigatonnes of carbon dioxide tossed into the atmosphere in 2021 than in the previous year, up to a total of forty, the world economy set a fresh record: never before had emissions risen as much in a single year.[5] One might imagine that such a thing would happen in the early stages of the crisis, before it had come fully into view. But by 2021, the world had seen at least 1.1°C of global warming, six IPCC reports, twenty-six COPs and immeasurable suffering for the most affected people and areas, and yet it generated the largest surge in absolute emissions – the input that directly determines the rate of warming – in recorded history: and on it went in 2022, with an increase of another 1 per cent.[6] Once again, emissions were higher than ever, and nothing suggested that they would not go higher still.

Early Seasons in Hell

These days were not lacking in vivid illustrations of the impact. To the contrary, the first three years of the third decade overflowed with extreme weather. Often relegated to the status of background noise, normalised as part of modern life and eclipsed by fresher events, this series is nonetheless worth recalling as, if nothing else, some portents of what was to come. It started in the early hours of 1 January 2020, when the heaviest rains in the local logbooks fell upon Jakarta, flooding the megacity-cum-capital apparently destined for drowning and slated for relocation.[7] On the opposite shores of the Indian Ocean, the flood was mirrored by fire. Australian bushfires burned an area the size of Great Britain – woodlands, heathlands, grasslands, farmlands – incinerated some 3 billion animals and drove a range of species towards extinction. They reached an apex around New Year's Eve, in the form of a smoke cloud three times larger than any hitherto registered, billowing into the stratosphere with the force of a volcano.[8] Then followed a season of disaster circumnavigating the globe. The double blow of a cyclone and an early monsoon put one third of Bangladesh under water.[9] On the other side of the world, it was fire: Pantanal, the planet's largest wetland, normally a lush delta fed by rivers from the highlands in the heart of Latin America, was enveloped in flames. Drought and heat had sucked moisture out of the biome and primed it for the worst ever burning, including of lands never burnt before.[10] The year ended with the most hyperactive hurricane season logged in the Atlantic – thirty named storms; within a fortnight, two hurricanes lacerating Nicaragua – and in mid-November, for the first time, a hurricane struck Somalia.[11] Thereby 2020 set the tone for the decade. Reinforcing prior trends set to spiral onwards, the year was typical in its atypicality, normal in the absence of anything normal, stable, average: weather so extreme that it was off the charts, either taking known phenomena to unknown heights (cities more deeply flooded) or introducing novelties (wetlands ablaze). All these events bore the mark of global warming.[12] All were amply documented in publicly available channels.

And it continued: on 14 August 2021, temperatures in Greenland measured 18°C higher than the seasonal average, and for the first time in known history, the peak of the ice sheet received rain.[13] The largest wildfire in California to date raged uncontrollably.[14] The tree-studded terraces around al-Quds/Jerusalem were covered by a thin layer of white, eerily snow-like ash.[15] Swathes of the southern coast of Turkey and Greece were aglow, while in the Chinese province of Henan, a year's worth of rain fell in three days – downpours 'unseen in the last 1,000 years' – but in southern Madagascar, drought forced eight in ten inhabitants to fill their stomachs with leaves, cacti and locusts.[16]

Summer had become a telltale season of disaster, each seemingly worse than the last. In 2022, the turn came, not for the first time, to Pakistan. A heatwave rolling over the city of Jacobabad pushed temperatures to 51°C, straining against the limits of livability.[17] Melting glaciers discharged an excess of water into rivers. Then came a freakish monsoon, dropping five times as much rain as the average for the past thirty years: by late August, one third of Pakistan was under water, arid areas that had never before seen boats navigable like inland oceans. Hundreds of bridges and tens of thousands of schools were washed away, 90 per cent of the crops in the national breadbasket Sindh ruined, some 10 million people – about twice the population of Denmark, losing the equivalent of its housing stock – displaced. The immediate floods killed nearly 2,000 people.[18] In the stagnant putrid water, diseases like malaria and dengue fever ran riot; before it began to subside, it was Nigeria's turn to be similarly deluged.[19] But the Horn of Africa suffered the worst drought on record.[20] The cradles of agricultural civilisation in Iraq were parched, crops shrivelling as prime farmland along the two rivers turned to desert.[21] Europe, too, had a taste of the dust. The warmest summer ever, the worst drought in 500 years of observation, the relentlessly scorching sun reduced the mightiest rivers of the continent to trickles: the Loire could be crossed on foot.[22] The Po was a pathetic dribble. From the bridges of Turin, it resembled water poured into a sandbox on a sunny day, the last puddles fated to dissolve and leave behind only a darker streak of dampness.[23]

On the Threshold

All this – a very limited sample – happened before the world had warmed by 1.5°C. Meteorological institutions put the temperature rise to 1.1°C or 1.2°C in these unsettled years.[24] The world, that is, was four or three fractions of a degree away from breaking one and a half degrees, identified as the boundary between tolerable and intolerable heating. We shall presently see how that identification came about and what it signifies. For now, we may simply note that a near consensus held 1.5°C to mark a step change in climate breakdown, that the planet rushed towards it at full throttle and that knowledge of this development was disseminated as through a tornado siren: in 2022, the World Meteorological Organisation (WMO) announced a one-in-two chance that any of the coming five years would breach 1.5°C. As recently as 2015, the probability had been close to zero. Between 2017 and 2021, it stood at 10 per cent, the event still exceedingly unlikely; but after 2022, each year would flip the coin.[25] And no one would be able to claim that people had not been forewarned.

If any one year would see the sum of temperature measurements around the globe average out to a rise of 1.5°C above the 'pre-industrial' baseline – that is, Holocene climate before the birth of fossil capital – that would not, however, necessarily mean the end of this matter.[26] One bad year could be a fluke. The scientific praxis was to even out warming levels across two or three decades, meaning that 1.5°C would be a done deal only when the average for at least twenty years adds up to it. On the other hand, by the dawn of the third decade, several parts of the Earth had already warmed that much over enough time to count – Australia, central Africa; the Arctic, by 1.9°C – and the fast approach of one year in exceedance of 1.5°C betokened a foreshortening of the schedule.[27] One year of that kind would be a springboard for the new average. The world might crash into or rather fly over the boundary. WMO predicted that mean warming between 2022 and 2026 could reach up to 1.7°C; come 2028 or 2030 and it could be higher still.[28]

Another method for gauging the proximity of the transgression was to count what remained of the carbon budget. Since warming is a function of the amount of emitted CO_2, one can estimate how much more of the gas may be released before hitting 1.5°C – namely, in the year 2022, some 380 gigatonnes. More precisely, to retain a 50 per cent likelihood of keeping warming to 1.5°C, no more than 380 additional gigatonnes could be piled on top of what was already deposited in the atmosphere. To be on the safer side of the limit, there would have to be less. At the velocity fossil fuels were being dug up and burnt in 2022, the entirety of this 380 gigatonnes budget would be gone in nine years.[29] If business as usual, that is, continued as in 2022 for another nine short years, at the end of that period, warming of 1.5°C might be prevented only by instant eradication of emissions, every furnace switched off from one day to the next. Such figures sum up something like the historical essence of the 2020s. At the beginning of that decade, staying below 1.5°C was still physically possible. At its end, it might no longer be.

To Respect the Boundary

What would it take to respect the boundary? The world was ushered into the decade with the injunction that emissions be halved until 2030. The IPCC had specified this as the 'pathway' to avoid warming above 1.5°C, and the message could scarcely have been more widely circulated.[30] In one calendar year, emissions would need to fall with the same magnitude seen in 2020 – the largest ever plunge – but not as a one-off, and not for accidental reasons of virus containment, but as an annual decrease repeated again and again in a sustained, linear contraction deliberately orchestrated by some agents in charge of society.[31] This had easily decoded implications for fossil fuels. There could be no new facilities for extracting or transporting or setting them on fire. Insiders understood as much. They learned it from perhaps their most trusted source. In May 2021, the International Energy Agency released a report translating the IPCC pathway to plainer language: as of that year, there must be 'no new oil and gas fields approved for development; no new coal mines or mine extensions' – an instantaneously executed moratorium, 'key milestone' on the path stopping before 1.5°C.[32] Old fields and mines, the Agency intimated, might remain open for some time more: but there could be no new ones.

Now this was not a wing of the climate movement or a revolutionary Marxist outfit speaking. Established after the 1973 oil crisis as a counterweight to OPEC, defending the interests of advanced capitalist countries in unperturbed flows of petroleum, the International Energy Agency was as respectable an institution as can be. It was to fossil fuels what FIFA was to football or al-Azhar to Sunni jurisprudence – or, at least, such was its standing until 2021. With the edict of that year, the Agency seemed prepared to kill the goose that laid the golden egg: 'this represents a clear threat to company earnings', it noted, somewhat bashfully.[33] The report 'sent shock waves through the industry'.[34] The Anglo-American business press covered it in extenso, the dismay palpable: 'investment in new fossil-fuel supply projects must immediately cease', the *Wall Street Journal* relayed; 'this is truly a knife into the fossil fuel industry', *Forbes* quoted one observer.[35] According to the latter outlet, the interest in the report was so massive that the Agency's website crashed.

Echoed by the United Nations, various think-tanks and scientific papers, the call for a moratorium on fossil fuel infrastructure reverberated through the years that followed: an unmistakable prerequisite for respecting 1.5°C.[36] A few months after the Agency report, a study in *Nature* determined that the bulk of known reserves of fossil fuels must be left untouched. When the clock strikes 2050, 89 per cent of the coal supplies staked out in 2018 will have to remain intact, 58 per cent of the oil, 56 per cent of the gas – figures that rested, the research team stressed, on conservative assumptions. For a better than 50 per cent likelihood of keeping below 1.5°C, even more of the mapped and claimed fuels would have to stay underground; perhaps virtually all of it, as of tomorrow. A peak in production should in any case occur 'now'.[37]

The ensuing steep and continuous decline could also be modelled with a measure of justice. It would be painful for a poor country with few other revenues to shut down fossil fuel production overnight, much less so for a prosperous economy with diversified assets to fall back upon. In the early 2020s, the nation of Equatorial Guinea derived 60 per cent of its GDP from oil and gas, while Norway, more saturated than most of the rich, tallied a mere 14. It stood to reason that countries in the former category should be given some more time to rid themselves of these fuels, but then, in keeping with a carbon budget for 1.5°C, the latter would have to do it correspondingly faster. One report from 2022 concluded that wealthy nations must halve their output of oil and gas in the following six years and bring it to zero by 2034. In other words, by 2034, countries like Norway and the UK and the US and Canada and Qatar and Australia can have no oil and gas – not a barrel, not a carrier – leaving their grounds.[38] Coal presented a similar picture. A country like South Africa, heavily dependent on coal for its electricity, should arguably be allowed to extend a phase-out, with North America and Europe picking up the slack. If so, the US would need to accept an assignment of annual emissions cuts sharpened from 10 per cent per year to nearly 17.[39]

While this pile of studies confirmed the 'knife' report from the International Energy Agency, the problem did not, in fact, end with future installations. It included existing ones as well. One team of researchers found that a pathway halting before 1.5°C necessitated foregoing 40 per cent of 'developed reserves' – or, letting nearly half of the fields and seams in which companies had initiated extraction lie untapped.[40] Not only could there be no more installations; many would need to be decommissioned before they had run their course. Indeed, it now appeared that the infrastructure already in place would take the world all the way to *two* degrees of warming, unless a chunk of it were dismantled.[41] A moratorium would be minimally necessary for averting 1.5°C *and* 2°C.

Wherever the exact figures ended up, whatever the methodological choices and modelling errors might be, there could be no mistaking where things stood when the decade dawned: fossil fuels needed to be stopped from reaching the fire. Their combustion would need to precipitously diminish. From enough corners to reach all the main players, the signal to retreat boomed.

A Bonanza of All Times

And yet on the ground, in actually existing capitalism, the exact opposite happened. Covid-19 put a damper on profits in oil and gas. As demand collapsed in April 2020 – aeroplanes grounded, roads empty – prices fell along a straight vertical line, to the point where American traders had so much oil on their hands that they paid buyers to take it away; for a brief eye-popping moment, oil prices turned negative.[42] 'No one is flying. No one is driving,' whined *Forbes*. Oil was trading scandalously low, and the fall looked set to continue: 'oil has no floor under it.' 'Oil is dying.'[43] But already in the following year, oil came roaring back, as consumers jumped into planes and cars and demand caught up with the supplies curbed during the pandemic. The months of global sickness left a legacy of bottlenecks to the quick recovery. By November 2021, the benchmark price for crude had returned to $85 and *Bloomberg* proclaimed 'the revenge of the fossil fuels', the dead energy refusing to stay buried; even coal was on a demand-driven upward curve.[44] And now profits suddenly looked healthier than usual. The turn of fortune was out of this world.[45] On 1 February 2022, ExxonMobil held an earnings call – those convivial occasions where owners and managers of a company chat with their investors and analysts about performance – and boasted about having weathered 'the worst conditions in this industry in living memory' and emerged on the other side, in 2021, with the largest cash flow in a decade.[46] For Chevron – its CEO answering a question from JPMorgan Chase – the last two quarters of that year were the best 'the company has ever seen'. The boss was confident that 'we're going to be more profitable', and he could not have known how right he was.[47] The bonanza was only just beginning.

When Russia invaded Ukraine on 24 February 2022, a war broke out between a giant of oil and gas and the West, which resolved to cut it out of the market. A most significant source of supply was placed under embargo. The trends already in motion since the post-pandemic recovery were thereby accentuated, prices soaring and profits skyrocketing, the latter (but not the former) hitting unprecedented stratospheric heights.[48] 2022 ended with oil and gas companies over the moon. The 'big five' – ExxonMobil, Chevron, Shell, BP, Total – all reported the largest profits in their histories.[49] ExxonMobil came out on top, with $56 billion, the finest result in 152 years of operations, the earnings call for the final quarter of 2022 a gala celebrating undisputed mastery of the private business booming everywhere.[50] BP was behind, but not in pain. 'The returns are pretty darn high', its chief financial officer mused, repeating to his audience of intimates that 'we've given you a lovely little chart' of the cash flow in late 2021 – just before the invasion – and admitting that 'we're getting more cash than we know what to do with'.[51] But none of these privately owned Western companies came close to the mightiest giant of them all: Saudi Aramco, profit of $48 billion for the second quarter of 2022. Between April and June, that is, this corporation made nearly as much money as ExxonMobil did in the entire year. This was said to be the largest quarterly profit ever posted by 'a listed company in global financial history', a candidate for the most extreme super-profit in the annals of the capitalist mode of production.[52] The results were publicised in August, just as Pakistan saw hundreds of its villages swept away.[53] The year's total ended at $161 billion in profits for Aramco, the most a fossil-fuel company, at the very least, had ever made.[54]

Belying the notion of capitalism having entered some ethereally virtual phase, oil and gas supermajors were again neck and neck with Apple and Microsoft in profits, the score that truly matters.[55] By late spring 2022, Aramco had overtaken Apple as the world's single most valuable company.[56] Even the fuel with the most obituaries written for it demonstrated its undying vitality: 'the world's largest coal mining companies tripled their profits in 2022 to reach a total of more than $97bn, defying expectations for an industry that was thought to be in terminal decline,' in the assessment of *Financial Times*.[57] Glencore made the most money. The world's largest commodity trader, this Swiss company was also its largest exporter of coal, handling a fourth of the fuel crossing borders, selling it to power stations in Germany, Japan and South Korea.[58] It raked in some $13 billion in profits from coal in 2022, tripling its haul over the previous year. But in relative terms, Glencore was outshone by Australian coal giant BHP, whose profits in 2022 grew by 3,200 per cent.[59] In Germany, still the world's largest producer of lignite – the brown type of coal known for generating the highest emissions per unit of energy – RWE, principal owners of the mines, feasted on cash.[60] To the north of Germany, there was Sweden, reputed to be greener and less dependent on fossil fuels than most; but even here, no man augmented his wealth as much as Torbjörn Törnqvist, owner of Gunvor, one of the world's largest traders in oil. With record profits, this individual made more than 2 billion dollars in a single year. He was an avid sailor. He lived in a luxury villa in Switzerland. He wore a restrained, confident smile in photographs.[61]

The year 2022 was a bonanza for fossil fuels across the board.

Total profits in this circuit of capital were estimated at 4 trillion dollars, about the same as the GDP of Japan, the third largest economy on Earth – but this being the sum of *profits*, not product or income, over the course of only twelve months.[62] Now what does a capitalist corporation do with a profit? It showers its owners – the shareholders – with the money it has made. But not all of it. Some is also poured back into expanded reproduction.

A Fossil Fuel Frenzy

Not one more pipeline could be built if the world were to avert warming by more than 1.5°C. But in 2022, there were 119 oil pipelines under development – planned, under construction, nearing completion – with a total length of some 350,000 km, enough to encircle the globe at the equator more than eight times. Not one more gas pipeline could be added: but in 2022, there were 477 in progress, with a combined length girdling this planet twenty-four times.[63] Over 300 liquified gas terminals were in the works.[64] Not one more coal mine or plant could burden the Earth, but underway were 432 new mines and 485 new plants.[65] These fossil fuel installations were in preparation already before the scale of the bonanza of 2022 became clear – expanded reproduction is the modus operandi – and with all the capital accumulated in that year, even more were spawned.[66]

'We will continue to invest in our advantaged projects to deliver profitable growth', ExxonMobil declared, an ambition so generic for a capitalist corporation as to be bland.[67] Here the new Golconda was Guyana, in whose territorial waters ExxonMobil first struck oil in 2015. One well after another then came gushing out of the seabed, accounting for one third of all discovery worldwide in the next seven years: abundant, cheaply produced crude, grabbed through hoodwinking and arm-twisting of the third poorest country in the Western hemisphere (after Haiti and Nicaragua).[68] And there was no end in sight. 'The resource base, as you know' – CEO Darren Woods turning to a gentleman from Wells Fargo – 'continues to grow. We continue to make discoveries. We continue to really optimize around these discoveries.'[69] Gas in Mozambique was another mother lode, and then there was, of course, the Permian Basin, Eldorado of fracking, where, as of 2022, ExxonMobil planned to ramp up production by some 70 per cent in five years.[70] In that year of historic profits, this company boosted its spending on new oil and gas projects by at least one fourth.[71] So did Chevron.[72] Together with ConocoPhillips, these were the companies most assertively writing up their investment plans, spearheading the expansion of the expansion – Aramco, far ahead in absolute terms, advancing one notch slower.[73]

It was as if the scenes playing out in Madagascar, Pakistan, Nigeria, the Horn of Africa, Iraq, even Australia and Italy belonged to a parallel universe. No feedback connected them to the calculi of oil and gas companies. The latter proceeded in the most studious conceivable disregard of the lives destroyed around them, not to mention the signal from the UN and scientific bodies and sundry other institutions to start winding production down. It was not heading downwards: production was on the up. Pouring more capital back into it implied output spiking in the years ahead. Even before 2022, ExxonMobil aimed at increasing oil and gas production by 8 per cent in 2027, Chevron by 16 per cent in 2026, Aramco by 16 in 2027, Total by 13 in 2030, Petrobras by 15 in 2027 . . . every company of status carrying a portfolio presuming demand, prices and lifetimes far outside the pathway leading to 1.5°C.[74] Indeed, a range of projects presupposed so much room for oil and gas that they could fit only into an envelope exceeding *two and a half* degrees. The jewel in the Guyanese crown, a sprawling complex of many dozens of wells, subsea cables and pipes connected with tankers; the deep-water fields off Libya picked up by Italian supermajor ENI; gas in the shallow waters of Angola (Chevron) and Malaysia (Shell); the sweeping exploitation of the oil reserves in and around Lake Albert launched by Total – these were some of the most far-out projects on the stocks.[75]

The oil and gas major that made the most aggressive move

beyond the 1.5°C pathway in 2022 was Total of France.[76] On 1 February of that year, the 'final investment decision' – the moment when capital is definitively dedicated to a venture – was ceremoniously announced for the East Africa Crude Oil Pipeline, or EACOP.[77] It would become the longest heated oil pipeline in the world. Stretching from the fields around Lake Albert on the border of the Democratic Republic of Congo, through Uganda and Tanzania to the coast, it was designed to cross 230 rivers, bisect 12 forest reserves, run through more than 400 villages and displace or otherwise severely affect the lives of around 100,000 people – many already ordered to cease growing crops and repairing their houses – all for the purpose of carrying 216,000 barrels per day to the world market. The resultant emissions would be twice those of Uganda and Tanzania combined.[78] Not satisfied, Total 'expressed an interest' in also accessing the oil stored in the peatlands of the Congo Basin.[79] And in May 2022, the Democratic Republic notified the industry that auctions would indeed be held for blocks in its rainforests, posting a slick video on Twitter – here was 'the new destination for oil investments' – tagging Total and Chevron.[80] Still not satisfied, Total was rummaging through Namibia, where it claimed to have found 'a potential new golden block', and Suriname; but the French aggressor clearly focused on the African frontiers.[81] So did ENI, from the old Italian haunts of Libya in the north to Angola in the south. The continent supplied more than half of the oil and gas produced by the company; likewise armed with record profits from 2022, it went out to find more.[82]

For majors like these, the acceleration of 2022 merely intensified a trend at work since the mid-2010s: growing capital expenditure on oil and gas (except for the pause of the pandemic); a growing share of earnings from their production; growing not shrinking reserves.[83] But the bonanza looked set to induce a change of gears. This could be discerned at the scale of countries as much as companies. The scramble for African hydrocarbons picked up speed after the invasion of Ukraine. Seeking gas supplies to supplant Russia, Europe shopped around in its former colonies and incited Mozambique, South Africa, Morocco and

Tanzania to embark on extensive construction of pipelines and terminals geared to the north.[84] The moment was even ripe for dusting off the four-decades-old idea of stitching together a pipeline taking gas from the Niger Delta through the Sahara all the way into the metropoles of Europe: in July 2022, a memorandum of agreement was signed by Nigeria, Niger and Algeria for this mother of all pipelines. Before the year's end, Nigerian welders were reportedly busy at work. 'European countries would like to see the project up and running within a maximum of two years,' one Algerian source explained the rush.[85] The state of Israel rode the same wave; in October, propelled by a deal with the EU, its flagship Karish gas field (in waters claimed by Lebanon) came online. It was the first time this state was elevated into a fossil fuel exporter of note.[86] From a Europe at war with Russia, the stimulus to erect brand new infrastructure spread to the four corners of the Earth.

Inside Europe itself, Germany, its powerhouse, went on a building spree, pipelines and floating terminals laid out along the coast to accommodate the imports. There was nothing temporary about the commitments. When Qatar and ConocoPhillips signed a contract with the Bundesrepublik for gas deliveries to start in 2027 and continue for fifteen years, Robert Habeck, 'minister for economic affairs and climate action' – representing the Greens – thought 'fifteen years is great'. Come to think of it, he says, 'I wouldn't have anything against twenty-year or even longer contracts', running towards and beyond 2050, that is.[87]

Unlike Germany, the UK had oil and gas fields of its own to resort to: it took a plunge deeper into the North Sea. In October 2022, the BBC matter-of-factly reported that Westminster had opened a new licensing round for companies. 'Nearly 900 locations are being offered for exploration, with as many as 100 licences set to be awarded. The decision is at odds with international climate scientists who say fossil fuel projects should be closed down, not expanded.'[88] Shell and BP jumped at the opportunities.[89] Shell had earlier withdrawn from the Cambo oil field outside the Shetland Islands, a target of protest activity; but this asset was snapped up by Ithaca Energy and scheduled to open in 2025 and peak in 2029 and carry on until 2053. (An up-and-coming actor in the North Sea, Ithaca Energy was based in Tel Aviv – one more sign of Israeli presence on the front – but entered the London stock exchange in 2022, in the largest floatation of that year.[90]) Another of the northern sea beasts to be awakened in the 2020s was – ironical intention unclear – named Tornado.[91] (An oil field in the Gulf of Mexico that went online in 2021 was also named Tornado.)[92] Westminster promised to 'max out' the reserves in its part of the sea.[93]

And then there was Norway. The biggest producer of oil and gas in Europe (Russia excluded) barrelled forward, always on the lookout for fields to get richer. During the decade leading up to February 2022, Norway awarded as many licenses for exploration (700) as in the preceding half-century, making this not only the biggest producer but the most aggressive hunter among European countries.[94] And in March of that year, the government invited bidders for another round of licences, 'including previously unexplored acreage in the Arctic' – how could it not?[95] In his letter, the energy minister clarified that 'access to new, attractive exploration acreage is a pillar in the government's policy for further development of the petroleum industry'; and when the round concluded with 47 new licences awarded to said industry, he praised it for contributing 'large revenues' and 'value creation'.[96]

The biggest producer of oil and gas in the world – the US – sped down the same road. It scurried to feed Europe with gas after the outbreak of the war. During the first half of 2022, after only six years in this race, the US overtook Qatar and Australia to secure a position as the world's foremost exporter of liquified fossil gas. It accounted for nearly half of export capacity under development.[97] In the early days of the invasion, the organisation lobbying on behalf of US companies in this business submitted a wish list to the Biden administration – more drilling on public lands; expedited approval of pipelines and terminals; the construction of 'virtual transatlantic gas pipelines' – and then jubilantly saw it come true within half a year.[98] The same administration handed out 307 leases for oil and gas exploration in the Gulf of Mexico in 2022 alone (Chevron bagged the most).[99] The Permian was booming again.[100] No other country in the world had more oil and gas reserves in the development stage – three times the amount in Qatar, four that in Saudi Arabia, six in Canada.[101] Of all the oil and gas expansion by then scheduled to take place until 2050, this single country accounted for *more than one third* (twenty-five times more than Saudi Arabia).[102] Here, more than ever, was the hegemon of hydrocarbons.

But as in any real hegemony, the US led a coalition of allies. One could imagine, as we have seen, that the most affluent producers would begin the descent first, as it would be easiest for them. Perhaps not so surprisingly, they did the opposite. Five countries in the global North – the US, Canada, Australia, the UK, Norway – together accounted for 51 per cent of the expansion from new oil and gas fields planned until mid-century.[103] (The share would be higher still if the plans of companies headquartered in these cores but active in peripheries were included – say, ExxonMobil in Guyana counted as American not Guyanese expansion). With the exception of Norway, these countries were also among the top ten historical contributors to global warming based on domestic emissions. Two centuries after fossil capital formed in Britain, it was still Anglo-America working hard to set the world on fire.[104]

What of coal? Here it was, of course, China above all others. 2022 witnessed a great leap forward of sorts: compared to 2021, a quadrupling of coal-fired power plant capacity greenlighted. Not in seven years had permits been handed out at that rate. On average, two large plants were given the go-ahead every week. Executives bragged about how fast they got them from drawing board to ground-breaking. 'Build first and reform later', ran the slogan in the province of Jiangsu; most of the spree, however, focused on Guangdong, home of the Chinese export miracle, while the build-out of mines to supply the southern chimneys remained heavily concentrated to a few distant northern provinces, led by Inner Mongolia and Xinjiang, the coal running on a conveyor belt from steppe and mountain to shore. The combustion capacity entering construction in China in 2022 was six times larger than that in the rest of the world combined.[105] But the rest of the world did not thereby sit still. In India, the

Modi government brandished the slogan 'unleash the power of coal' and superintended an unbroken expansion.[106] Part of the 'unleashing' consisted in transferring coal resources to private hands, notably those of Gautam Adani, who in 2022 rose to a position as the richest man in Asia and third in the world thanks to his mining fortunes, often seized through the eviction of villages and destruction of their forests.[107] The tentacles of his Adani Group extended to the seams of Australia, most infamously the Carmichael mine in Queensland, from which the first shipments of coal left in the first days of 2022.[108] Measured by province – and the geography of coal might be better understood at that scale – this one had more coal under development than any other in the world. Coal in the 2020s was largely a question of Queensland. In fact, Australia, the nation that exported most of the substance, lagged only slightly behind China in its plans for future expansion.[109]

If the centres of coal fell within an eastern arc running from Inner Mongolia to Queensland, the original western hubs also showed signs of renewal. In the summer and autumn of 2022, Germany reopened at least five mothballed plants burning lignite. Anthracite installations were plugged into the grid too. RWE enjoyed the good times.[110] Austria and the Netherlands moved in the same direction, while the birthplace of fossil capital, the UK, saw its first coal mine authorised since the 1980s. From pits sunk 500 metres into the ground of Cumbria, the Woodhouse Colliery would produce coking coal for export – a minor additive to the market, but symbolic proof of the undeadness of fossil fuels.[111]

Investment in coal stayed consistently high worldwide for the last two decades of the twentieth century and the first two of the twenty-first.[112] In 2021, it rose by some 10 per cent; in 2022, by another 10 per cent, most of the projected mines being 'greenfield' – that is, mines where none had existed before. Actors like Glencore and BHP and the Adani Group jostled for more stuff to extract.[113] Meanwhile, in the oil and gas departments, the mood was bullish. Leading consultancy Rystad Energy rounded off the year 2022 with a reference to Genesis: 'global oil and gas suppliers look set to echo the Biblical story about the Egyptian pharaoh's dream of seven years of feast and seven years of famine – only in the opposite order. All signs point towards 2022 being the start of another super cycle for the energy services sector.'[114] In other words, the period 2015–22 would have represented seven years of famine with profits mediocre and investment lacklustre: what then commenced would be an equally long feast. The priorities of the men at the helm were not in doubt. 'We are underinvesting as an industry', Darren Woods of ExxonMobil made clear. The CEO of Halliburton, the unrivalled giant of refinery, pipeline, fracking rig and chemical plant construction – footprint stretching from Karish to the Permian – looked forward to 'multiple years of increased investment.'[115] In this widely shared assessment, the bonanza marked the beginning of a long boom in the primitive accumulation of fossil capital.[116] But this required engagement from the world of finance.

It would not play hard to get. Between 2016 and 2021, the world's sixty largest banks poured nearly 5 trillion dollars into fossil fuel projects, the sums bigger at the end of this half-decade than at its beginning.[117] The world's largest bank, JPMorgan, was also the most generous. ('It is not against climate for America to boost oil and gas', said the CEO, and demanded 'immediate approval for additional oil leases and gas pipelines'.[118]) Perhaps briefly, JPMorgan was bested by the Royal Bank of Canada as the top financier in 2022. Closely behind were Wells Fargo, Bank of America, Citigroup, Barclays and BNP Paribas, all busy pumping money into projects from the Amazon to the Arctic, not to forget Aramco.[119] The lust for coal was just as strong: JPMorgan, in 2021, nearly tripled its financing of coal production, while Crédit Agricole poured money into Glencore and BNP Paribas into RWE.[120] In the midst of the bonanza, fossil fuel producers were in lesser need of borrowed money than usual, as they had 'more cash than we know what to do with'; yet the banks continued to disburse money to them, and when expanded reproduction took on a new scale in 2022, expectations were that borrowing would commensurately grow: more money would be needed to finance the next generation of fossil fuel projects.[121] It would take years for these projectiles to rain down on Earth.

Out of Control

Seven words from Theodor Adorno summed up the situation: 'society is not in control of itself.'[122] Three years into the third decade, things were out of control. Things were completely, infernally, demoniacally out of control: the classes ruling the planet seemed bent on burning it as fast as physically possible, and nothing – nothing – had yet reined them in to even the most minimal degree. Capitalist society enacted a dissociation between the reality of climate breakdown and the interior drives of accumulation. In the physiology of this highly peculiar type of society, no synapses connected the information and experience of the catastrophe unfolding with the centres of decision-making – above all, concerning what energy sources to use.[123] It did not matter what people went through along the Indus or the Po and certainly not in Somalia, nor did it matter what the International Energy Agency said. What about the secretary-general of the UN? António Guterres did not mince words: 'high-emitting governments and corporations are not just turning a blind eye; they are adding fuel to the flames. They are choking our planet, based on their vested interest' – moreover, 'investing in new fossil fuels infrastructure is moral and economic madness.'[124] But by the same token, Guterres came across as an impotent Cassandra or prophet Amos, for the wicked men he denounced could shrug him off and laugh all the way to the bank. It did not matter what he said.

What all of this amounted to was a fossil fuel frenzy, instigated right on the threshold of $1.5\,°C$ of global warming. All kinds of capitalist social formations partook in it, from Norway to the Democratic Republic of Congo – governments social-democratic and conservative, centrist and far-right; even cabinets with Greens in them. It obviously did not matter how solemnly they had sworn themselves to the $1.5\,°C$ limit. The bombs planted under their supervision would blow the budget for that limit sky-high. One way to measure this was to home in on 'carbon bombs', defined as proposed or existing fossil fuel extraction projects that would generate more than 1 gigatonne of CO_2 if allowed to run their full course: and one list – not exhaustive – identified 425 such bombs. Coal mines in China made up the largest batch; on the list were items ranging from Canadian tar sands to lignite mines in Germany. All in all, they would emit twice as much as could fit into the remaining budget for 1.5°C.[125]

These were figures preceding the fillip from the profit bonanza. Already in 2021, governments were planning for a growth in fossil fuel production until 2030, which would bring it to double the quantity consistent with 1.5°C; it would also deviate by 45 per cent from the pathway to 2°C.[126] All the new mines planned as of 2021 would take coal output at the end of the decade to a size four times larger than could be squeezed into 1.5°C, three times that into 2°C.[127] One comprehensive database yielded the prediction that proven fossil fuel reserves had the potential to blow the former budget seven times over – or, to emit more CO_2 than in the entire period from the Industrial Revolution up to 2022.[128] Whichever way you measured it, there was no sign of the sustained emissions decline needed, any more than there were signs of a classless society or a free Palestine.

And as the curves still pointed in the wrong direction while the world was rapidly heating up, there appeared a series of absurd spectacles. In Germany anno 2022, the RWE dismantled a wind farm to make room for an expanding lignite mine; but during the summer drought, the water levels in rivers fell so low that barges carrying coal to the newly opened plants could not move.[129] Just as the Pakistani floods were nearing their peak, the country signed an agreement with neighbouring Afghanistan to send it more coal. The European shopping spree had pushed the price of gas to levels where countries in the global South could no longer afford it: hence the turn to coal; and barely had the monsoons ended before a Saudi delegation arrived at the scene.[130] Led by the energy minister, it headed to Balochistan, the poorest and normally driest province, accounting for 60 per cent of the houses lost to the floods. The minister came to announce that the Kingdom had finally settled on pouring $12 billion – equivalent to one fourth of Aramco's summer profit – into the construction of a colossal new 'oil city' on the Balochi coast. Centred around a refinery four times larger than the largest then operating in Pakistan, it would also include a sprawl of petrochemical plants, all nourished by oil from the Kingdom, which, the minister graciously explained, 'wants to make Pakistan's economic development stable'.[131]

And why did China escalate its coal expansion so spectacularly in 2022? Perhaps the most forceful move in the carbon bombs arms race, it had little if anything to do with the war in Ukraine. China scrambled to open more mines and plants in no time because of the heatwave of that summer. As it continued for several months, covering an area of continental scope, breaking record after record – according to one weather historian, no heatwave in any archive compared to it – people put their air conditioners into overdrive. Sixty-six rivers having dried out completely, hydropower was down. It was to make up for these swings in demand and supply that Chinese authorities lost their inhibitions on coal. And what was the world capitalism built? An overheated and dried-out Guangdong province, serving shopping malls on all continents from its factories, choking on itself, grasping for even more coal to keep breathing a while longer.[132]

Mad Capital

We write these words in the year 2023. What the rest of this decade will hold, no one can know. But it is evident that the first three years bequeathed, to put it mildly, some problems to the near and indeed long-term future. They provided an object lesson – the starkest so far, but starker may come – in how capitalism has related to the climate crisis: by intensifying it, and by intensifying it more the more intense it gets. Whatever follows next in the crisis will inevitably spring from this fundamental condition: the obduracy of business as usual; its extreme inertia in the face of consequences; the practical refusal to heed the calls to stand down or even just moderate a little bit. Covid-19 was but a blip. The invasion of Ukraine, on the other hand, cranked up business as usual to a higher volume, from which it might come down – or, some other war may have a similar effect.

One lesson from 2022 was that in wartime, even the most presumably green-minded Western nations push climate mitigation to the bottom of the agenda and go frenzied for fossil fuels, if these are perceived as serving immediate interests. And the rest of this century does not seem short on geopolitical conflict. New pandemics or other shocks with a reductive profile may very well be in store too, of course; but one lesson from 2020 was that fossil capital has the power to shake off the cold water from such an incident like a dog getting out of a lake. In fact, if Covid-19 had any enduring effect on the fossil economy, it was to generate that bottleneck sending prices and profits bouncing back as the recovery set in, spurring a rush to reinvest in fossil fuel infrastructure subsequently sped up by the war. The two shocks were linked by the upwards curve of 2021. Here, then, was a general lesson from those first three years: the articulation of the climate crisis with any number of other crises has, if anything, a tendency to aggravate the former. However this overdetermined multiplicity is conceptualised – as a 'polycrisis', 'cascading crises', 'organic crisis', 'chronic emergency' or some other term for a world lurching from one disaster to the next – business as usual itself has, as of this writing, remained the steady default trajectory.[133]

Already here we may notice a psychic dimension to developments. The object lesson concerns a monumental failure to adjust to reality. If we encounter a person who is not in control of himself and incapable of adjusting to reality, we recognise someone with mental health issues. Sigmund Freud considered this the hallmark of neurosis and psychosis, the two classical disorders: both are 'the expression of a rebellion on the part of the id against the external world, of its unwillingness – or, if one prefers, its incapacity – to adapt itself to the exigencies of reality, to Ἀνάγκη [Necessity].'[134] We may then abbreviate the secretary-general's diagnosis. Investing in new fossil fuel infrastructure in the third decade of the millennium represented not merely moral madness: it was madness without qualifier; madness in the original, clinical sense of the term. As for the economy, there was, as we have already gathered and will see more of below, nothing mad about such investment; rather it was perfectly sensible. Precisely because of the way the economy was constituted, some immoral, clinical madness held the future of the planet in its hands. Did this also mean that it was now too late to defend such a future?

2

When Is It Too Late?

As 40,000 people descended on the Egyptian Red Sea resort of Sharm el-Sheikh in November 2022 for yet another international climate summit – the twenty-seventh addition to a long list of unremarkable achievements – one question was on everyone's lips: could warming still be limited to 1.5°C? Was the target – an unusual condition for a target, but such was the language – still alive? The official message coming out from Sharm el-Sheikh was that yes, indeed, the target was alive: but only just so. When Guterres ascended his podium, it was 'on life support – and the machines are rattling'. As negotiators gathered in noisy, air-conditioned meeting halls on the outskirts of the resort, regularly interrupted by a steady stream of aeroplanes on their descent to the nearby airport, the secretary-general sought to instil a sense of purpose and urgency in the audience. If only countries rose to the occasion and put their self-interests aside *this* time around, the patient could be saved and the worst dangers averted. 'Humanity has a choice: cooperate or perish,' Guterres boomed across the room. 'It is either a Climate Solidarity Pact – or a Collective Suicide Pact.'[1] The gathered dignitaries applauded politely and then went on their way, spending the next two weeks bickering about who, if anyone, should pay for climate damage, whether the final document should mention all fossil fuels or only coal (they decided on only coal) and how to understand the eye-wateringly complex new carbon market they had summoned into being at COP26 in Glasgow the previous year.[2]

After twenty-seven iterations of this show, anyone following the climate debate would be forgiven for collapsing into bouts of cynicism by this point. Indeed, many onlookers outside the media-trained echo chambers of the United Nations Framework Convention on Climate Change (UNFCCC) secretariat were not buying it anymore. Grown tired of the political hyperbole, they started calling the bluff. Just as the summit got underway, the *Economist* sought to sink it by declaring the patient already deceased. In an obituary entitled 'Goodbye 1.5°C', the magazine's editors concluded that 'there is no way Earth can now avoid a temperature rise of more than 1.5°C.' Citing the unrelenting emissions growth, the equally unrelenting shrinkage of the carbon budget and the presumably obvious fact that 'fossil fuels will not be abandoned overnight', the hebdomadal of choice for the Atlantic liberal intelligentsia pronounced that time had come for 'a dose of realism'.[3] Necrophiliac attachment to the 'lost cause' would not merely recreate false hopes; it would distract from the other actions that should be pursued instead. What were those? We shall inspect the favoured shortlist later.

At this juncture, the position enunciated by the *Economist* rapidly gained traction, but not only among bourgeois commentators with a parti pris: scientists and activists took it up too. Indeed, parts of the research community seemed to be holding an undeclared wake for the 1.5°C target. A 2021 survey by *Nature* asked IPCC authors what temperature rise they thought most likely by 2100: only 4 per cent of the ninety-two respondents believed the world would limit warming to 1.5°C. A full 60 per cent bet on 3°C or more.[4] Even scientists posing on the barricades were furling up the flag. Scientist Rebellion, an academic offshoot of Extinction Rebellion, or XR, raised eyebrows in late 2022 with an open letter demanding that friends of the climate come clean on 'the inevitability of missing the 1.5°C goal'. Citing the familiar trends, the *Nature* survey and the XR tagline 'tell the truth', the several hundred more or less illustrious signatories called for that goal to be formally buried.[5] In their view, it would clarify the next phase of the battle; clinging to 1.5°C would merely keep people high on 'hopium'.[6] The starting point for a realistic climate movement must be a lucid registration of historical defeat. Call this position *one-and-a-half-degree defeatism*.

But far from everyone subscribed to it. To the contrary, the question of the fate of 1.5°C remained as unsettled as any in climate science, even at this late date. On the commentary pages of newspapers, in Twitter threads and climate blogs, the academic equivalent of a pub brawl was fought over it. Opponents of defeatism argued that no geophysical dynamics had yet condemned the world to as much warming. To pick two, H. Damon Matthews – a pioneer of carbon budget research – and a colleague of his reaffirmed that it was still – technically, physically, in any strict sense of the term – possible to avoid warming by 1.5°C, on the condition that CO_2 and other greenhouse gases be 'aggressively mitigated over the coming decades'. That conclusion was not derived from some roseate assessment of where things stood. Writing in *Science*, these two scientists began by asserting that warming had already reached 1.25°C and that one more decade of ongoing trends would smash the budget: and yet they stuck to the possibility of one last-ditch effort. As for the real sources of the problem, they pointed to 'current sociopolitical systems'.[7] Such are amenable to disruption. Forceful enough, it could break the trajectory. Call this position *one-and-a-half-degree voluntarism*.

Defeatists like those from Scientist Rebellion claimed that the phrase '1.5°C is still alive' had become 'a fig leaf for business as usual', sustaining illusions, fostering a complacent belief that it would all be fine in the end.[8] Voluntarists threw the charge back. Declaring 1.5°C unavoidable before it had been realised constituted a surrender to business as usual, rather like burying a child alive because someone is force-feeding her with drops of poison. Resignation like that reflected in the *Nature* survey stemmed from a failure to imagine political ruptures.[9] Even someone like Fatih Birol, head of the International Energy Agency – a former petroleum pal who by now sounded almost like a climate activist – lashed out at the '1.5°C is dead' criers, called their arguments 'factually incorrect' and 'unhelpful' and accused them of playing 'into the hands of fossil fuel proponents'. The latter would be the sole 'beneficiaries if the obituary of 1.5°C is written'.[10]

But oddly, this faith in the lifeforce of 1.5°C was reaffirmed from the highest quarters too. For a while, it looked like the target would not be mentioned in the final document from Sharm el-Sheikh: but this triggered indignation from the 'We Mean Business Coalition'. The primary front for companies at COPs, starring business leaders like Sir Richard Branson, CEO of Virgin, the Coalition drafted a letter of protest, declaring that 'there can be no excuses for backsliding on the commitments made a year ago'. It assured the world that corporations 'are committed to doing everything in our power to limit global warming to 1.5°C' and called on governments to join them in this noble cause.[11] Even former climate denialist and recently ousted UK prime minister Boris Johnson made a cameo appearance in Sharm el-Sheikh to reiterate the need to 'keep 1.5°C alive', a phrase he had pushed to breaking point at the previous COP.[12] During the final days of the negotiations, G20 convened a parallel summit in Bali, and in a ceremony befitting of the cause, all twenty leaders – Biden of the US, Macron of France, Scholz of Germany, Modi of India and the rest – dressed up in white and grabbed hacks to plant saplings in a mangrove, looking mildly uncomfortable in the heat, while confirming their profound determination 'to limit the temperature increase to 1.5°C'. The saplings were planted to form the letter G and the numeral 20 in green.[13] This establishment had its way: 1.5°C stayed in the text from Sharm el-Sheikh. An interesting moment it was, when the likes of Branson and Johnson rushed to the defence of an ambition that activist scholars wanted to lay to rest.

Somewhat crudely, we can then say that there was a left version and a right version of one-and-a-half-degree defeatism, mirrored by the same two poles on the other side. In this matrix, capitalists most sanguine about their system and activists most sceptical about it could end up anywhere. But there was a logic to the confusion. And not only did the debate have antecedents almost four decades old: it was attached to an arrow of time set to fly for as many decades or more. What was at stake in it? To understand this, we need to wind back the clock nearly half a century, for any notion of it being too late is, of course, a product of change over time.

Mostafa Tolba Outlines the Limits

When people realised that the world was in the process of heating up, due not to some chance event or natural destiny but to fossil fuel combustion, and when they understood that very high temperatures would entail very considerable dangers, some of them came to ask: at what point do we have to stop? Where is the limit beyond which the process must not be allowed to run? The first attempt to formalise such deliberations took place in the mid-1980s, on the initiative of one Mostafa Tolba. Born in a village in the northern Nile Delta, with a PhD in plant pathology, he rose to a post as minister for higher education under Gamal Abdel Nasser, the heyday of anti-colonial politics shaping his outlook.[14] (As a Nasserist diplomat, he might well have attended the conference of the Non-Aligned Movement held in 1964 at Sharm el-Sheikh.) In 1972, he was dispatched to Stockholm to represent Egypt at the UN conference on the environment, where the United Nations Environment Programme (UNEP) was set up; between 1975 and 1992, spanning the period when ecological issues rose to the surface of world diplomacy, it was Tolba who ran it. If he is remembered today, it is as the architect of the 1987 Montreal Protocol. From his chair at UNEP, he corralled the nations into an agreement for phasing out substances depleting the ozone layer, following a superbly straightforward model: we have a shared problem with 'global emissions'; the objective must be their 'elimination'; along the way, the signing parties shall cut the level of production by a fixed share between certain dates, until it hits zero – absolute zero, that is. The text of the Protocol is written in a language of limits, control, regulation, strict allowances, imperatives. The word 'shall' appears sixty-nine times in its ten pages.[15] Every country ratifying it would be legally bound to comply. With the ozone layer steadily healing, the Montreal Protocol is, we now know, the most resounding success of environmental protection the modern world has produced. Could it be emulated elsewhere? Fresh from having negotiated its precursor in 1985, Tolba turned to another problem that worried him and prepared to treat it with the same model: global warming.[16]

His many speeches from the 1980s were redolent with Third Worldist rhetoric. He railed against the exploitation of the poor by the rich, who laid the resources of the former to waste. He repeated the calls for a New International Economic Order. The path to recovery passed through 'the overturn of a global economic system that is not only grossly unfair' but also destructive to nature, including the climate.[17] Having convened a symposium in the Austrian town of Villach in 1985, Mostafa Tolba stressed the urgency of this particular matter: 'we run the risk of being overtaken by events, and of having to deal with a global warming' when 'it is already too late to do anything about it or to deal with its impacts'.[18] The science had been clear enough since the 1960s; now was the time to impose restrictions on fossil fuels, or else temperatures might rise by '2°C or thereabouts' by the year 2030.[19]

Tolba proceeded to form an Advisory Group on Greenhouse Gases, a synod of seven scientific experts. (One was his fellow botanist and compatriot Mohamed Kassas: the Egyptian presence at this moment of birth for international climate politics is remarkable, a distant memory when the country hosted its first COP in 2022).[20] At workshops in Villach and the Italian resort town of Bellagio in 1987, the Group came up with the principle of 'targets' – specified limits for global warming, the transgression of which would be impermissible. Their utility lay in their being clear-cut; once adopted, progress towards meeting them would be 'quantifiable and unambiguous'.[21] No fudging on targets. But targets of what kind? Global warming could be measured from plenty of angles, and so the Group toyed with ideas for sea level rise (maximum 50 mm per decade: in the 2010s, it was 40), atmospheric concentration of CO_2 (maximum 400 ppm: in May 2022, it was 421), rate of warming (maximum 0.1°C per decade: in the early millennium, it was 0.2°C).[22] It also considered absolute rise in average temperatures (2°C came to mind). So did 1°C. 'Temperature increases beyond 1.0°C may elicit rapid, unpredictable, and non-linear responses that could lead to extensive ecosystem damage', the Group foresaw, while recognising that warming of such magnitude may be 'unavoidable' due to emissions already made. And even if, by some miracle, it stopped at 1.0°C, there would still be 'adverse impacts to ecosystems and human systems'.[23] In some sense, then, it was already too late.

The US did not like Mostafa Tolba. He was a thorn in their side, an annoying holdover from the age of anti-colonialism, who had alienated allied regimes in Latin America during the negotiations over the Montreal Protocol. The White House would not let him capture climate policy as he had ozone policy. His messages to the world and the deliberations of his Group intimated that something like the Protocol – limits: reductions: elimination – was in the offing for fossil fuels, a wholly different matter than chlorofluorocarbons.[24] The Group even had the temerity of singling out the disproportionate use of such fuels in the US.[25] Too activist for American taste, it was sidestepped and consigned to oblivion – Tolba himself leaving UNEP in 1992 – its role transferred to the IPCC, which, initially, did not engage in target practice.[26] Nor was there any mention of a particular limit to warming in the text of the UNFCCC. The objective of that Convention – adopted at the UN conference on the environment in Rio in 1992, the foundational document for all the COPs ever since – was, according to the famous article 2, to stabilise 'greenhouse gas concentrations in the atmosphere at a level that would prevent dangerous anthropogenic interference with the climate system'.[27] A level – but which one? The UNFCCC evaded the question Tolba and his Group had raised.[28] But it would impose itself on climate negotiations, unworkable without a notion of where truly intolerable dangers would set in, and then the original struggle would resume: between a US averse to unequivocal limits and the global South.

One to Two and Back to One and a Half

One of the many questions laid down by the UNFCCC for the COPs to ponder every autumn was precisely the level that demarcated the onset of danger; but few advocated 1.49°C or 2.3°C as the most adequate response. 'Two degrees' had the appeal of a round number. In the early 1990s, scientists affiliated with Tolba's Group encapsulated their approach in the image of a traffic light, where green represented modest damage to ecosystems and people, shifting to amber – extensive damage – at 1°C and to red – unacceptable harm – at 2°C; business as usual was a car that needed to stop when amber flashed. Continuing to drive after red would be the height of recklessness. Early IPCC reports and COPs stayed silent on the question, but in 1996, the European Union interpreted the available science as advising a 2°C stop sign. This was the era of the Kyoto Protocol, which, again, shied away from formally endorsing any one temperature degree; yet Kyoto became associated with 2°C, as the most commonly envisioned upper limit at the time.[29]

In the two decades around the turn of the millennium, this was also the banner under which the environmental movement in the global North would march. For Greenpeace and Friends of the Earth and Stop Climate Chaos, a coalition of NGOs gathering tens of thousands of demonstrators in London in the dying days of Kyoto, it was 2°C that must not happen. In perhaps the finest book to come out of this cycle of protest, *Heat: How to Stop the Planet From Burning* (2006), George Monbiot averred: 'our aim must be to stop global average temperatures from rising to more than 2°C above pre-industrial levels', only to immediately throw it into doubt. Given the amount of carbon in the atmosphere, we might 'already be committed to 2°C. But I am writing this book in a spirit of optimism, so I refuse to believe in it' – two-degree voluntarism, as it were, backed up by spiritual faith.[30]

The bridge leading from Kyoto to another era was erected during COP15 in Copenhagen in 2009, and here, for the first time, 2°C entered the final text as statutory target, against the wishes of the US.[31] Copenhagen thereby marked the crossing into official target chasing. But it was no less significant in another key respect. During the Kyoto era, negotiations were conducted on the assumption that if countries shouldered emissions reductions in an agreement, these would be legally binding, their implementation mandatory, noncompliance subject to sanctions in one form or other, just as in the Montreal Protocol.[32] Indeed, this common-sense principle was enshrined in the Kyoto Protocol itself, the advanced capitalist countries of 'Annex B' obliged to cut emissions by 5 per cent – a paltry burden, but a compulsory one.[33] It was not levied on the global South. Such distribution stemmed from another formula in the UNFCCC, namely 'common but differentiated responsibilities and respective capabilities' – an acknowledgement that some had done more than others to cause global warming and were better equipped to remake their economies.[34] The US and Germany would have to carry that burden, Madagascar and Pakistan not so.

A second assumption followed. Negotiating teams from the global South and their allies expected any agreement succeeding Kyoto to revolve around some specified mechanism for fairly and equitably sharing the reductions: some would have to cut deeper than others. As the Protocol approached its expiration date – 2012 – a flurry of schemes was floated for ensuring the unity of justice and survival.[35] The most popular was 'Contraction and Convergence', a model that posited, as the self-evident starting point, a total cap on emissions. The cap would gradually be pressed down towards zero. But before that endpoint, rich countries would have to reduce their emissions so fast that poor countries could temporarily increase theirs, so that they all converged on the same level of per capita emissions – Americans and Pakistanis meeting in equality – and then reached the finishing line as a humanity unified in average fossil-free plenitude.[36] 'Contraction and Convergence' was radicalised in the 'Greenhouse Development Rights', a framework that also took into account class distinctions within nations: poor everywhere would carry no obligations, but rich Pakistanis would have to tighten their belts like their American counterparts.[37] There was a moment when expectations rode high that something like this would follow Kyoto. People who have engaged with climate politics for several decades might remember these as the days of a youthfully innocent idealism. In hindsight, that is how they appear. But the conjuncture is better understood as a long afterglow of the anti-colonial revolutions. In the 1990s and very early 2000s, the global South still retained a combativeness in the UN arena, putting up a united front and insisting on rights vis-à-vis the imperialist core, and only for this reason – not out of some naïvety – justice was in the air.

COP15 in Copenhagen broke the front. Towards the end of the drawn-out, sour negotiations at that summit, Barack Obama presented the world with a draft text that disposed of commitments: henceforth, it said, it should be up to each country to do what it wanted, without duties or risk of punishment. In a private meeting with the leaders of China, India, Brazil and South Africa, Obama ensured their support for the ultra-liberal initiative. All others from the South baulked at it. They regarded it as a blatant attempt by the supreme mega-emitter of the world to duck out on any limiting of emissions. If the US had its way, there would be no future for a just distribution of cuts, since it presupposed a cap of legal standing; and so countries from the South – minus the defectors siding with Obama – refused to sign the draft.[38] COP15 ended in an unsettled anti-climax. Paradoxically, a temperature target was proclaimed, just as the very premises on which negotiations had proceeded until that point went up in the air.[39]

But COP15 was more paradoxical still, for it was at this summit that voices were first raised for a sharpening of the target: to 1.5°C. The vanguards of this demand were the small island states. Allied in a negotiating bloc, they considered 2°C tantamount to collective drowning, as warming of such magnitude would raise sea levels above the lands on which their peoples lived. From Villach/Bellagio to Copenhagen, sea level rise was perhaps the facet of global warming that most agitated the South. In an op-ed from 1988, 'For a World Campaign to Limit Climate Change', Mostafa Tolba anticipated that a temperature rise of more than 1.5°C would displace some 10 million people from his native Nile Delta.[40] At the very first COP in Berlin in 1995, one Bangladeshi representative warned the North that 'if climatic change makes our country uninhabitable, we will march with our wet feet into your living rooms.'[41] By the time of Copenhagen, the small island states – the Maldives, the Marshall Islands, Dominica, Jamaica, to mention only four – had concluded that the difference between 1.5°C and 2°C was that between survival and annihilation, and with their mightiest efforts, teaming up with the bloc known as 'the least developed countries', they managed to squeeze in a reference to the former number in the final text. While 2°C was the target of the day, it should, the document stated, be reviewed and reconsidered in relation to 1.5°C.[42] In this respect too, Copenhagen was a transitional town.

During this period, scientific debates on targets were all over the place.[43] In the first article in a top journal to take on the question, published in 1991, two scholars coming out of the Villach and Bellagio workshops contended that warming should be held 'below 1°C' (a limit crossed circa 2015) while also defining 2°C as 'the "red" zone'. That zone, they predicted, would be entered in 2025, in the absence of radical emissions cuts feasible in principle but 'extremely difficult to achieve'.[44] Most research in the 1990s chose 3°C or more as the future to study.[45] Around Copenhagen, scientists gravitated towards 2°C, but some thought the goal too lenient while others claimed it was slipping out of reach: half a year before COP15, the *Guardian* polled some two hundred climate scientists and found that a tiny minority thought warming would be limited to 2°C by 2100. All but eighteen put their bets somewhere between 3°C and 6°C. Most recognised that staying below 2°C was possible – it was just not where the world was heading, and so a 'continued focus on an unrealistic' goal would 'undermine essential efforts' to prepare for what was coming.[46] Others charged that 2°C was 'unattainable', its utility that of a fig leaf for business as usual.[47]

Two-degree defeatism, then, was rife. The *Economist* wrote an obituary for 2°C, copied almost verbatim in its later obituary for 1.5°C: given the persistence of emissions growth and the speed of the decarbonisation needed, the former goal had become a 'wishful dream'. Already in 2010, this magazine announced that 'the fight to limit global warming to easily tolerated levels is thus over.'[48] Clearly, this repeated throwing in of the towel had a performative function: the more often it was repeated, the truer the impossibility of doing things differently became. It is an act that has accompanied business as usual at every station through which it has rushed on the way into the abyss, likely not for the final time; it is not contingent on any particular limit or year or deadline, but repeatable *ad infinitum*. But it is worthy of note that two-degree defeatism, by the time of COP15 – which could easily reappear by the time of, say, COP35 or 45 – was followed by a reversal of the parameters of the debate. As emissions merely continued to rise after Copenhagen, some expected that 2°C would be formally buried and replaced with a more permissive target.[49] Instead, the exact opposite happened.

1.5°C to Stay Alive

After Copenhagen, climate negotiations deepened their schizoid trends: towards a sharper temperature target on the one hand; towards ever softer instruments for treating emissions on the other. At COP16 in Cancun in 2010, the reference to a coming consideration of 1.5°C was reiterated, while most of the global South succumbed to overwhelming pressure from the US to give up on legally binding commitments.[50] With a mixture of carrots and sticks – including holding back aid – American diplomats used the year between Copenhagen and Cancun to break most of the resistance against the ultra-liberal turn.[51] The path was cleared for the next milestone in the negotiations, the city synonymous with the second era of climate politics: Paris.

But the South had not lost all punching power. On the road to Paris, the small island states assembled a front of more than 100 countries to demand 1.5°C as the absolute maximum.[52] The climate movement, including the more established NGOs that frequented the corridors of the COPs, swung behind it. Whereas 2°C was a target for the rich, '1.5°C to stay alive' was the cry of the poor.[53] The North would not have it. Just before COP21 would begin in late November 2015, President François Hollande summoned diplomats from the Pacific islands to the Élysée Palace to inform them that 2°C it would have to be.[54] The US resisted even that, calling for any mention of a target to be kept out of the central treaty text.[55] When it became clear, however, that more than 100 governments present inside the negotiation halls backed the equation 1.5°C = survival, in an alliance with NGOs shuttling between these halls and the streets, as well as the movements that marched through Paris, the northern front caved in. A rejection of the demand risked coming across as cruel.[56] Even more importantly, as we shall see, the North had just acquired a special asset by which it could cushion this blow from the South. In a famous article 2 to match that of the UNFCCC, the final text offered a two-part response to the long-standing question: the world had agreed to 'holding the increase in the global average temperature to well below 2°C above pre-industrial levels and pursuing efforts to limit the temperature increase to 1.5°C'.[57] The second part resounded louder than the first. The message heard from Paris was 1.5°C, an ambition that governments from the global South and their movement allies had pushed into the text against the desires of the North – a genuine victory, and one that would shape climate politics for years to come, if not exactly in the way intended.

The small island states still sought a legal status for emissions cuts.[58] But on this most decisive issue, the southern front had already splintered. The US made it perfectly clear that it would not countenance any mandatory aspect of anything at all coming out of Paris. And no one could stand up to Obama: 'commitments' became 'contributions'; more precisely, 'nationally determined contributions' – it would now be up to each nation to decide what it wished to do.[59] A state could say that it meant to cut emissions by a favoured amount until a date of its own choosing, or just reduce the emissions intensity of GDP, or undertake some other mitigation effort to its taste – basically, throw any kind of hat into the ring, without the risk of being held to account for a failure to honour the pledges. The one thing signatories to the Paris Agreement would be obliged to do was to *say something* about some kind of aspirations.[60] Moreover, they would be called back to say something new and (preferably) more aspirational every five years.[61] Beyond that, no regulations or imperatives applied. In the sort of petulant picking on words that characterised much of the negotiations at COPs, when it came to emissions reductions, the US objected to the word 'shall' and had it replaced with 'should', a verb with a more elastic ring to it; 'fulfil' was likewise deleted as too overbearing, and instead the parties were asked to have 'the aim of achieving the objectives' of their contributions.[62] Here was 'a laissez-faire accord' of the highest rank.[63]

Advocates of the Paris Agreement – and there was no shortage of those in the late 2010s – claimed that this voluntary system would be the most effective in getting countries to buckle down. On the same view, more taxes would be collected from the rich if they were free to choose whether to pay up or not and by how much; if obliged only to offer *something*, and then left to conduct any actual transfer of money at their own discretion, the tax havens would run empty. In reality, of course, chances that countries live up to their words increase if an agreement imposes some penalties on those that do not.[64] And justice presupposes limits. In the Paris Agreement, there was no cap on emissions, no aggregate sum to be shrunk, and so there could be no fair distribution of the necessary cuts; without any common dates, quotas or other parameters, what remained was a free-for-all. If there was equality in Paris, it came in the form of a 'shared unaccountability: the agreement required that no one was required to act at any certain level'.[65] In one fell swoop, Paris chased away the thoughts of Contraction and Convergence or Greenhouse Development Rights; they have not been heard of since.

The South, then, won the battle over targets but lost the war over commitments. For the North, the former concession was not all that painful in light of the latter triumph. In substance, Paris favoured, as one scholar noticed, 'developed countries of the North, who won most of the key battles', the agreement being 'least fair to the African Group and other Least Developed Countries'.[66] But it was the mention of 1.5 °C that provided it with a saving grace, the reason Paris received so much love even from progressive forces. The agreement had the nature of an unstable compromise. On the one hand, the limit for warming was tightened to 1.5°C; on the other, the very concept of limits was hollowed out, towards a point where limits would no longer be limits at all. Now what do you get when a seemingly strict target is combined with such lax rules? You get overshoot.

Towards Limits of Human Livability

The impetus for 1.5°C came not from academia, but from countries on the frontlines of sea level rise and their confederates. The former was rather taken by surprise. As with 2°C in and around Copenhagen, scientists had generally assumed that 1.5°C was way off the charts of political possibility. One observer in Paris commented that the agreement was seen as 'so ambitious that many climate analysts are rolling their eyes'.[67] But it needed their signature. Overriding opposition from the US – here partnering with Iran, climate negotiations ever a parade of naked interests – COP21 tasked the IPCC with compiling a special report on the impacts of 1.5°C and how they would differ from higher temperatures.[68] History folded back on itself. Here was, for all practical purposes, a southern front dictating to the scientific body that had succeeded Tolba's Group what research to commission and collate. Before Paris, studies on 1.5°C had been exceedingly scant, because interest in it was strongest in the South, where academic infrastructure was weak, and institutions in the North would not bother investigating such an impracticable proposition.[69] The task scandalised some scientists: should we do the bidding of obviously self-interested actors like small island states and 'a new generation of civic activists'?[70] Overall, however, the research community generously responded with an explosion of studies on 1.5°C.[71] Thanks to the (partial) southern offensive in Paris, the world learned what warming beyond that limit might entail.

The Special Report on 1.5°C duly appeared in 2018. The question it set out to answer was essentially the following:

what difference does it make if warming stops at one and a half degrees, compared to two (or higher)? Is 1.5°C something more than a number on paper? Does it correspond to a real break in natural systems, for which it is a convenient shorthand? The IPCC answered in the affirmative, roughly in line with '1.5°C to stay alive'. In its examination of the 'avoided impacts' if warming were to be capped at that level rather than 2°C, the Panel found markedly lower risks: destructive downpours, protracted droughts, crop failures, water scarcity – all would be less frequent and devastating.[72] Some of this writing had a platitudinous quality to it. If you plunge a knife into someone's abdomen, it is less bad to stop at five centimetres than at ten; similarly, global warming will be less deadly at earlier stages of progression. This goes for all its aspects, with one significant exception: if the warming is mild and slow, bats, rodents, monkeys, and other mammals will have time to move and keep track of their preferred climate, migrating in droves, crossing paths, exchanging the parasites and pathogens that ride on them. They will also brush by human settlements and shed some of their viruses, massively increasing the risk for zoonotic spillover of the kind that caused Covid-19. But if warming is brutal and swift, these mammal populations will simply die off.[73] On this one count – the exposure of humans to viruses from animals on the move – things would be worse at a plateau of 1.5°C than in an ascent to 2°C, which is to say, conversely, that the latter would be all the worse for biodiversity: doubling the extinction rate for plants, insects and vertebrates, as their habitats would be swept away for good.[74] The Special Report identified a particularly stark before and after for coral reefs. In a 2°C world, heat stress would kill 99 per cent; staying at 1.5°C would save at least one and possibly three tenths of the corals.[75] Some would say that should be reason enough.

The IPCC judged it unlikely that the Arctic Ocean would be denuded of ice at 1.5°C, but likely at 2°C.[76] Subsequent research showed that this part of the world was heating up nearly four times faster than the rest – not twice as fast, as previously believed – lowering the odds of the former event, further underlining that an early brake would be the only chance to save some essential features of Arctic ecosystems.[77] What of sea level rise, the original spur to the target swap? The IPCC and others estimated that 1.5°C might possibly mean half as much of it as 2°C, the rise perhaps peaking at one and a half metres instead of three, in the best case allowing small island states to survive; but much of this forecasting described the sliding, gliding movement of a knife pushed deeper, and much of it rested on underestimations.[78]

With the Special Report, the discourse around 1.5°C rather moved away from sea level rise. Another danger rose to the fore: lethal heatwaves. Nothing so clearly separated 1.5°C from 2°C as the avoided impacts of extreme heat. The IPCC cited a fresh study – one of many prompted by the call from Paris – calculating that nearly 2 billion people could be saved from the fate of severe heatwaves rolling over them every fifth year.[79] At 2°C, cities like Karachi and Kolkata would cook every summer. As a line between life and death in the heat, this average half of a degree ran deepest through the tropics.[80] But not only there: appearing on the back of a searing heatwave in the Northern hemisphere – the summer of 2018 memorable for wildfires in places like Sweden – the Special Report heightened general awareness of this prototypical danger.[81] It was real already before 1.5°C.

As for one category of particularly lethal heatwaves, the science, it should be noted, is a product of the second decade of the millennium. Before that point, there was little reason to worry. But in 2010, there appeared a paper outlining the contours of a furnace under construction. A human body has a core temperature of 37°C, and if the environs are hot – in the 20s or 30s, depending on tolerance – it starts to sweat, the normal way for inner heat to exit the frame. The heat flows outwards only because the skin is cooler than the core, at 35°C. But if the surroundings become so hot and humid as to match the skin, the sweating will bring about no cooling and no relief, for the second law of thermodynamics prohibits an object from losing heat to an environment as hot as itself. Evaporation no longer works. At this point, the human body cannot eject heat: it accumulates inside, until heat stroke sets in and the person in question expires. The metric for pinpointing this limit to survivability is 'wet-bulb temperature', a compound of heat and humidity, measured with a thermometer covered in water-soaked cloth. If the wet-bulb temperature reaches 35°C, it takes at most six hours before a human body – however healthy and fit, however deep the shade – is finished. The 2010 paper noted that no such levels had yet been observed on Earth: the wet-bulb temperature 'never exceeds 31°C', let alone 35. Only 30 had been detected. But the authors warned that heatwaves climbing to the lethal heights would 'begin to occur with global mean warming of 7°C' – seven degrees, seemingly at a safe distance.[82]

Climate science has a way of revising itself to adjust to the onrush of impacts, coming faster, harder, earlier than expected. In 2016, one paper reported that wet-bulb temperatures exceeding 31°C had in fact been observed in the Persian Gulf, though not yet 35; the latter, this modelling exercise suggested, would be common in places like southern Iraq, southern Iran and northern Oman towards the end of the century.[83] The region was considered so exposed to the ordeal because it combines a fierce sun with the water of the Gulfs. Come 2020 and scientists reported that, for several moments in the summer of 2017, 35°C had in fact been measured at a couple of weather stations in this very region; for a few hours, not days or weeks, and yet above the critical threshold.[84]

Soon after, one study submitted a more comprehensive prognosis. Above an average warming of 1.5°C, waves of unsurvivable mugginess would begin to roll through the tropics – not above seven degrees, but one and a half. 'These results suggest that limiting global warming to 1.5°C will prevent most of the tropics from reaching a TW [wet-bulb temperature] of 35°C, the limit of human adaptation.'[85] The *Guardian*, the indisputable leader of climate journalism, duly reported this piece of news. The headline said: 'Global Heating Pushes Tropical Regions towards Limits of Human Livability'.[86] But not even the editors of the *Guardian* could resist the compulsion to post this report below the latest headlines from the Meghan and Harry drama. Tropical regions, where are they anyway? In a band circling the globe, between a northern line running through Mexico, Libya, India and a southern through Brazil, Madagascar, Australia: home to about four tenths of humanity. This zone, pushed towards the limit of human livability. In a sane society, the news would be 'plastered on every lamppost and stop sign in America, no less the world'.[87]

That did not happen, but climate science humbly continued its reappraisals and produced a still more extensive study reporting blips above 35°C from six stations in the Persian Gulf plus the Indus valley and presenting new simulations that buttressed 1.5°C as a life-and-death guardrail. At this global average, the killer heatwaves would begin to last for three or four days in said places. At 2°C, twice as many humans would be subjected to such waves. At 3°C, in the year 2100, seventy-five times more humans would contend with them than if warming is kept to 1.5°C, and so on: a scourge coming first for people living between the Euphrates and Indus, then west to the Amazon and east to the Irrawaddy and Mekong.[88] But then again, south Asia might suffer the lethal spells already at 1.5°C; 2°C would just double the hazard.[89] It would come first for certain classes. Agricultural workers in fields, construction workers on scaffoldings, manufacturing workers in poorly ventilated factories – these would be the least shielded, together with the hundreds of millions living in informal settlements.[90] Weighing the feasibility or desirability of 1.5°C in the early 2020s was an exercise concerning their lives.

The Special Report, then, was not the conclusion of research on 1.5°C, but rather a belated starting signal; the closer to the line the world treaded, the more knowledge of its significance was amassed. It extended to tipping points once thought remote. At a certain threshold, a system might undergo qualitative change, abruptly shifting into another regime or mode in a self-perpetuating loop that can no longer be stopped. In 2022, *Science* published a paper by a team led by researchers from the University of Exeter – call it 'the Exeter paper' – updating the forecasts for tipping points and concluding that at least four would be activated above 1.5°C: the coral reefs extirpated; the Greenland Ice Sheet doomed to complete meltdown, generating seven metres of sea level rise over a millennium or more; the West Antarctic Ice Sheet likewise condemned, yielding three and a half; thawing of the boreal permafrost, currently holding some 3,000 gigatonnes of CO_2 in its frozen stores. For these systems, it would then be too late. Few if any small island states would survive. All glaciers would go around 2°C. Other tipping points still seemed several additional degrees away – the East Antarctic Ice Sheet holding out longest, to 7.5°C – but this was more than enough to corroborate the hard character of the 1.5°C target.[91] We shall return to the problem of non-linear climate breakdown in further detail in the second instalment of this study.

The southerners and their crowds in Paris got it right. In the light of all the research conducted at their prompting, the subalterns of the global South, led by the small island states, came to resemble 'a sphere which has a universal character because of its universal suffering and which lays claim to no *particular right* because the wrong it suffers is not a *particular wrong* but *wrong in general*', a sphere 'which is, in a word, the *total loss* of humanity and which can therefore redeem itself only through the *total redemption of humanity*'.[92] The Special Report put a seal of scientific legitimacy on their struggle for survival. It became the most influential publication by the IPCC to date. In the continuing dialectic between science and activism, this success was constituted by the wave of climate mobilisations surging forth in 2018: when Greta Thunberg reached a point of desperation after the heatwave of that summer, sat down outside the Swedish parliament for her school strike and set off Fridays for Future, she constantly cited the Special Report and demanded that politicians 'listen to the science'.[93] The same report inspired the Sunrise Movement in the US to embrace the Green New Deal.[94] XR chimed in too. Between the autumn of 2018 and that of 2019, science and activism – perhaps more deeply interlaced than ever before – induced the most intense worldwide media coverage of the climate issue in a decade; in the US, it reached an all-time high.[95] Few public actors could have missed the gist of the matter. But what then happened on the ground was, as we have seen, another story entirely.

The Revolutionary Imperative

None of this registered in the only place that mattered. The graph tracking CO_2 concentrations in the atmosphere continued its inexorable upwards march. From the beginning, 1.5°C had been haunted by questions of feasibility, the push in Paris rebuffing the sceptics and expanding the realm of the possible, at least for a brief moment. Warming above 1.5°C was 'not geophysically unavoidable: whether it will occur depends on future rates of emissions reductions', the Special Report avowed.[96] But after three years of the third decade, the sheen of such happy possibilism had faded. Those 'future rates of emissions reductions' had been altogether null. The disputes over 1.5°C resumed, this time with a darker resonance, based on some rather obvious conditions. Climate politics had become revolutionary politics. More precisely, from now on – if not earlier – any attempt at meaningful mitigation of the crisis would have to waylay the dominant classes with a force and confrontational resolve unlike anything in the common memory or imagination.

This was not a communist delirium. It was a mathematical certainty of rare absoluteness. Like 1 from 100–99, it resulted from the cumulative character of the problem: the more CO_2 dumped in the atmosphere, the warmer the Earth will be, in virtually exact proportion; if a lot has already been dumped, not much more can be so before some crucial line is crossed; but if more CO_2 is still at that point being dumped, and if the dumping even increases in volume, then the only option is to violently slam the brakes.[97] Else the line will be crossed. The below graph neatly captured the logic.

We can then adapt a phrase from Ernest Mandel and say that *revolution is born out of the historical delay of mitigation* – not necessarily as actuality, but as logical necessity.[99] Revolution in what sense? We shall explore the substance of it in part II of this book; for the moment, we shall retain the term 'revolution' in its fuzzy, intuitive meaning and note that climate scientists with no left-sectarian credentials were being compelled to use it. Among them was Kevin Anderson, the much-respected professor in

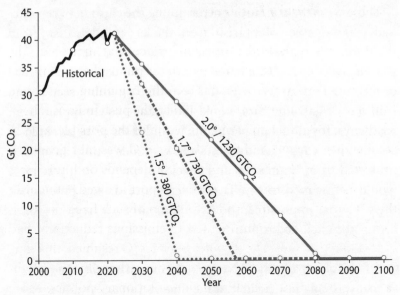

Figure 2 Global CO_2 pathways using IPCC AR6 remaining carbon budgets
Source: Pierre Friedlingstein, 'Global Carbon Budget: Presentation',
Global Carbon Project, 11 November 2022.[98]

Manchester and Uppsala, here writing in *Nature*: 'in the 1990s, technocratic approaches could have reduced emissions', but the opportunity was squandered, and 'climate change is a problem of cumulative emissions. Ongoing failure to mitigate emissions has pushed the challenge from a moderate change in the economic system to a revolutionary overhaul of the system.'[100] Or, in the words of Inger Andersen, who in 2022 held Tolba's old post as chair of the UNEP: 'we had our chance to make incremental changes, but that time is over. Only a root-and-branch transformation of our economies and societies can save us from accelerating climate disaster.'[101] Note that this was, again, formulated at a high level of abstraction. There was no Petrograd Revolutionary Military Committee awaiting the signal to strike: this was revolution as theorem. The very same UNEP estimated that policies in place would bring the world to 2.8°C before the end of the century.[102] It knew that there was 'no credible pathway to 1.5°C in place'.[103] Its adoption of quasi-revolutionary rhetoric was born not of some millenarian euphoria or successful organising of the garrisons, but of disillusion and simple arithmetic.

These were the conditions determining the clash between one-and-a-half-degree voluntarism from the left and defeatism from the right. When *Scientific American* published a piece about the passing away of 1.5°C, it based its case on the 'colossal' challenge of meeting that target: 'it would be unlike anything seen so far. Millions of gasoline cars would likely have to disappear from roadways, fossil fuel power plants would close or be adapted to confine their carbon, and forests and wetlands would have to be protected from chain saws and development.'[104] Well, yes, some would be inclined to say. Twelve years after it had proclaimed the 2°C goal impossible, the *Economist* argued from the same law of the excluded rupture:

> An emissions pathway with a 50/50 chance of meeting the 1.5°C goal was only just credible at the time of Paris. Seven intervening years of rising emissions mean such pathways are now firmly in the realm of the incredible. The collapse of civilisation might bring it about; so might a comet strike or some other highly unlikely and horrific natural perturbation. Emissions-reduction policies will not, however bravely intended.[105]

That statement won the 2022 award for best rendering of the maxim 'it's easier to imagine the end of the world than the end of capitalism'.

Now much of this was not exactly news. In the Special Report, the IPCC spoke straight into the ears of 'policymakers': meeting the 1.5°C target would necessitate 'systems transitions' of an 'unprecedented' scale.[106] They could come about only through 'more planning and stronger institutions' than observed in the past.[107] The Panel called for 'effective planning' in cities, 'coordinated planning' of agricultural sectors and, cognizant of trends, 'a marked shift in investment patterns' – sotto voce, a transitional programme of sorts.[108] The proto-revolutionary implications of the 1.5°C target were evident from the beginning: the obverse of its presumed infeasibility. They merely came into sharper view at 1.1°C and 1.2°C.

Seeing no other options left, parts of the research community then went further down the road of questioning the prevailing order. In its sixth assessment report from 2022, the IPCC included, for the first time, buried in its many thousands of pages of text, references to eco-Marxist thinkers and a nod to the validity of their insights about 'the centrality of fossil fuels' to processes of 'economic growth and capital accumulation'.[109] In the same year, *Nature Climate Change*, the top journal of climate science, published a wide-ranging essay by a team of researchers ascertaining that 'conventional interventions are simply not fast nor deep enough to slow climate change' – 'instead radical interventions are required', 'radical' as in attacking the 'root drivers of a problem rather than its proximate causes and symptomatic effects'. What root drivers? 'Capitalism', first of all, followed by 'asymmetrical power relations', 'colonialism', 'global inequality', 'exploitation'. With no discernible link to any revolutionary mass organisations, this team claimed to know that 'there is now serious appetite for deep radical intervention, as we define it here, and substantial likelihood of broad-scale acceptance.'[110] Not music to the ears of the *Economist* and its ilk.

Here, then, was the argument of one-and-a-half-degree voluntarism from the left: abandoning 1.5°C would really be about escaping revolution. More concretely, the strongest case for that position was the fact that every single fossil fuel installation in progress in the 2020s, as in any other decade, was politically constituted. Either states owned them directly or they acted as their handmaidens.[111] Every pipeline, every terminal, every mine, every plant could see the light of day only if one state apparatus or other issued a whole string of permits. Every piece of infrastructure already in operation could have its permits revoked. It could be shuttered. Many of the carbon bombs listed in the early 2020s were not yet ignited; had they been fixed in position, they could still be defused.[112] In this sense, capitalist society was in excellent control of itself: it knew full well what it was doing, every investment decision passing through the offices of 'policymakers' for a mindful seal of approval. And this also meant, as per logical truism, that fresh installations *could be blocked* – if not by states, then by non-state actors with a determination to struggle and win. Such installations did not in fact arrive on Earth the way stars move in orbit or waves lap with the wind. There was nothing immutably natural about them.

One of the most noteworthy papers in a leading journal of climate science from those years, and perhaps the single most inspiring, analysed a database of nearly 400 cases of 'place-based resistance' against fossil fuel projects from all corners of the world: people marching, petitioning, litigating, divesting, sitting in, blockading and engaging in other direct actions to stop them from going through. In 15 per cent of the cases, pipelines meeting such resistance were cancelled, suspended or had their investments withdrawn. For fracking projects, the share was 26 per cent; for other types of oil and gas, 18; for fossil fuels on average, coal included, it stood at 25.[113] Now one in four is not a success rate good enough for resistance. But it was sufficient proof that projects could be thwarted, and that even if the enemy had *mostly* not ceased to be victorious, it was by no means invincible.

A handful of cases could be mentioned in brief. A groundswell of place-based resistance across the Philippines stood in the way of new coal plants, to the extent that the government had to announce an actual moratorium in late 2020; two years later, the Philippine Movement for Climate Justice claimed responsibility for having stalled eight gas projects in five months.[114] In 2015, there were fifteen fossil fuel projects afoot in Portugal: five years later, Climáximo, an uncommonly effective chapter of the European climate movement, celebrated the defeat of each.[115] Persistent campaigning from communities and climate groups in Uganda, teaming up with 350.org and other international networks, had by late 2022 convinced several major banks and insurers to drop their funding of EACOP. There were three banks to go.[116] After prospective clients faced an outcry, the Democratic Republic of Congo postponed the auctioning of oil blocks in its peatlands.[117] The trans-Saharan gas pipeline had long been held up by various rebellions, with a penchant for pipeline sabotage, along the prospective route from the Delta to the Maghreb.[118] In Mozambique, Exxon and Total turned their gas projects off and on as an armed insurgency swept through the fields, fuelled by anger over these companies displacing and leaving them destitute.[119] An in-depth discussion of the prospects of such militant action is outside the scope of this book. Suffice it to say that investments flagrantly incompatible with 1.5°C were not invulnerable: through a diversity of tactics, some had been thrown off the rails. Anyone with a serious interest in limiting global warming would then, presumably, seek to join this resistance and ratchet up its success rate. The feasibility of 1.5°C was essentially a question of subject formation: whether or not an agent for ramming through the necessary changes could be formed.[120] One-and-a-half-degree defeatism intervened on the other side to broadcast the futility of trying. If this was a prophecy, it would be of the self-fulfilling kind.

On the other hand, however, there was also a real case for defeatism from the left. The scalar mismatch between any actually existing subject of resistance and the task at hand remained dizzying.[121] No revolutionary subject was ever built in a day. We have seen just how perilously close the world was to 1.5°C a couple of years into the decade. In a tragic irony, two events in autonomous nature conspired to beef up the effects of fossil fuel combustion: the underwater Tonga volcano lofting enormous quantities of heat-trapping water vapour into the stratosphere in January 2022 and the expected onset of an El Niño in 2023, possibly dispatching warm water through the Pacific Ocean for years to come. Both led scientists in 2022 to raise the probability of the breach even further.[122] It would then be irresponsible, bordering on escapism, to maintain that 1.5°C was as possible as ever, if only everyone threw themselves into the resistance and upgraded it into a world revolution.

Does that mean that it was too late already in, say, 2022? It would not have been the first time in history for the question to arise:

> But first we must answer here another question, what does 'too late' mean in the given instance? Must this be understood to mean that even the boldest about-face on the road of revolutionary

policy is no longer capable of radically changing the relationship of forces? Or does it mean there is neither the possibility nor the hope of achieving the necessary turn? These are two different questions.[123]

That was Leon Trotsky, writing about the rise of Nazism on 5 February 1933, six days after Hitler was appointed chancellor – a moment when, in some basic sense, the Rubicon had been crossed. Yet even at this point, he rang the tocsin and called on the workers of Germany to rise up. 'It is perfectly clear that in its tactical estimates the proletariat must proceed in terms of *very little time.*'[124] Such sentences were written in a frantic effort to cobble together an anti-fascist subject out of the splintered and paralysed forces of the working class, a united front capable of beating back the catastrophe: a pursuit to which Trotsky had devoted himself for years, to no avail. What, if anything, might be gleaned from this? We shall have reason to return to the question of what makes up meaningful politics when catastrophe is already a fact; for now, we shall only underline some distinctive aspects of the temporality of this one.

As is true of every possible analogue, the rise of Nazism does not exactly conform to global warming. One difference – one of many – is that the latter has no unique 30 January 1933. There is not *one* date in the calendar at which the warming seizes power. It cannot be excluded that some climatic blast rips through the Earth and comes to be regarded as a caesura in time; but even if that were to happen, there would have been a sequence of fractal iterations leading up to it (and more beyond). Nor is there one single average rise in temperature that distinguishes harmless from harmful warming. If there were, it would not be 1.5°C: just ask the peasants and workers of Pakistan whose lives were shattered by the floods.[125] 'The Earth may have left a safe climate state beyond 1°C global warming,' the Exeter paper suspected, recognising that the tipping point for the West Antarctic Ice Sheet might have already been triggered, confirming the lower proposal of the Tolba Group.[126] At 1°C, it was clearly too late to undo some damages already done. And if by 'too late' we mean too late to prevent a process from kicking in, then it was too late from day one of warming. Wrote Tolba in 1988: 'climate change cannot be prevented altogether, but it may be possible to limit its extent, delay its onset and reduce the effects' – a statement that could be recycled four decades later, at a higher pitch.[127] Moreover, if capitalism is the root driver of this problem and its ultimate cause, like the quake in the seabed that causes the tsunami, then it might have been too late ever since 1918–23, when the historical chance to deal this mode of production a mortal blow in its European heartlands was missed; or perhaps since the centuries following the Black Death, when the English peasantry failed to prevent it from striking roots in the land; or even, if we follow the regress to its end, since the fourth millennium before Christ, when the hunter-gatherers of Egypt were herded into the first centralised class society, from which all else followed.[128]

And yet if by 'too late' we mean an absolute, objective, technical-physical impossibility of preventing things from getting even worse, it might never be too late with climate, at least as long as some substantial human population and scrap of the biosphere exist. After 1.5°C, there will be a struggle over 1.7°C, the limit 'well below 2°C' next in line (the one Scientist Rebellion seemed to aim for, after its espousal of one-and-a-half-degree defeatism). There would then be one-point-seven-degree defeatism and voluntarism, but with higher stakes, even larger impacts to avoid, business as usual having proved itself obdurate in extremis. And then there would be 2°C . . . At every point, the revolution will be reborn out of the historical delay of mitigation – not automatically as actuality, but automatically as ever more inescapable necessity; and if the latter does not translate into the former, if the necessary never becomes the actual, then the equation will push the course of events in other directions, into channels formed by the revolution that did not happen.

At every point, the conditions and debates of the early 2020s would then be restaged. We have seen how the dichotomy between defeatism and voluntarism was prefigured already in the discussions about 2°C in the late Kyoto era, and if the lower boundaries are overstepped in the next few decades, because fossil capital is so formidably powerful, they will reignite with a vengeance. Climate politics has a way of moving in circles, or perhaps rather in a spiral, old themes, initiatives, 'solutions', contradictions re-emerging in the higher coils. Defeatism and voluntarism seem poised to be recurring, almost timeless positions.[129] At each stage, they would either affirm or deny the possibility of a revolutionary rupture. Those who repeat before every limit that it is too late to avoid its crossing would point the way into Hades. But voluntarism, on the other hand, must be chastened with the realisation that *this* particular limit might in fact be crossed, the cause for now lost, only to reappear in a moment. In the staggering warming condition of relative too-lateness – always-already too late, never too late; in the years ahead, another deadline looming – the historical task would remain exactly the same if 1.5°C were to be passed. It would only become even more pressing.

But the terrain of climate politics does undergo a qualitative shift on the cusp of 1.5°C: that limit is – was? – no trifle. (No climatic or other planetary boundary is.) Developments in the Paris era indicated that a reordering of the fronts was underway. It centred on the phenomenon of overshoot.

3

The Rise of
Overshoot Ideology

To locate the genesis of overshoot, we must, again, return to
the Kyoto era and, more particularly, to its computer models. It
was late in the first decade of the millennium. The Protocol had
come into force, but the US had withdrawn from it, considering
the rules too stringent and inimical to the American economy,
leaving the EU as the main northern anchor of climate governance. At this point, the EU was preparing for what would come
after the first phase of Kyoto. It was, as we have seen, calling for
2°C as the appropriate target. But it faced pushback, and critics
pounced on the supposedly arbitrary and even authoritarian
nature of 2°C; and, admittedly, the science backing it up was
rather thin.[1] Only a handful of studies had tried to chart a path
to this limit.[2] What would be required to get there? It was to this
question the EU needed an answer, and so it turned to scientists
and asked them to switch on the favoured crystal ball of late
capitalism: the computer.[3]

The tool considered best suited for the task would become
known as 'integrated assessment models', or IAMs. (A word on
acronyms is in order. Nowhere was official climate policy and
science as productive as in the department of acronyms. The
lack of tangible results in dealing with the actual problem stood
in proportion to the fabulously prolific invention of ever more
acronyms for talking about it: a very short shortlist would include
AGCM, CBDR-RC, CCUS, CDM, CDR, ETS, GCF, GHGs, GWP,

LULUCF, NDCs, NETs, REDD+, SAI, SCC, SLR, SRES . . .[4]
This correlation between poverty of substance and richness of
alphabetic characters in combination was arguably more than
coincidental. A laywomen or layman who sought to approach a
session of climate negotiations or a document of climate science
would slam into a wall of language ugly, soporific, esoteric, as
if a notice had been hung on the threshold to this abode saying
'no admittance except on business' – the acronymic logorrhoea,
in other words, had an ideological function. It sealed off governance behind a semblance of objectivity. It raised the bar for
popular participation. It represented the internment of the climate
question in a factory of technicalities, which allowed the outside
world of actually existing capitalism to further speed up planetary
destruction – so much linguistic litter churned out onto the beach,
seemingly destined to be swept away by the rising seas.[5] Here
we shall use acronyms as sparingly as possible. Only when they
are fairly well-established, or when the alternative is even duller
and clunkier, shall we opt for them. On account of the latter, we
shall speak of 'IAMs' rather than 'integrated assessment models'.)

Such models existed in computer labs. They should not be
confused with climate models. In these, scientists could simulate
how a rise in CO_2 or methane or some other greenhouse gas,
would drive up temperatures or the acidity of oceans or rate of
glacier melt – all in the domain of nature, society left out of the
picture.[6] The mechanisms here represented on a screen derived
from laws of physics not of human making. A glacier would melt
above 0°C independently of what people felt about it. IAMs,
however, were *integrated* models, in the basic sense of spanning
the divide between nature and society: a series of equations
representing the Earth coupled with another standing in for the
economy, the two domains unified in one digital visualisation.[7]
But how could an economy of distinctly human making be captured in the same model as something like glaciers? In the IAMs,
the trick was to render the economy lawlike on the assumptions
of neoclassical theory: individuals are rational agents. They
choose the cheapest options and maximise the sum of welfare.
Perfect information is fed into their brains. Their foresight is

equally perfect; they do not hesitate or make mistakes or invest in suboptimal assets.[8] 'The models', in the words of the IPCC, 'typically assume fully functioning markets and competitive market behavior, meaning that factors such as non-market transactions, information asymmetries, and market power influencing decisions are not effectively represented' – an economy purged of human impurities.[9] It could be easily pointed out that this was a distortion, for not even northern consumers respond to price signals the way an icicle reacts to heat. Had that been the case, there would, for example, have been only small and cheap cars on their roads.[10] SUVs would not have existed.

But concessions to such realities had no room in IAMs, which persevered on the postulate that the economy is inhabited by one 'representative agent'. No flesh or blood, this is a portrait of the utility-maximising subject, whose discernment is unclouded by any particular allegiances, interests, outside influences or other idiosyncrasies: pure homogenous reason. No classes tear it into halves. No relations of race or gender or core and periphery mark it with tensions or fissures.[11] With the table thus cleared, the IAMs could process the question: what is the optimal policy for dealing with climate change? This will be a matter of lowest cost. Choices about mitigation will be made on the criterion of price, for so are all choices. Or, by the way, is that true? William Nordhaus – maven of bourgeois climate economics, the only economist winning the Nobel prize for work on the issue; also on his CV, father of IAMs – briefly turned to this question in a paper from 1991, tellingly titled 'To Slow or Not to Slow: The Economics of the Greenhouse Effect'. He there acknowledged that a 'variety of non-marketed goods and services escape the net' of cost accounting. 'Among the areas of importance are human health, biological diversity, amenity values of everyday life and leisure, and environmental quality. I am aware of no studies that point to major costs', and since no quantifiable costs were attached to these areas, they could be omitted from the equations without further ado.[12]

Only some things can be counted, so count only them.[13] What Nordhaus claimed to be able to measure was the monetary cost that emissions reductions would impose on the economy. He could then weigh this against a 'greenhouse damage function', meaning the costs of climate change itself – the price tag for lower 'crop yields, land lost to ocean, and so forth'.[14] Out would come the most efficient optimum, the least cost for avoiding any costly damage. In this early trend-setting paper, Nordhaus concluded that it would be imprudent to try to slow down global warming to any measurable extent, for only 'a modest reduction of green-house gases can be obtained at low cost'.[15] He would later derive more precise measurements by speaking to his own IAM, called the Dynamic Integrated Climate-Economy Model, charmingly abbreviated DICE; in one paper from 2007, he put the optimal cuts to 25 per cent of business as usual by 2050 – not much of a slowdown, mid-century.[16] In many ways, DICE was in a category of its own – older, less sophisticated, originally designed for a different purpose – but the other IAMs were built on the same analytical foundations. They depicted a future of rational choices in a world slowly heating up. A price will be attached to carbon, and as it rises, the signal transmitted across markets, agents will catch on and seek to avoid emissions. This is, however, a burden on economies: it subtracts from GDP. Growth is desirable and necessary, but a steep carbon price weighs it down and causes financial pain. The version of mitigation that carries the day is then axiomatically the cheapest one. IAMs were constructed to select the dose of emissions cuts least painful for agents respond-ing to small, marginal changes on a market in general equilibrium. A policy would come out on top of the algorithms not by dint of being superior in some other respect – say, ecologically or ethically or aesthetically preferable – but exclusively on its cost-minimising merits.[17]

And cost was assumed to fall over time. Because growth will make future generations richer than the present, a price hike of one dollar will mean less to them than to us; their wallets will be bigger than ours, and so they can better afford any given burden of mitigation. Costs close to the present must therefore be weighted higher than in the future. With this practice of 'discounting' embedded in their code, the IAMs were tilted to postpone emissions cuts: most cost-efficient would be to place them in a wealthier hereafter. And because technologies will also continue to get better and better, costs will fall deeper still, the postponement extended.[18] Ruled out were fast, deep, sweeping cuts starting immediately – 'a revolutionary overhaul of the system', in Anderson's parlance, could not be fitted into these models any more than a mass demonstration into a bedroom.[19] Some studies reporting IAM modelling results were forthright on this point. One important paper was put out by a team of Dutch researchers, who chose the poetic acronym IMAGE for their model – the first and most influential IAM on European territory – and described, in characteristically passive voice, how 'a maximum reduction rate was assumed reflecting the technical (and political) inertia that limits emissions reductions. *Fast reduction rates would require the early retirement of existing fossil-fuel-based capital stock*, and this may involve high costs.'[20] The obduracy of business as usual was built into the model as an axiom about how the economy works and mitigation toned down for the express purpose of *protecting fossil capital from losses*. In the next chapter, we shall see how the latter is the crux of the matter.

In the same spirit, the IMAGE model 'spread out' emissions cuts 'over time as far as possible' and made sure they were 'only allowed to change slowly'.[21] You can't just march into the master bedroom and throw the lord of the estate out, so any transformation must be gradual and gentle: shocks of any kind would be unmodellable. Neither revolutions nor cataclysmic effects of warming that slash through the economy could be grasped by IAMs.[22] Since the latter were left out, benefits from avoiding truly large and bad warming were also missed.[23] From the beginning, this practice blended into denial. To minimise the 'greenhouse damage function', Nordhaus employed some highly fanciful arguments: because humans currently thrive in all sorts of climes, from Arabia to Alaska, climatic variables have negligible effects on productivity (as if future warming would not throw every region off-kilter).[24] And because most economic activity takes place indoors – think 'cardiovascular surgery or microprocessor fabrication in "clean rooms"' – most output is insulated from the climate and won't be ruffled by the temperatures; more precisely, 87 per cent of it (as if what goes on indoors happens on another planet, the cardiovascular surgeons living not off food and water but enclosed air).[25] DICE would then tell its creator that climate change in the twenty-first century is unlikely to cause 'substantial net economic damages'.[26] Those 3°C might at most shave off a couple of per cent from GDP. Another prominent bourgeois climate economist and IAM inspirator, Richard Tol, took this line of reasoning to new heights when pressed on the question of whether *ten* degrees of warming might take a toll on the economy. 'We'd move indoors, much like the Saudis have,' was his carefree response.[27] Far from all IAMs threw up results of this kind, of course, but built into them from the very start was a *tenuous relation to reality* which, as we shall see, produced some real, troubling consequences.

The models, in other words, had a way of substituting themselves for the external world. Above all, this applied to feasibility, generally defined as 'a function of model solvability': that which can be housed within the computations can be done.[28] So as not to be mistaken for pessimism, it should be stressed that this stemmed from an upbeat view of how the economy works, always close to an optimum, with no mass unemployment or resource waste (hence policy interventions necessarily being burdensome).[29] Every human artifact has a dimension of ideology. Climate models were not exempted from this rule. But IAMs were positively drenched in non-innocent ideological positions, of which we can quickly list a few: rationalism (human agents behave rationally), economism (mitigation is a matter of cost), presentism (current generations should be spared the onus), conservatism (incumbent capital must be saved from losses), gradualism (any changes will have to be incremental) and optimism (we live in the best of all possible economies).[30] Together, they made ambitious climate goals – the ones later identified as in line with 1.5°C or 2°C – seem all but unimaginable.

IAMs became possible only at a certain level of development of the productive forces. Much like other types of information technology, they came of age in the 1990s and then boomed and blossomed in the first two decades of the millennium, stimulated by novel software, the power of the internet, access to immense databases and exploding computational capacity. The IAM became, so it seemed, an increasingly trustworthy orbuculum. Plainly a northern phenomenon, the most impressive models were kept in a few hubs – apart from IMAGE of the Dutch Environmental Assessment Agency, places like the Institute for Applied Systems Analysis in Vienna, the Potsdam Institute in Berlin and the Pacific Northwest National Laboratory in the state of Washington – together making up the 'IAM community', also likened to a league of football teams.[31] It was from them Brussels sought an answer to the question of the late Kyoto era. If we want a limit on global warming, such as 2°C, how can we best make it happen?

Some Shall Be Freed by Overshoot

Up to this point, models had mostly been run from the present into the future, like digital ships sent out to explore, in more or less open-ended fashion, various scenarios for how climate change might unfold and what routes might be optimal for navigating it.[32] But on the initiative of the IPCC, modelers were now asked to also flip things around. Scenarios should start from a handful of possible, predefined levels of warming in 2100 and then work backwards to map out the processes that might eventually lead up to those destinations.[33] Along the way, they would assess the costs and benefits associated with each of these. Warming levels in 2100 were expressed in terms of radiative forcing – essentially the amount of energy, measured in watts per square meter, added to the atmosphere when a quantum of greenhouse gases is pumped into it. One concentration yields one level of radiative forcing gives one rise in temperature on Earth. Scientists baptised their new scenarios 'representative concentration pathways' (yes, another acronym: RCP) to reflect the idea that they should be 'representative' of the available literature and focus on 'concentration' of gases and, of course, plot a number of 'pathways'.[34] The EU shopped around for something that would lead to 2°C, and it found IMAGE. Only the IMAGE team came up with scenarios that stationed the planet at 2.6 watts per square meter in 2100, roughly corresponding to 2°C. All others ran with concentrations taking Earth to 3°C or more.[35] Within the corridors of the IPCC, the EU now enthusiastically promoted the IMAGE scenarios as just the nautical charts needed for navigating to the goal, which so many others regarded as unreachable.[36] For the IMAGE team had found an ingenious way around the dire straits.

What if one could run global warming in reverse? The terminus was dated to 2100, but that did not preclude flexibility in early decades. If the winds are on your side, you might sail far away from the island you have in sight and still make it back on time. There was no logical reason why temperatures could not briefly *exceed* the limit of 2°C, as long as there was a way to bring them down again by the end of the century. Could such a way be found? The IMAGE team had pondered this question and come upon a solution of a kind. Their model had long included a standard function for sucking CO_2 out of the atmosphere, because that is an integral part of the active carbon cycle. Plants incorporate CO_2. The atmosphere is like an open clay pit, from which all that grows extracts carbon dioxide and, through the process of photosynthesis, fixes it as carbon in green organs. When those organs or bodies die, microbes break them down like so many brittle bricks, feed on the crumbs for their own modest metabolism, hook the loose carbon up with oxygen and shovel it back into the atmosphere, renewing the supplies so that 'consumption is also immediately production' – gas consumed is vegetation produced: vegetation consumed is gas produced – the one moment realised in the other and leaving an imprint in the sky, most beautifully captured in the graph of atmospheric CO_2 concentration measured on the Hawaiian volcano of Mauna Loa.[37] The graph zigzags its way upwards in a symmetrical pattern. It drops a little during spring and summer in the Northern hemisphere, when green bodies expand across the largest landmass; in autumn, when leaves fall to the ground and fungi and bacteria have their season, it spikes.[38] What if you could put one moment of this process on steroids?

Some wizards place things inside their crystal balls: into their IAM, the IMAGE team inserted curves and grids for what they called 'carbon plantations'.[39] If humans plant crops on the land, these will capture CO_2. The challenge, if this is to be more than a seasonal flux, is to prevent the carbon from swinging back into the air: there must be no decomposition. The CO_2 has to be locked out of the active cycle and into the passive underground. Might this be possible? Here the IMAGE team stood on the shoulders of a group of researchers from Sweden, who had no IAM but a simpler model from which to pull out the following recipe: cover land with sugarcane or willows or poplars modified to grow so fast the eye can almost see it. Let those plants absorb CO_2. Harvest them before they shed any leaves and carry them off to a power plant, burn the matter, get electricity in the process – it could substitute for coal – but make sure to put a filter on the chimney, so that the CO_2 molecules are grabbed from the column of smoke. You now have a pure stream of CO_2 in your hands. Compress it further and inject it into a pipeline or a tanker and send it off to a depleted oil or gas field or a deep aquifer or some other cavity where it can be stored forever. Now repeat and repeat again and again and you have created a pump for removing carbon dioxide from the atmosphere – inevitably, christened with an acronym. This would be Bio-Energy Carbon Capture and Storage: say BECCS. The Swedes discovered that if BECCS were to happen in the future, CO_2 emissions might well be allowed to continue in the meantime, a greater total sum permitted for any given ultimate target, because *gas dumped in the atmosphere could be later removed.*[40] It was on this idea the players in the IMAGE team seized, feeding it into their more advanced digital loops and returning to the EU and other governments with an astonishing finding: even very low limits are eminently feasible. You just have to give yourself the freedom to first go beyond them.[41]

They called it 'overshoot', meaning that any particular concentration serving as target could be missed, emissions let loose and the extra carbon subsequently swept up for reinternment in chambers. Temperatures may well go above 2°C and then back down to where they ought to be. There would be no need to crash the models with radical interventions, no 'early retirement of existing fossil-fuel-based capital stock'. If you want the CO_2 concentration to stop at 450 ppm, you simply let it rise to 510 ppm and put your BECCS pump into action and get back to 450 ppm 'before 2200. This overshoot is justified by reference to present concentration levels, which are already substantial [as of that writing: 383 ppm], and the *attempt to avoid drastic sudden reductions'* – overshoot, in plain language, as the alternative to revolution.[42] The EU listened attentively. It could apparently have the 2°C cake and eat it too. Indeed, the IMAGE team hinted at the possibility of overshoot for any limit, high or low.[43] It was an idea with a long future ahead of it.

One Solution, Overshoot

The first paper to present the case for overshoot appears to have been one commissioned by the OECD – the governments of the US, Canada, Germany chipping in extra money – and authored by Tom Wigley, a climate scientist of prominence, in 2003.[44] 'In principle it may be advantageous to follow an "overshoot" pathway for *any* target,' Wigley meditated.[45] The advantage would lay in lower mitigation cost. And money is time: thanks to overshoot, 'an immediate reduction in emissions is not required', which would 'give us time to develop the infrastructure changes' that have to be carried through one day, some day.[46] Overshoot would save the economy from the tribulations of early, deep cuts; there would be no rush, no stress – benefits especially alluring in light of the recent American walkout from Kyoto. Perhaps even a George Bush Jr. could be won over to the climate cause if overshoot was on the table. Wigley had the prudence to notice a higher risk of bad things happening in the climate system under overshoot and therefore stopped short of embracing it as the royal road to prosperity – nor did he specify by what kind of pump CO_2 would be removed – but given the very considerable advantages, he found it 'clear' that scenarios unified by this fresh principle 'warrant further investigation'.[47] And the IAM community stood ready for just this.

We can see how perfectly overshoot harmonised with the functionality of these models: it cuts the cost of mitigation.[48] The inertia of business as usual does not come under attack.[49] Something like BECCS can be counted upon to materialise later in the twenty-first century, because when the carbon price starts creeping upwards, the ability of a technology to remove the substance will stand revealed.[50] It will be cheaper for agents to start capturing it than to keep releasing it at a price; rational as they are, they will be storing the carbon in slabs. And then there is the discounting: because the removal happens decades hence, it will be a bargain, compared to any mitigation here and now that would in fact keep warming to a limit. If the modellers merely discounted a little less – that is, ascribed a little higher monetary value to future projects like BECCS – most of the presumed cost-efficiency would evaporate.[51] But that would be another unwarranted concession.

Overshoot, then, tallied with rationalism, economism, presentism, conservatism, gradualism, optimism, and so it became ubiquitous in the IAMs, just as Kyoto segued into Paris. The IPCC lent its aura of legitimacy to it. The Panel sided with the EU on IMAGE; at a meeting in 2007 where all of this was discussed, it dubbed the Dutch scenarios 'scientifically interesting' and welcomed overshoot as a 'novel concept that the climate community has not thoroughly explored to date'. Including it in future assessments would generate insights 'regarding "reversibility" of climate changes and impacts' – the Panel opening, that is, for the possibility of viewing the climate of the Earth as that of an apartment.[52] If it becomes uncomfortably hot, you find a way to dial the thermostat down. The Panel asked other IAM teams to replicate the Dutch scenarios; for upcoming tournaments, they should 'employ their standard assumptions' and include BECCS but 'avoid non-traditional assumptions' such as 'dramatic dietary changes' or 'severe economic collapse'.[53] Creativity and innovation were encouraged, in some directions.

The IAM community duly complied. By the time the IPCC published its fifth assessment report in 2014, the number of scenarios that landed at 2.6 watts per square meter in 2100 had expanded to 114, spanning different models. The EU had got its 2°C in Copenhagen. The governments of the world had signed up for it. Not everyone may have realised, however, that *all* of those 114 scenarios leading to that favoured future now assumed some degree of overshoot.[54] In a wider ensemble of 400 scenarios retaining at least a fifty-fifty chance of hitting 2°C, 344 postulated massive rollout of some technology for carbon dioxide removal (while the rest presupposed global emissions peaking in 2010 – that is, 'an ability to change the past').[55] Like the proverbial army of robots, the IAMs had taken over the zone where climate science and policy met. The fifth assessment report, the last of the Kyoto era, had those computer models as its 'backbone'.[56] The Panel accepted the logic in principle: 'overshoot scenarios entail less mitigation today in exchange for greater reductions later.'[57] BECCS would provide redress. It was by these means 2°C came to look officially palatable and the stage was set from the northern side for the next act: Paris.

The fifth assessment report appeared in 2014, when all eyes were trained on the epochal COP to come; plenty of presidents, prime ministers, diplomats, expert advisors, lobbyists and others heading for the French capital would have picked up the key lines from the IPCC or at least caught wind of the opportunities opened up. In the late Kyoto era, overshoot had made 2°C look doable from a northern perspective – then why not also 1.5°C?[58] One would simply have to administer even more steroids to the process of drawdown. Surely IAMs could model this too.[59] The south could have its way on the question of targets, because the north had armed itself with a strategy for robbing it of the victory. This did not become immediately apparent during the rapturous final days in Paris, but it did so during the process of hammering out the meaning of 1.5°C in the Special Report. The IAM community, at first taken aback by the audacity of the goal, conjured a new representative concentration pathway into being – finishing at 1.9 watts per square metre, roughly corresponding to 1.5°C – and promptly served the IPCC with a fleet of scenarios that would get there through the by now established detours.[60] In the first draft of the Report, 191 out of 191 scenarios ending up at the right limit in 2100 broke it en route. But this is not what we agreed on! Locked up in the Panel deliberations, the ideological class struggle recommenced: delegations from the South, critical reviewers and movement allies protested vehemently at the total dominance of overshoot. 'This is not what we would define as a 1.5°C degree target as we [will] have water up to our necks by then', one objection went.[61]

Some crumbs had to be swept down from the computer tables, and the IAM community was not impervious to criticism. A fraction took it to heart. Members of the IMAGE team broke ranks by developing a 'lifestyle change scenario' that tested 'a radical value shift towards more environmentally friendly behaviour' – less meat, less solo driving of cars – as an alternative to the default deployment of late twenty-first-century removal and found that yes, indeed, such a shift would make BECCS almost redundant. But even in this most nonconformist model run, one measure could still not be pictured: 'rapid forced closure of fossil-fuelled power plants'.[62] Another team staked out a path to 1.5°C that relied on reduced energy demand rather than removal.[63] Studies like these were rushed off to the peer-reviewed journals to meet the deadline for the final version of the Special Report. They became the exceptions proving the rule: out of the 578 scenarios included in the compendium – the authoritative bundle of maps to a 1.5°C world – 10 did not pass through overshoot. The other 568 did.[64] In line with the schizoid trends running since at least Copenhagen, the Report, then, did two things simultaneously: it described just how dangerous an exceedance of 1.5°C might be, and it consecrated the notion of overshoot.[65]

The first chapter contained a curious definition. It classified two pathways as '1.5°C-consistent': those in which warming actually 'remains below 1.5°C throughout the twenty-first century, and pathways in which warming temporarily exceeds ("overshoots") 1.5°C and then returns to 1.5°C either before or soon after 2100'.[66] This was a logical somersault in plain sight. A was redefined as also encompassing -A, at least for a time. It was a bit like saying that the speed limit on this road is 60 kilometres per hours, so there are two speeds consistent with it: speeds up to 60 kilometres, and then also speeds up to 120, provided the drivers take a break after three or four hours. Or, the limit for allowed alcohol levels in your blood is 0.08 per cent, so you can have 0.08 when you drive or 0.2 if you make sure to sober up afterwards. Like a very Janus-faced or torn policeman waving on the cars, the IPCC then laid out the new grammar of climate politics. Overshoot 'allows initially slower or delayed emission reductions', the overaccumulation of CO_2 to be undone by technologies for drawdown further down the road; while some scenarios banked on reforestation and afforestation – letting forests grow back or planting new ones – the vast bulk went with BECCS.[67] Or else, 'large, immediate, and unprecedented global efforts to mitigate greenhouse gases are required'.[68] Avoiding overshoot-and-later removal 'can only be achieved if global CO_2 emissions start to decline well before 2030'.[69] The IPCC spelled out the two options henceforth facing the world: overshoot or revolution. It would then be up to the 'policymakers' to choose.

On the Way to Three Degrees

Most did not dwell long on that choice. After 2018, one country after another responded to the Special Report and the wave of movement mobilisations by announcing 'net zero' goals – promises of a future date when any remaining emissions would be 'balanced' by removals, the sum of both equalling zero. (Those late emissions were often referred as 'residual' or 'hard to abate' or even 'unavoidable' – very flexible categories all.[70]) As of this writing, 151 countries have made such promises, some of them on paper, enshrined in law.[71] If emissions were to fall in a straight line from today to net zero, the goals would have to be reached in 2040 – the most recent estimate puts the date earlier still, at 2035 – to avoid an overshoot of 1.5°C.[72] Any later date would imply a high likelihood of flying past it.[73] If we also apply a modicum of climate justice – reviving at least the basic principles of Contraction and Convergence – northern countries would need to place their targets earlier still, while those in the South could push them back a few years. But no such ambition was forthcoming from the signatories of the Paris Agreement. The vast majority rallied behind the date that the Special Report had put forth, namely 2050, a date that had overshoot sewed into its seams. Some, like China and India, asserted their right to fossil-fuelled development and put their arrival at net zero further away still, to 2060 and 2070. And so, as the third decade was well under way, no country had officially embraced overshoot as the route to the target; but nearly all of them had done so de facto, merely by virtue of their deference to the calculus of the Special Report and the IAMs that underpinned it. Unbeknownst to many, perhaps including those in various ministries engaged in setting the dates, overshoot had sneaked out of the models and into the formulation of climate policies, where it quickly asserted its place.

But on another level, of course, it had asserted itself already the moment countries put in place the new voluntary architecture of climate politics in 2015. Promises were, after all, just that, and most were not backed up by measures that could take them out of the realm of 'blah blah blah', to use Greta Thunberg's phrase.[74] When in 2023, eight years after Paris, one scratched away at the net zero façade and summated all the *actual* efforts – not promises – in place across the globe, the sobering result was that the world was on track for 2.7°C; or, rounding the number, 3°C, meaning a warming twice as large as that which the global South had insisted on to stay alive.[75] And we know that the warming does not produce a linear rise in damages: 3°C would be something far worse than just a doubling of the impacts at 1.5°C. But deep into the Paris era, this is where the world was heading. Keeping to the limits was never part of the intention to begin with.

Hatching the Anti-revolution

There was a reason that programmatic overshoot came to the fore in the 2010s: things were about to spin out of control. In one of its papers, the IMAGE team began by observing that emissions ought to shrink fast, but 'in contrast, historical trends so-far have resulted in an almost continuous increase'. Worse, the COPs themselves – this was written in 2013, when the debacle of Copenhagen still weighed heavily on minds – seemed to have descended into futility. 'There is also no evidence that ongoing negotiations for policies beyond 2020 will be more successful, although decision-makers proposed a new round of negotiations in Durban [the forgettable COP17]. A possible way to deal with this is the consideration of so-called overshoot scenarios' – an acquiescence masked as saving, through a syllogism that would come to inform the case.[76] The situation is now so dire that something like a revolution is needed. We cannot have a revolution. Therefore, we must do something else. Or, *emissions are out of control and we have no project for bringing them under control and so we must try different ventures than classical mitigation* – the core doctrine of what we shall call *overshoot ideology*. In the case of this IMAGE paper, conforming with the early Paris era, the liquid poured into the vessel happened to be BECCS. But it could be replaced with almost anything, as long as the substance stayed true to the ideological form and steered clear of 'a revolutionary overhaul of the system'. One should never underestimate the creativity of the anti-revolution. It can, as we shall see in our second instalment, throw the most astounding innovations into the world.

We must be careful to distinguish this from *counter*-revolution. Such a thing presupposes a revolutionary situation – that is, one of those moments when the ruled no longer put up with being ruled and the rulers have lost their ability to go on ruling, and so the two arrange themselves into contending blocs: a popular source of power-seeking to supplant the beleaguered old regime. Society is suspended in an awkward position that must be straightened out one way or another. It cannot be ruled by soviets and provisional government at the same time: one must go. *Now* the counter-revolution intervenes, to close the situation down 'on terms favourable to the old order'.[77]

Overshoot, obviously, was conceived before any situation of this kind had developed. In climate politics, it would be marked by some popular subject launching a challenge to fossil capital so profound as to express a generalised refusal to abide by it any longer, while this same capital and its guardians would be so shaken in their capacity to rule as to lose their grip: dual power putting society on a knife's edge. Transition would then be a real prospect. 'The essence of a revolutionary situation is that profound transformation, social or political, goes from being something that couldn't happen to something that very well might.'[78] While resistance was budding in a thousand places in the early millennium, no revolutionary situation thus defined was in sight, and so no counter-revolution was called for; but the logic inherent in the planetary situation was such that an *anti*-revolution had to be thought up.

> Anti-revolution tries to prevent an imaginary revolution, whose terrible spectre constantly pursues the ruling powers and heralds their demise. This approaching imaginary revolution does not have obvious roots in society, and it lacks an apparent strong-willed political subject – in fact, its potential future participants know nothing about it. But this imagined revolution lives a full life of its own in the consciousness of state authorities, and it has been outlined by experts in dozens of documents.

– words formulated for another context, but perfectly matching the climate.[79]

It was the rapid depletion of the carbon budget that made the anti-revolution of overshoot seem like the only way out. By counting on BECCS, that budget could be artificially extended: allowances made for many hundreds or thousands more giga-tonnes, as they could be subtracted at a later date.[80] In a retrospect written for its sixth assessment report from 2022, the IPCC explained the career of the idea: 'overshoot was particularly allowed for low concentration and temperature targets *as many models could not find a solution otherwise'*, presupposing that one kind of solution had been forbidden a priori.[81] For a scientist like Kevin Anderson, this was a betrayal of the budget concept itself. IAMs represented the cutting edge of an 'almost global-scale cognitive dissonance in acknowledging' the 'revolutionary implications' of what was now basic climate science (and, we might add, experience).[82] An irreducibly cognitive, ideational, psychic phenomenon, the anti-revolution was hatched inside models – but the warming did not happen there, of course, but very much in the real world: which meant that sooner or later, in the continued absence of revolution, overshoot would have to take on *material force.* Then it would appear not as minus signs in a calculation but as layers of facts on the ground, on top of all the pre-existing fossil fuel infrastructure. At that point, the basic coordinates of climate politics would be rearranged and the fronts at least partly reordered. But long before this point had been reached, it was clear that, as ideas go, programmatic overshoot was a very bad idea indeed.

Let the Poison Flow

'Overshoot' can mean simply 'missing the target', but in the guise we have encountered it so far, the idea comes with a grander meaning: the target can be missed and then hit after the event, if efforts are properly retrained on it; more particularly, the emissions that caused the missing can be eliminated through removal – or, as it is also referred to, 'negative emissions'. Too much of the positive can be balanced out by an equal or larger amount of the negative.[83] The way this idea developed alongside, and indeed intertwined with, the science of the severity of warming above $1.5\,°C$ bears some resemblance to the following scenario. Imagine a facility close to a major city that is poisoning its inhabitants. A lot of money is made from it. Shutting the facility down would therefore cause economic trouble to the owners, but voices are raised from inside the city demanding that this happens regardless, since the lives of the residents are at stake. What to do?

A report is submitted for consideration by the city authorities, based on some advanced modelling, which concludes that the facility, hazardous as it is, may well continue to operate, because a pharmaceutical innovation will soon be able to suck the poison out of the bodies when it reaches life-threatening elevations. Just before patients die, they can be cleaned of the substance. What would we think of the ethical status of such a recommendation, not to mention the authorities that acted on it? Presumably, most would consider it a reckless code of conduct, perhaps also a form of cruelty: letting the poison continue to flow with the promise of eventual, last-minute healing would amount to the deliberate infliction of harm and the gambling with the lives of multitudes. To choose such a policy on the premise that the results could later be annulled and the worst outcomes avoided would be blameworthy, not to say impermissible, particularly if combined with or even based on knowledge of the risks accepted.[84] As is the nature of hypothetical scenarios, this one reflects only some aspects of overshoot; there is, for example, no reason the innovation would have to intervene before death. One might as well consider a bet on fancy de-extinction technologies applied to the residents at some point in the future and bringing those who have died back to life.

In overshoot, the equivalent of the city authorities and the owners of the poisonous facility may never come to see its victims – or see them only as children – because these would belong to a later generation. The recommendation is here to defer mitigation and removal decades into the future. One of the first papers to bring the idea to a wider scientific audience, a short letter published in *Science* in 2007, made this clear:

> In this scenario, emissions would be reduced less severely in the short term, but more severely later on (possibly using carbon capture technology), when compared to a nonovershooting scenario. As such, 'overshoot' could be a conscious policy that removes some of the burden of mitigation from the present

generations while protecting future generations from exposure to the most severe impacts.

– 'protecting' here somewhat of a euphemism, along the lines of protecting the poisoned people from the most severe impacts by promising them late antidotes.[85] The virtue of this course of action, for which it would be consciously chosen, would not lay in the most excellent protection it would afford future generations, but in the lifting of a 'burden' on certain members of present ones. It would not be an act of altruism. It would be about letting some currently living people off the hook. 'The situation changes substantially if negative emissions are assumed to be possible: the urgency for near term emission reductions becomes much less,' in the honest words of the IMAGE team.[86] It takes no sixth sense to discern what kind of appeal such an argument might have.

There is, however, as we have seen, a practice that purports to lend legitimacy to this procedure, namely that of discounting. Nordhaus made the case for it in 1997, in an article in which he warned against any ambition to stay below 3°C as too costly and deduced that 'optimal climate-change policy reduces long-run global warming from 6.6°C to 6.2°C.'[87] Growth, he argued, has been 'more or less continual' for more than a century. It has raised standards of living. All indications are that this success story will extend into the future. Therefore, we ought to delegate to coming generations the task of carrying a yoke that will be lighter for them than for us, 'just as high-income people pay a larger fraction of their income in income taxes' – Robin Hood dumping burdens in the maternity ward.[88]

Critics have struck this argument so hard as to leave it with more holes than a riot fence: we may mention less than a dozen. Suppose, to begin with, that it were true. Suppose that humanity twenty years from now will in fact be many times richer than in, say, 2020. If it is advisable for the generation of 2020 to leave mitigation to their richer descendants of 2040, then why should not the same also apply to *that* generation, insofar as its offspring is placed on the same growth curve, the discounting and postponement continuing in perpetuity. The practice lends itself to endless deferral. Even if all of this were true, leaving the warming unmitigated – or just *less* mitigated – would then still be wrong. If the climate is wrecked, because fossil fuel combustion was not discontinued in time and warming overshot 1.5°C, there will be pain and suffering, and then it matters little if there is also more money in the banks.[89] But why take for granted that the latter will even be the case? There could be world wars. There could be recessions.[90] The rate of profit might fall so low as to induce the terminal collapse of the capitalist mode of production – who can claim to know the opposite? The world might not look like the American 1950s forever. Climate change itself will bite into growth rates: only privileged, white, middle-aged economists cocooned in the virtual reality of money signs could imagine that living standards will be unaffected by lost harvests, inundated lands, blasting storms or indeed summers too hot to permit any work outdoors; rather, the material conditions of possibility for any kind of economy will be stripped away one after the other. To leave it to later generations to deal with this mess because we presume they will be richer is to misunderstand what kind of a mess it will be.[91]

Moreover, if the past two centuries of capitalist growth are anything to go by, a future continuation will produce winners and losers, and the losers might well end up poorer than the rich today, and no average rise in per capita GDP – a statistical artifact – will be able to cancel out that poverty: the poorest of the future may be more numerous than their most fortunate coevals. It is on them, we know, that the effects of global warming will disproportionately fall. There can then be no justification for abstaining from mitigation today, on the assumption that future people will be wealthier in the aggregate and therefore cope well; to the contrary, any such abstention would be tantamount to sending grievous harm down the generations, among which it will primarily hit the poor. The redistributive pretension of the Nordhausian argument is the most disingenuous. If one wants to make sure wealthy people carry the burden of mitigation, it makes rather more sense to place it on wealthy people who actually exist – and who happen to be responsible for exacerbating the mess on a daily basis – than to any notional nabobs of the future.[92] But, as we have seen, class relations were effaced from IAMs in the Nordhausian family line, leaving them by definition blind to questions about who is causing the crisis and who is suffering from it and who might be fairly or unfairly targeted by this or that policy: the models were designed to remove redistribution from the agenda.[93] The ersatz taking from rich descendants and giving to poor contemporaries could not make up for this. Indeed, it inverted the signs of intergenerational justice too.

For the central point, of course, is that *any* delay in mitigation will, because of the cumulative character of the problem, make it worse: more CO_2 will be emitted than if the cuts had been sharp; temperatures will rise faster and higher, handing down greater harm to whatever people come next.[94] In *A Perfect Moral Storm: The Ethical Tragedy of Climate Change*, the greatest classic of climate ethics, Stephen M. Gardiner anatomises this ongoing mistreatment of coming generations. Because warming is 'backloaded', so that emissions from day one only strike home with their full impacts at a later day twenty or one hundred, it is 'hard to grasp the connection between causes and effects' and tempting to pass the buck.[95] Easy as it may come, this constitutes a form of exploitation or domination of later generations.[96] They are used and abused for the purposes of present gain, with no regard for their well-being. As broad as the consensus is on discounting among bourgeois economists, as pervasive is the agreement among moral philosophers of all stripes that a grave wrong is here being committed – emissions cuts in fact *being* kicked down the road, the COPs the annual festivals for the renewed kicking.[97]

Nordhaus built his career and the entire subdiscipline of bourgeois climate economics around the case for postponing mitigation to a future when it is cheaper: the intellectual bedrock of the IAMs. Philosophers played catch-up and pointed out the analytical errors. But in the idea of overshoot, the intergenerational buck-passing was elevated from blind practise to explicit credo – with the added proviso, of course, that the mess will be efficiently dealt with later, when removal technologies such as

BECCS have materialised. When Gardiner published *A Perfect Moral Storm* in 2011, this ideology was still in its infancy; since then, it has made the storm worse.[98] On the surface, the idea would seem benevolent, since it includes a clause about abetment of the problem, and since it (so far) speaks about returning to 1.5°C (or 2°C . . .) rather than trimming the warming from 6.6°C to 6.2°C; but transferring responsibility for this task still remains unfair. It is a bit like hosting a massive party in which you let the guests set the roof on fire, flood the basement and spoil the garden, only to then move out and leave the repair bill on the table for your children and grandchildren to take care of.[99]

And that bill will grow: philosophers understand that the burden of mitigation itself will *become heavier* if the problem is allowed to fester.[100] But the mistakes go beyond both ethics and economics. They concern how history works. In *The Pivotal Generation: Why We Have a Moral Responsibility to Slow Climate Change Right Now*, the first sustained engagement with overshoot from a leading philosopher, Henry Shue is drawn to the irresistible analogy of Nazism. A retrospective overshoot argument would say: would it not have been better to leave the struggle against Nazi Germany to the 1950s or 1960s, when people were more affluent and better armed? But the task would likely have been harder two or three decades after 30 January 1933, the Nazi regime so much deeper entrenched. Conversely, whatever good conditions prevailed in those decades were dependent on the Third Reich *first* having been vanquished. We should, Shue argues, be grateful to the poorer, less high-tech generation of the 1940s for rising to the occasion.[101]

The deeper error here is historiographical. It lies in smoothing out history and conceiving of it as identical time units succeeding each other in uniform, mechanical fashion, as in a digital clock – only in this image world can mitigation be reshuffled to posterity, as if the task would remain identical to itself, like an alarm signal put on snooze. Real history has an 'integral fabric' (Shue), dense and textured, fractured by breaks, composed of processes with distinct rhythms 'coiled and slotted inside one another, like circles within circles, determining the enigmatic patterns of historical time, which is the time of politics' (Daniel Bensaïd); and in this real history, some threats can build a momentum of their own.[102] The rise of Nazism in Germany is indeed a case in point. When it had progressed far, some of its enemies advocated something like a benevolent overshoot policy – namely, the leadership of the KPD.

> The sense of the theory is the following: fascism is growing unrestrainedly; its victory is inevitable in any case; instead of 'blindly' throwing ourselves into the struggle and permitting ourselves to be crushed, it is better to retreat cautiously and to allow fascism to seize power and to compromise itself. Then – oh then – we will show ourselves,

as summarised by Trotsky in November 1931, in one of his desperate missives to the German proletariat, imploring its two wings to put their differences aside and intervene in time to prevent the catastrophe from being consummated. If the battles were delayed to some point 'not *before* the seizure of power by the fascists but *after* it', they would be incomparably more gruesome, taking place 'under conditions ten times more favourable for fascism than those of today'.[103] Trotsky's argument against overshoot fell on deaf ears.

Clearly it would have been better had it been otherwise. European Jewry and the German working-class movement and tens of millions of Soviet citizens, to mention only some, would have been spared destruction, had an anti-fascist subject formed and picked the decisive battle before, not after, the *Machtergreifung*. The KPD theory that Nazism could be conveniently knocked over after the seizure of power proved delusional.[104] The sacrifices required to roll it back can hardly be overestimated. Again, the analogy is incomplete: the position of the KPD in relation to the NSDAP, for a start, is not commensurate to that of advocates of overshoot vis-à-vis fossil capital. But analogies can help us grapple with questions even in something as singular as the climate crisis. Here, two distinct ones are at stake: whether overshoot can ever be a wise policy in the face of a catastrophe gathering steam; and whether resistance *post festum* can ever be meaningful. Real history suggests 'no' to the former question, but 'yes' to the latter.

Dance around the Carbon Unicorn

If these were several ethical objections and one historiographical, most damning for programmatic overshoot, however, might be a psychological observation. The idea was born out of the rejection of systemic change, on the argument that it was a pipe dream only starry-eyed environmentalists could possibly entertain. Yet what its architects substituted for it was, if anything, even more a figment of the imagination: when climate policy and science swerved towards it, BECCS did not exist. At the time of Paris, there was exactly one pilot plant in the whole world, in Illinois.[105] If BECCS had any existence to speak of, it was in the *minds* of modellers, in the fantasy futures of the IAMs; in real life, the technology had zero proven efficiency.[106] What the friends of overshoot were asking of the world was to stake its future on a force of fictional quality. Rather than undertaking immediate mitigation along principles fully understood, proven and available – shutting down fossil fuels and replacing what was needed with renewables – it would here gamble on an entity still foreign to the physical world. If the tech never made it there, what would then happen?[107]

Critics were not slow in pointing this out, leading some modellers to retort that 'large-scale inclusion of BECCS in ambitious pathways is justifiable because IAM work does not seek to produce direct representations of reality', which merely conceded the point: BECCS plants had the ontological status of 'carbon unicorns'.[108] They were brandished 'like a light saber, incredible but not real'.[109] They represented 'a kind of magical thinking'.[110] Beyond the question of BECCS, the broader syndrome was that of a tenuous relation to reality, as some anonymously interviewed members of the IAM community readily and unhappily acknowledged. 'I'm concerned that everything becomes so focused on modelling results that are totally theoretical and detached from reality,' said one.[111] A deeper investigation of the processes at work here must take psychopathology into account. Indeed, one might be tempted to see in the acronyms the modellers picked for their creations – IMAGE, DICE, WITCH – parapraxes bespeaking their manner of relating to the external, material world. If this manner was innate to the IAMs, it was intensified in overshoot. In the anti-revolution, we may hypothesise, allegiance to the status quo is maintained at the cost of a break with reality.

Astrochickens and Other Great Ideas

The idea of overshoot did not succeed on its own merits qua idea. Barely had they been assembled before the crystal balls were picked apart. In the wake of the oil crisis of 1973, American universities, led by Stanford, created the first proto-IAMs to model how oil demand might develop in the future, only for attentive scholars to lambast their 'unsubstantiated assumptions', sensitivity to arbitrary inputs, unrealistic rationalism, fidelity to fossil fuels, naturalisation of existing institutions, blindness to disruption, blithe expectation of universalised growth and faux objectivity.[112] It was all there from the start. The models were prone to substitute themselves for external reality: 'our scenarios are comprehensive and allow for no escape', the main study of the period declared, eerily.[113]

It was also in this historical moment that the idea of removal first appeared, likewise with recognisable features. In his very first paper on climate, 'Can We Control Carbon Dioxide?' from 1975, Nordhaus responded positively: 'it is possible to remove carbon dioxide from the atmosphere by running combustion in reverse.' He had in mind the growing of trees or the injection of carbon into the deep oceans or removal 'by an industrial process' not further specified. The two former options, he claimed to know, would be extraordinarily cheap ways of controlling CO_2, while early mitigation would entail difficult losses: it would make 'coal and oil shale royalties fall to zero'.[114] Better to wait at least half a century before any emissions be cut.

These speculations and back-of-the-envelope calculations were developed in dialogue with Cesare Marchetti, an Italian physicist based at the Viennese institute that would go on to become one of the main IAM hubs: he favoured the solution of scrubbing CO_2 from power stations and blast furnaces, pumping it through pipelines, piling it on barges and sinking it in exhausted gas fields, all for a song.[115] The focus on trees came from another man, the eccentric American theoretical physicist Freeman Dyson. In an article from 1977, he adumbrated a logic of removal as anti-revolution. 'In spite of various dire warnings, it seems inevitable that we shall continue for many decades to burn fossil fuels', possibly leading to an emergency. 'Suppose that with the rising level of CO_2 we run into an acute ecological disaster. Would it then be possible for us to halt or reverse the rise in CO_2 within a few years by means less drastic than the shutdown of industrial civilization?' Here too, the response was positive: it would indeed be possible to plant water hyacinths and sugarcanes – perhaps also doubling as energy crops – and draw the carbon down for burial in 'artificial peat-bogs' and thereby 'reverse the growth of atmospheric CO_2 within a few years'.[116] Dyson added the caveat that he simplified things. 'The purpose of this paper', he acknowledged, the psychic dimension worn on the sleeve, 'is to begin a process of *mental preparation* which may enable us to have realistic plans ready if ever the danger of catastrophe from CO_2 becomes acute' – mental preparation, that is, for doing something else than shutting down the source of the problem.[117]

Freeman Dyson had many fancies. He believed it possible to send unmanned missions to Mars and other celestial bodies to plant seeds, growing into trees genetically engineered to build their own greenhouse atmospheres the way turtles do their shells – 'the greenhouse would consist of a thick skin providing thermal insulation, with small transparent windows to admit sunlight.' Such trees would then huddle together into habitats hospitable for humans. 'The private settlement of pilgrims all over the solar system will begin.'[118] Dyson came up with a scheme for 'astrochickens', a species of artificially intelligent, fully auto-mated spacecraft weighing less than a kilogramme, launched from the Earth to collect 'nutrients' from planets and lay eggs that would hatch into new astrochickens, and so on, opening up the solar system for endless exploration.[119] Dyson thought he could overcome heat death. Through a series of equations, he claimed to demonstrate that before the second law of thermodynamics brings the universe to an end, there might well be 'a technological fix that would burst it open', intelligent beings escaping the limitations of the old universe and taking their stored energy into the next.[120] Dyson died in 2020 as an unrepentant climate denialist.[121] Something about his mind, it seems, made him strive to detach from tellurian reality.

From such intellectual soil did programmatic overshoot sprout: patently the idea did not make its way through the world because of its sheer brilliance and persuasive power. It required some connection to the base.

A Bankruptcy Petition of Consciousness

When modelling how glaciers react to heat, the results and their dissemination do not in and of themselves change that reaction. But when the modelling concerns human responses to warming, this can, due to a *differentia specifica* of the social sciences – the object of inquiry also being a subject – happen: in the early Paris era, the IAMs themselves came to condition 'policymaking'. Some two dozen models in the global North defined the spectrum of possibilities. They mass-produced scenarios then packaged and bundled by the IPCC and passed on for consumption by governments. They were like leviathan computers, with inner workings inscrutable to outsiders. They had to be taken at their words, which allowed them to operate as '"Trojan horses" for undeclared interests', most effectively through the work of exclusion: the IAMs would not touch degrowth, the Green New Deal, rewilding, nationalisation of fossil fuel companies, a state-led shift to 100 per cent renewable energy (of which more below), half-earth socialism, ecological war communism or other proposals that veered from the middle of the road.[122] They worked to 'shut down alternative imaginations'.[123] In the early years of the third decade, these machines determined the bounds of acceptable climate politics, above all through the one theme that increasingly dominated it: carbon dioxide removal.[124]

Perceptive critics of IAMs characterised them as 'performative'; anything but neutral mirror images of the future, they brought some types of future into being.[125] They were 'political machines' in possession of 'world-making power'.[126] Under the influence of Latourian theory, some even designated them 'actors in their own right' holding 'a powerful position' as such.[127] But this was to ascribe to them a mysterious protagonism. Did these strings of computer code rule on their own initiative? Had they plotted to take over the making of climate policy? As easy as it might be to blame the IAMs (or the modellers behind them) for the rise of overshoot ideology, there are good reasons not to: for a start, as we have seen, the ideas they thrust into the mainstream were not all that bright, which only deepens the mystery – were governments gullible fools? Had computers somehow hypnotised them? Critics would shred them to pieces and conclude that 'the IAM Emperor has "no clothes"' and still it remained on its throne.[128] But this emperor had not staged a coup or anointed itself. It had been summoned by state apparatuses, the EU in particular. The IAMs were in fact not an army of robots, with intentions and goals and other properties of agency, but rather servants called up to ideological duty.[129] Their mission was to square the contradiction of *practical loyalty to business as usual and nominal fealty to temperature targets*, first 2°C, then 1.5°C. They did what they were asked to do.

The alternative, then, is to understand IAMs and programmatic overshoot with the help of some Marxist ideology theory: they could rise to prominence only because they fit the material interests of dominant classes – namely, and most simply put, their investment in the maintenance of the status quo *and* its legitimacy at the same time. 'An ideology as such can really emerge only where the base is articulated enough to provide a close motivational context to which the superstructure then corresponds,' in the words of Adorno.[130] The early Paris era brought about precisely such a context of compelling motivation. Adorno, as so many others, voiced discomfort with the metaphor of base and superstructure, largely because he regarded technologies themselves as imbued with ideology. Rather than being added to the material base, as a gloss or sanctification dispensed by priests, ideology, in this view, is *in* the machines (like Christianity is in the cathedral). 'Critics might use every industrial administration building and every airport to show to what extent the base has become its own superstructure.'[131] The role of professors, teachers, editorialists, artists and other old-fashioned intellectuals, who once had to work overtime to justify the status quo, has become superfluous or at least of diminished stature, since the productive forces have now 'acquired a kind of halo or suggestive power of their own'.[132]

The analysis could have been thought up with IAMs in mind. Does it also obviate the distinction between base and superstructure? Only if the base is equated with the purely physical productive forces, along the lines of orthodox determinism; but a different reading of Marx would, instead, define it as the *relations between classes* in the process of production. The base is where the dominant class ties the direct producers in subordination and controls the metabolic exchange with the rest of nature.[133] Ideas can form a superstructure to it only in this sense – or, as Marx and Engels put it in *The German Ideology*:

> The ideas of the ruling class are in every epoch the ruling ideas: i.e., the class which is the ruling material force of society is at the same time its ruling intellectual force. The class which has the means of material production at its disposal, consequently also controls the means of mental production . . . The ruling ideas are nothing more than *the ideal expression of the dominant material relations*, the dominant material relations grasped as ideas.[134]

In this *locus classicus* of ideology theory, not a word on productive forces: it is class power that dresses up in ideas. Once we dispense with determinism and adopt a constructivist Marxism, we can comfortably follow Adorno's intuition and expect machines to also be ideational entities.[135] The technologies characteristic for the age of overshoot might well traverse the realms of the material and the ideological and the psychic. But they would do so with the duty of keeping the *base of class power* intact.

This is not to suggest that 'base' and 'superstructure' are the last words in the quest for a metaphor capturing the linkages between classes and ideas; rather, fresh efforts in creativity are always called for. One could, for example, think of the whole panoply of ideas produced in any given moment as so many gloves sewed in a workshop. What ideas will be picked up and gain efficacy in the real world will depend not on their intrinsic beauty or symmetry, but on whether they fit the hands of the dominant classes. Conversely, other ideas might be selected by subaltern forces: and so the ideological class struggle can begin. *'Ideologies in class societies always bear the mark of a class'*, in Louis Althusser's phrasing, 'with the dominant tendency of ideology representing the dominant class's interests' – quite possibly in mechanical or digital garb.[136] This appears to be what happened in the case of the IAMs, when they were accorded the dignity of being a toolbox for climate policy and the notion of overshoot was weaponised in ways the modellers might not ever have imagined.

The IAMs could be the easiest example of machines located in the superstructure, since their output consisted not in commodities but in ideas. As we have seen, several vintage motifs of bourgeois ideology were inscribed in their codes. 'The exchange principle', as Adorno would have it, dictates that everything be given a monetary value or no value at all; 'it occurs to nobody that there might be services that are not expressible in terms of exchange value', and if it occurs, as it did to Nordhaus, then the thought is shooed off like a housefly.[137] The 'representative agent' is the picture of the bourgeois subject. He is like Robinson Crusoe on the island, detached from all bonds, preoccupied with his bookkeeping and stocktaking, alone.[138] He prefers more to less. He is incapable of 'kindness, generosity, or compassion towards others – not only because there are no others, but also because if there were others, he would not derive any "satisfaction" from these activities.'[139] He is the bourgeois *ratio* consummated.

As skewed and twisted as this image of humanity might be, however, it is a distillation of relations that structure capitalist society in its depths: in the process of exchange, under the law of value, everything really is swapped for everything else, the qualitative aspects of things dissolved in the acid of the universal equivalent, the abstraction renewed each millisecond of market activity.[140] An agent who wants to make his way in this world must adopt the 'calculatory equation' as his mindset. 'Every businessman who calculates has to act according to this fetish. If he does not calculate in this way, he goes broke.'[141] The IAMs were at their most ideological when programmed to treat this way of life as a given – not as something that once popped into history and might well pop out of it again, but natural and eternal, like the fact that ice melts in heat. 'Positivism is so blinded by society that it regards second nature as first nature and identifies the data of society with the data of natural science': the integration of the IAMs.[142] In an age when data in large numbers on a luminous screen was perceived as the quintessence of credibility, the models did attain a dazzling authority.[143] But in the last instance, it was a function of a correspondence between superstructure and base.

Once we see the IAMs from this angle, their triumphal procession through the juncture of climate science and policy is no longer mysterious, any more than the preceding success of Nordhausian economics, however wretched its analytical poverty might be. As much as in the nineteenth century, the categories of bourgeois economics were '*socially* valid, and therefore objective, for the relations of production belonging to this historically determined mode', Nordhaus continuing the lineage of a Ricardo and a Say.[144] The thinking he pioneered wielded an enormous influence during the Kyoto era, especially in the US. When he received the Nobel prize in 2018, it was no exaggeration to say that 'the failure of the world's governments to pursue aggressive climate action over the past few decades is in part due to arguments that Nordhaus has advanced.'[145] Among the many factors contributing to the historical delay of mitigation, this one was, in the realm of ideology, vital.

The idea of overshoot took the same principles into the age of consequences. If it was now too late to avoid 1.5°C (or 2°C) of warming, negative emissions would come to the rescue, since a positive amount could always be neutralised by the same amount of the opposite sign, an addition undone by a subtraction. The premise here, of course, was that one quantum of CO_2 released in 2020 would equal the same quantum of CO_2 removed in 2120. History would not intervene in the meantime. The exchange principle was extrapolated decades or centuries into the future, as the solution to the problem to which it had itself contributed: 'exchange is the rational form of mythical ever-sameness. In the like-for-like of every act of exchange, the one act revokes the other; the balance of account is null.'[146] The IAMs of the Paris era did not invent this notion of exchanging one emitted sum of CO_2 for one captured; such equivalence was foundational already for the Kyoto era, then often referred to as 'offsetting'. It merely took on a new life in the age of overshoot, often through the shibboleth of 'net zero'. What set overshoot apart from its prefigurations was the context of ideological contradictions in need of resolution when the limits of endurable warming were about to be breached.

How do you let things run out of control and still pretend to be steering them in the right direction? You invent something like overshoot: the idea that business as usual can continue for another while and *then* we will set the course straight by reversing it.[147] Through this sleight of hand, any given target could be both missed and met and any missing rationalised as part of the journey to meeting it, like Schrödinger's cat simultaneously alive and dead.[148] At 1.6°C or 1.7°C or even higher, the world might still be heading for 1.5°C: before the breach, 'policymakers' had bought themselves the freedom to commit it. Come the day and only the breach will, of course, be empirically verifiable, but the hypothetical detour and return may absolve dominant classes from judgement, their failure impossible to prove, as there will always be the possibility of going back.[149]

In the early years of the third decade, overshoot subsumed one-and-a-half-degree defeatism and voluntarism from the right. But the latter current might have been noisier, and it explained why hardcore climate activists could find themselves in the company of men like Branson and Johnson: when the 'We Mean Business Coalition' insisted that 1.5°C was still alive, what it meant was overshoot. (And a forceful argument for defeatism from the left was the need to puncture such phony optimism.) Just like the positions of defeatism and voluntarism, overshoot was then poised to accompany climate politics through many decades to come, potentially far beyond 1.5°C or even 2°C, as the idea could always be resurrected, its escape clause perennial: for every breach, a promise of its healing. No one could say that it would ever be too late. Or, with this attitude, it would inevitably always become too late, for just that reason. And so overshoot had something surreal about it. The IAMs producing it seemingly fulfilled another of Adorno's recondite dicta:

> The computer – which thinking wants to make its own equal and to whose greater glory it would like nothing better than to eliminate itself – is *the bankruptcy petition of consciousness in the face of a reality* which at the present state is not given visually but functionally, an abstraction in itself.[150]

In this regard, in this bankrupt relation to reality, overshoot was part of a wider turn in climate governance launched to much fanfare in Paris.[151]

The Irreal Turn in Climate Governance

'I believe this moment can be a turning point for the world', said Barack Obama from the White House on 14 December 2015. It was a Monday, less than forty-eight hours after the conclusion of COP21. Establishing 'the enduring framework the world needs to solve the climate crisis', creating 'the mechanism, the architecture, for us to continually tackle this problem in an effective way', the agreement just signed was, the president solemnly avouched, his intonation and gesturing as inspirational as ever, 'the best chance we have to save the one planet that we've got'.[152] Four days later, on the Friday, he put his signature to a bill that repealed the forty-year-old ban on exporting crude oil from the US.[153] Sixteen oil companies had banded together to lobby for the reform. It was expected to open the last sluices to the twenty-first-century American boom in fossil fuels, and indeed, by 2022, this country – not shipping off a single barrel before the day the Paris Agreement was finalised – had become the third largest oil exporter in the world.[154] If this was an enduring material legacy of Obama, his infectious optimism had some staying power too.

He made the entire world talk about Paris in glowing terms.[155] Indeed, faith in the goodness of outcomes might have been key to Paris happening at all. One of the protagonists of the summit, Christiana Figueres, then executive secretary of the UNFCCC, wrote a reminiscence in *Nature* in 2020, recalling how she had walked into the negotiations with a firm yes-we-can resolve, without which they would have foundered. 'When the Paris agreement was achieved, the optimism that people felt about the future was palpable – but, in fact, optimism had been the primary input.'[156] Optimism was the auto-renewable fuel of the new era. It replaced the traditional drawing up of mandatory commitments. In the absence of legislative rules, climate governance came to hinge on the participants displaying their willpower and abounding good cheer; instead of a text with defined rules, a performance that had to be restaged at every ensuing COP.

'Incantatory governance', Stefan Aykut and his colleagues have labelled it.[157] Starting in Paris, the annual summits became occasions for the incantation of optimism about a win-win low-carbon transition already underway.[158] If anything obligatory remained, it was the continuous transmission of positive signals, from COPs as well as other events: Figueres chose to address the World Economic Forum in Davos in early 2020 and called on the attendees to commit to 'stubborn optimism'. There were grounds for it, she claimed. 'Leaders in the oil and gas industries have told me privately that shareholder and public pressure, plus questions from their own children, have prompted them to change their practices.'[159] Her own enterprise was named 'Global Optimism'; in a TED talk from 2021, she explained that such an approach to life 'makes you jump out of bed in the morning because you feel challenged and hopeful at the same time'.[160] At COP27, she joined We Mean Business in devotion to the feasibility of 1.5°C. This 'hurrah-optimism', this 'frantic optimism', this 'constantly enforced insistence that everybody should admit that everything will turn out well' amounted to a kind of superstition – 'the insidious bourgeois superstition that one should not talk of the devil but look on the bright side. "The gentleman does not find the world to his liking? Then let him go and look for a better one"' (Adorno).[161]

The medium for this ideology was theatre. If Adorno smashed the idol of optimism, it was, of course, Althusser and his followers who emphasised the theatrical forms of ideology: it takes place through 'rituals, rites and ceremonies'.[162] In an essay incidentally published just in time for COP21, Étienne Balibar stressed that speakers who perform the act of interpellation always appear on a stage, literally or figuratively, preferably with a degree of 'pomp and ceremony', combining 'a machinery and a show' – a theory of *'politics as theatre'*.[163] Climate politics in the official register was then deep into the process of becoming just that sort of politics, the main raison d'être of the summits now the enactment of a pageantry of world leaders taking care of the planet and its inhabitants.[164] Every November or December, people were interpellated as the subjects of a capitalist society still sailing towards sustainability. They were asked to buy into it, the COP a kind of environmental counterpart or sideshow to Black Friday (coincidentally, taking place around the same time, and spreading across the world during the same post-Paris years). If ideology was built into the IAMs – invisible, inaudible – here it was communicated in the opposite style: overt, histrionic, forced, as when G20 leaders planted saplings in a mangrove to certify their commitment to 1.5°C.[165] What mattered was not what anyone did, but what these leaders appeared to be doing, as seen from the audience floor.[166]

At a certain point, this optimism bid farewell to the reality of the crisis. Aykut and colleagues have pointed to a 'virtualisa-tion' of climate governance, and if we consider another couple of factors, we may push their analysis one step further.[167] The Paris Agreement did not contain one single mention of 'fossil fuels' or 'energy'. The key texts of international climate diplo-macy carried out a redaction of this determinant of the reality in question: the UNFCCC used the word 'energy' six times and 'fossil' four, the Kyoto Protocol 'energy' seven times and 'fossil' none, the Paris Agreement 'energy' three times and 'fossil' again none.[168] The last was the most craven in its refusal to mention the problem by name. It was like a peace treaty that does not give the names of the conflicting parties. (Or, an analogy from the real world, the documents of the 'peace process' never referring to 'the occupation': and if root causes are not even named they are guaranteed to make things worse). Add to this the centrality of carbon unicorns, and we can infer that climate governance took an *irreal turn* in Paris. Overshoot was the only way the optimism could be sustained while the carbon budget neared depletion. Between the IPCC's fourth assessment report in 2007, in which most scenarios projected the world solidly into a 3°C future, and the Special Report in 2018, an additional 300 gigatonnes had been poured into the atmosphere and no peak in emissions was in sight, and yet a low ceiling on temperature was made to seem far more reachable and cost-effective at the latter point.[169] The gravity of what was happening on the ground and its implications for climate action did not quite register.

Through this irreal turn, climate governance increasingly dis-connected from reality. Temperature targets were not all that real to begin with. In the early third decade, then, the pressing question was not – or should not have been – whether 1.5°C was dead or alive, but why it had never been allowed to gestate in the first place. At the moment of its supposed birth, it had been kept stillborn. In the mainstream iterations of science and policy, it was entirely conditional on the idea of some degree of overshoot to resurrect it from the dead. Overshoot denied 1.5°C any real life by flying away from an objectively revolutionary reality: it marked the synthesis of laissez-faire and temperature targets in the Paris era, on the unifying principle of limits not being limits, in perfect correspondence with developments in the base. It was fully established that as long as more CO_2 is added to the atmosphere, the warming will continue and the crisis will get worse; and in the early third decade, *no limits* had yet been imposed on this process of addition. In Shue's words, 'we have failed to place any outer limit on the severity of climate change', meaning that 'we are currently on course to do however much damage to the climate of our own planet humans are capable of doing.' Only real limits could alter this trajectory. 'Each kind of disaster is possible and can be reached easily from the route we are now on, until limits make it impossible.'[170] But limits were the one thing capitalist society could not bring itself to institute and respect, which meant that, in the fullness of time, it would have to deal with the consequences in some other way.

Adapt, Remove, Engineer

A tragedy of overshoot is its self-fulfilling character: if it becomes too hot, the world really will have to do something else – or more – than trying to cut emissions. But this does not have to mean carbon dioxide removal solely. If one cannot stop global warming, the immediate alternative is to live with it. In its first obituary for the idea of limiting the process, anno 2010, the *Economist* frankly stated as much.[171] In one of the earliest high-profile articles on overshoot, published in *Nature* in the run-up to COP15, with Martin Parry – co-chair of the fourth assessment report – as lead author, the schema was 'overshoot, adapt and recover'. Emissions would need to peak in 2015 and then contract by 3 per cent year on year, Parry and his co-authors argued. But if this best-case scenario played out, there would still be a high risk of exceeding 2°C. Bending the curve in 2015 would be a tall order, 'because it would require substantial reductions in fossil fuel use and deforestation', and so the peak might well come to be postponed to 2025 or even 2035, with correspondingly growing risks of shooting past targets – what to do then? Overshoot, in whatever quantity, would mandate 'massive investment in adaptation'. What cannot be prevented must be better endured. 'We should be planning to adapt to at least 4°C of warming', Parry et al. advised.[172]

Or, if overshoot cannot be prevented, it might be whisked away by some very unconventional means. Tom Wigley, progenitor of the former idea, penned a piece in *Science* in 2006 proposing that cumbersome mitigation be traded for a much lighter load: geoengineering. Putting some substances or objects into the stratosphere would block a portion of the incoming sunlight. Temperatures on Earth would drop. Wigley spun three scenarios: one with no climate policy, one with mitigation only, one with overshoot treated with geoengineering. The latter had the great attraction of allowing 'much larger CO_2 emissions and a much slower departure' from business as usual – several decades extra; so much more time to 'phase out' technologies at the source. Respite could be extended to capital invested in fossil fuels by nullifying their climatic effects. The overshoot would not even have to register on thermometers: planes in the sky could just shoot out material to cool the Earth. Wigley went for sulphate aerosols, the most popular candidate; in his modelling, the stuff that would permit overshoot without any noticeable warming. 'A relatively modest geoengineering investment', he concluded, IAM-style, 'could reduce the burden on mitigation substantially, by deferring the need for immediate or near-future cuts in CO_2 emissions', while making overshoot a non-problem.[173]

Both adaptation and geoengineering had long prehistories. Both were familiar to the cognoscente of climate politics. In his 'To Slow or Not to Slow' article, to take but one example, Nordhaus placed mitigation side by side with these two options. Adaptation 'could take place gradually on a decentralized basis through the automatic response of people, institutions, and markets as the climate warms and the oceans rise': if certain zones of human inhabitation become 'unproductive', the labour will simply migrate to more 'productive' ones. If the seas rise, settlements will 'gradually retreat upland' or be protected by walls. Perhaps governments might also escort resources towards more adaptive locations. And then there is geoengineering – 'shooting particulate matter into the stratosphere', quite possibly at bargain-basement prices.[174] If these ideas had been around for some time, and adaptation in the 'automatic' sense very much already practised, a novel circumstance kicked in during the early third decade: developments in actually existing capitalism were pushing the first premise of the overshoot idea towards actualisation. The limit was about to be crossed in real life and the historical alternatives to mitigation ipso facto knocking on its door.

Ironically, or not so ironically, dominant classes intensified their preparations for this new stage in Paris. In January 2018, French president Emmanuel Macron announced the formation of the 'Paris Peace Forum', a match for the World Economic Forum of Davos, more focused on politics and 'governance', less directly on money-making.[175] As president of this Forum, no lesser a dignitary was appointed than Pascal Lamy: between 2005 and 2013, director-general of the WTO; before that, a commissioner for trade in Brussels; with stints in banks and other businesses, as reliable and weighty as any neoliberal technocrat on European soil. In April 2022, Lamy and his Peace Forum unveiled the fresh initiative of a commission for managing overshoot.[176] A dozen luminaries were handpicked to ready the world for the all-but-inevitable. They included Laurence Tubiana, the French diplomat in charge during COP21 who would later often be named *the* architect of the Paris Agreement, and Jamshyd Godrej, CEO of the Indian conglomerate Godrej & Boyce, assets ranging from aerospace to auto.[177] Setting itself up as an informal shadow Panel, the 'Climate Overshoot Commission' entered deliberations in June 2022: as if to spit on the memory of Mostafa Tolba, the first meeting was held in Bellagio, where his Group had laid down the principle of 'unambiguous' targets.[178] For the fourth meeting, in sinking Jakarta in February 2023, Bill Gates sent a 'supportive video' to the Commission.[179] It was looking into three 'approaches to reduce risks beyond what emissions cuts alone can achieve': adaptation, removal, geoengineering.[180] Outside the presumably air-conditioned halls where its meetings were held, wheels were now rolling along those tracks.

Since the sources continue to vent their heat, not much else can be done: adapting to the impacts and removing the carbon and engineering the amount of sunlight fairly fill out the menu of remaining options. In the early third decade, this triplet jelled as the fill-in for the emissions cuts that refused to transpire. In its obituary for 1.5°C, the *Economist* departed from the unbreakable bond between the world economy and fossil fuels and inferred that 'greater efforts must be made to adapt' to flood, drought, storm, fire; thankfully, 'a lot of adaptation is affordable.' On top of this, 'having admitted that the planet will grow dangerously hot, policymakers need to consider more radical ways to cool it' – radical in an utterly unradical sense of the term. The magazine was thinking of removal and geoengineering.[181] In the many versions of this argument circulating by the early third decade, business as usual would drive past 1.5°C (or 2°C, or indeed any other limit), but adaptation would soften the impacts, removal take out the excess carbon and geoengineering 'shave off the peak' – as the saying went – of the temperature rise; all in all, making the overshoot manageable. The second instalment of this study will delve into each of the three measures. But in the rest of this volume, we shall stay with the basics.

The Rationalist-Optimist View of Overshoot

Overshoot ideology is the ideology of anti-revolution in a warming world. It posits that an extended time above any given temperature limit is now unavoidable, because fossil fuels cannot be phased out at the required speed. But beneath this veneer of descriptive statements, normative dimensions lurk: some futures are more worthy of consideration than others; some things are just too valuable to smash. By the third year of the third decade, the ideology had become prevalent, to the extent that an average article in *Nature Climate Change* now discussed whether the overshoot of 1.5°C would last for sixty-seven years or forty-seven. The notion of 'limit' had migrated from pathways that limit the warming as such to those that 'limit the degree of temperature overshoot' – a matter of keeping the transgression within some bounds.[182] The central wager was that removal technologies would, ultimately, roll it back. But the removal was flanked by adaptation and geoengineering, in an ideology branching out into several types of technologies, united in the blunting of the effects of business as usual; and in its most programmatic form, this ideology still maintained that everything would turn out well.

It said: agents are rational. They will deploy adaptation and removal and geoengineering in such a way that harm is minimised. They will realise that the cuts so long deferred must at some point be undertaken, alongside these auxiliary efforts; they will adapt, remove, engineer but also reduce, so that a return to 1.5°C (or 2°C, or . . .) can eventuate. The three interim measures will merely *buy time* – the phrase so expressive of purchasing power and priorities – for the mitigation that must, one of these days, ensue, or else the overshoot will go on forever. And because agents are rational, they will not renege on this obligation. Removal will be their main return ticket, complemented with adaptation and geoengineering, redeemed with the overdue cuts: talk not of the devil but look on the bright side. We shall call this *rationalism-optimism*, a central component of overshoot ideology, carrying over the elements of the IAMs in which it was incubated.

Against it, a suspicion immediately arises. Each rather seems susceptible to capture by fossil capital and likely to prolong business as usual even further, the promise of cuts receding into the distance, the excuses to go on redoubled, the return ever more elusive, until all that is left is the catastrophe unmitigated. How reasonable is this suspicion? A look around the world in the early third decade might give a prima facie answer; we shall offer further considerations below.

The Not Quite Genocidal Logic of Overshoot

If rationalism-optimism marked much theory and some practise of the three types of overshoot management, there was also a darker side to the ideology. Having explained why we must abandon the idea of quitting fossil fuels overnight, the *Economist* admitted that doing so can 'feel, to those who care, like giving up on the poorest, who will suffer more than any others after the threshold is breached. But the truth needs to be faced, and its implications explored.'[183] The truth in question was that one must, in fact, give up on the poorest. 'Overshooting 1.5°C does not doom the planet. But it is *a death sentence for some people, ways of life, ecosystems, even countries.*'[184]

Such words ought to have sent chills down the spine of readers, if there were any who cared. This magazine of note did not shy away from passing a mass death sentence in the service of the sanctity of fossil capital: because 'a revolutionary overhaul of the system', to again speak with Anderson, was unthinkable, entire peoples must die. It did not quite meet the definition of genocide, because the intention was not to destroy any particular group of people and hunt down its members for elimination. The passing of ways of lives and countries into history was just a price to be paid, with or without pangs of remorse.

Counter-revolution in the classical era was, of course, the fountainhead of truly genocidal violence. Perhaps the anti-revolution could afford to be less discriminating about its victims. It did not train its guns on a designated subhuman race, but on anyone who happened to go unshielded from the bullets. Perhaps we should call it paupericide.

Into the Overshoot Conjuncture

Because overshoot ideology is an ideal expression of the dominant material relations, the uncontrolled raging of these relations grasped in ideational form, we might also speak of an overshoot *conjuncture*. It is the period in time when the dominant classes demonstrate their constitutional inability to shut down the drivers of global warming in the face of fast approaching limits. The inability is then converted into other projects for managing the fallout. At the moment of this writing, we seem to be entering this precise conjuncture – indeed, as we have seen, the years between 2018 and 2022 formed its opening act – and it looks set to last for a potentially very long time indeed. Already here, we may hypothesise that it entails a multiplication of the fronts of climate politics. The climate movement and its allies will have not less to do, but more. There will be not fewer struggles but more of them, even if under conditions ten times less favourable than at 0.5°C or 1°C: struggles over adaptation and removal and geoengineering. But all will remain defined by and drawn back into the unfinished struggle over mitigation.

What got us into this conjuncture will not stay behind like some passive launching pad. As long as a 'revolutionary overhaul of the system' is not part of the programme, the fundamental forces thrusting humanity into overshoot, driving beyond declared temperature limits, will rather continue to determine what unfolds. So what are they?

PART II

Fossil Capital Is a Demon

The first place to look would be the base, not the superstructure. Investment flows are a prime suspect. If you buy an SUV today, you do not plan on sending it to the scrapheap tomorrow. Even if you have a taste for fast fashion in automobiles, you will mean to drive this car for more than a day or two, because it has cost you a goodly sum – even if, as is likely, you have a lot of money. You will not spit it out like a piece of chewing gum. Now if you have a lot of money, and instead of consuming 'durable goods' such as swanky oversized cars, invest it in heavy fixed capital, you will be less inclined still to relinquish your purchase, sworn to keeping it in operation for more than a few years. If, for instance, you elect to underwrite an oilfield deep under the waters of some remote nation, you do not intend to use the crude for your own benefit: you plan to have it sold and get your money back with an increment, also known as profit, or else you might as well have kept the money in the bank; and for this to happen, the crude must be sold for a decent amount of time. A deep-water field – think Exxon in Guyana – is not cheap to develop. For any sizeable oil or gas field or coal mine, it takes about ten years to reach 'break even', the point at which the initial outlays have been recuperated.[1] Deep under water, it might take between twelve and twenty years.[2] Only after that day does the phase of profit-making kick in. If you count yourself among the owners, it is now that your good times begin, and you will want to extend them years and decades into the future, squeeze every drop of profit out of the installation, which, in those later phases, can be considered practically 'costless and its further use as free': a pure source of money, as long as the fuel fetches a price on the market.[3] You are committed to the endurance of the installation like a holy man to that of his temple. Then what happens if there is a burst of investment in fossil fuels just as a limit to the rise in temperatures is fast approaching?

4

The Political Economy of Asset Stranding (or, Blood and Gore Come to Wall Street)

There is a glut of such commitments to future production. We have seen that Germany in 2022 signed a contract with Qatar and ConocoPhillips for gas deliveries to start five years later and run for fifteen, although Robert Habeck, the climate minister from the Greens, would have preferred more than twenty, imports continuing into the second half of the century. We have seen how the Cambo oil field was planned for operations until 2053. EACOP had an expected lifetime of twenty-five years, the trans-Saharan gas pipeline a 'minimal' span of thirty.[1] In Guyana, ExxonMobil was 'taking steps to ensure' that it could maintain production in the newly opened fields 'for as long as possible', counting in the decades.[2] Similar time horizons stretched from the gas fields of Mozambique.[3] ConocoPhillips used the record earnings of 2022 to add 'high-quality strategic projects to enhance our global portfolio for decades to come', oil in Alaska the latest frontier. It received an extension of the license for extraction from the Norwegian field of Ekofisk until 2048: by that date, it will have been in operation for seventy-nine years since the same corporation discovered it.[4] This would take us to mid-century, at the least. But that would be early in some calendars.

One of the main players in the Norwegian oil and gas industry, Aker BP, announced that fresh stimuli from the state, rebating much of the capital expenditures, and the ongoing profit bonanza motivated new projects not to be unplugged before 2080.[5] Coal plants were counted on to stay in business for at least fifty years, possibly seventy-five and beyond.[6] Plants opened in China in 2022 would, in other words, reach the end of their line somewhere between 2072 and 2097 or later. Mines generally lasted shorter – maybe twenty-five years, maybe fifty; but these were lower bounds: the Adani owners of the Carmichael mine in Queensland calculated it would last for sixty years, or to 2082.[7] EMR Capital, the private equity firm based in the tax haven of the Cayman Islands that owned the Woodhouse Colliery in Cumbria – where coal extraction based on capitalist principles had been going on since the Elizabethan leap of the late sixteenth century – settled for a more modest fifty years, or the early 2070s.[8] But if there were still oil to extract in the fields of Aker BP in 2080, or coal in Carmichael or Cumbria around that time, there would certainly be an interest in going on for longer, the licences extended further, as in the case of Ekofisk – all perfectly rational from the standpoint of capital.

What would it take to stay below 1.5°C in a world where the primitive accumulation of fossil capital proceeds apace in this fashion – or 2°C, or 3 . . . ?[9] It would require that investors suffer the blow they dread the most: installations just funded, or just finalised, or just inaugurated, or just about to reach break even or starting to yield a profit would have to be sealed and locked up for good. More broadly, this problem is known as that of 'stranded assets'. The dry standard definition says that stranded assets are assets that 'suffer from unanticipated or premature write-offs, downward revaluations, or conversion to liabilities' – in everyday language, milch cows killed before their time is up.[10] Apart from being dry, the standard definition is also deficient, as we shall shortly see. For the overshoot conjuncture, it is nonetheless of utmost importance that a discourse about stranded assets emerged in the second decade of the millennium – 'discourse' in the vernacular sense of something talked about, discussed, indeed hotly disputed and debated; in this case, analogous to that of 1.5°C, on the initiative of subaltern forces.[11] This discourse marked out the threat posed by mitigation to investors. It bolstered their motivation to cling to business as usual and deflect climate politics onto secondary fields. It defined a mission on the primary, central front yet to be accomplished.

Talk of Keeping It in the Ground

Three interrelated moves engendered the discourse. First, popular opposition to northern oil companies active in countries of the South – notably Nigeria and Ecuador – jumped scale and became a cause célèbre beyond their borders, entering the deliberations of courts and governments (as in the various lawsuits against Shell for its destruction of the Niger Delta and the Yasuní-ITT initiative for foregoing oil extraction in an Ecuadorian rainforest). The slogan 'keep it in the ground!' travelled with the winds and passed over into proliferating struggles against fracking.[12] Place-based resistance, one might say, strove to convert certain assets into worthless jetsam. Second, the climate movement in the North turned away from the COPs in disgust after the fiasco of Copenhagen. Instead, it trained whatever fire it could muster directly on the producers of oil and gas and coal. The campaign against the Keystone XL pipeline in the US and Ende Gelände in Germany became emblematic. Focus fell on supply, not demand; on profits from the exploitation of fossil fuel reserves rather than the epiphenomenal emissions.[13] Third, in a move that would enter movement lore, in 2012 Bill McKibben wrote an essay for *Rolling Stone* on the 'terrifying new math' of global warming: on their books, fossil fuel companies had reserves five times larger than what could be contained within a carbon budget for 2°C. If they had them extracted and burnt, the planet would go up in flames. McKibben declared this industry 'Public Enemy Number One' and proceeded to launch a movement for divesting from it, activists marching through campuses, churches, pension funds, philanthropic foundations and other institutions and trying to shame them into withdrawing their money from this line of business.[14]

In the waning years of the Kyoto era, 'stranded assets' worked their way up to the top of the climate policy agenda. Indeed, already the year before McKibben's *Rolling Stone* essay, Nicholas Stern – a mild, progressive lord to match Nordhaus, the Genghis Khan of bourgeois climate economics – took to *Financial Times* to warn about a 'profound contradiction' between the stated goals of mitigation and the reserves fossil fuel companies had earmarked.[15] In 2013, Al Gore joined forces with an investment manager by the name of David Blood to advise the businessfolk reading *Wall Street Journal* to sell such assets off. These would soon be stranded anyway, because of regulatory efforts ensuring 2°C, or even just 'grass-roots protests and changing public opinion'.[16] (We do not need to resist the temptation to see some meaning in the authors Blood and Gore coming to Wall Street.) In the US, the push for asset stranding reached peak proximity to power (so far) in 2015, when Bernie Sanders introduced the 'Keep It in the Ground Act', which would have banned all new fossil fuel projects on federal land and called off drilling in the Gulf of Mexico and the Atlantic, Pacific and Arctic oceans, then still in early stages.[17] By the time of Paris, the resistance stemming from the banks of the Niger and the Napo had launched the discourse into the metropolitan mainstream.[18] It should be noted that, very crucially, the case made in these years by Stern, Gore, Blood et al. did not, unlike the IAMs in parallel development, relax the carbon budget by means of removal technologies.[19] The stranding would happen in their absence.

The climate movement of the second decade did not attain any of its loftier goals, but it did succeed in this one regard: a cloud of uncertainty was hung above fossil fuel properties. 'Much about the fossil-fuel status quo can no longer be taken for granted,' three scholars summed up the situation in 2020.[20] A business, until then routine and mundane and largely inoculated against the presumptions of mitigation, was called into question. Or, more accurately, the calling into question of this business – a corollary of the discovery of global warming, feared and fought as such for half a century – came closer to concretion. This achievement at the level of discourse was underpinned by local victories: every cancellation of a pipeline, every decommissioning of a coal mine inflicted losses on their proprietors. Between 2017 and 2019, to take but one instance, resistance caused the scrapping of fossil fuel projects in Canada to the tune of 100 billion dollars.[21] Such blows stayed at a limited scale, evidently. They might be considered a form of prefigurative asset stranding.

What would it mean to scale them up? What sort of crisis would that induce, with what kind of consequences? Would this mode of production be able to live with asset stranding, or would it rather pose a mortal threat, beyond the precincts of drillers and diggers and into the vaults of all kinds of capital? Can it fit into existing theories of crisis, or are we here dealing with something without precedent? To ask these questions is to size up the forces that cannot stop at temperature limits.

The Inertia and Exposure of Fixed Capital

If asset stranding is indeed defined as 'unanticipated or premature write-offs, downward revaluations, or conversion to liabilities', we must conclude that this would fall well within the framework of capitalist normality. That much is part of the game of accumulation. At the most humdrum level, it proceeds by discarding old stuff for new: in a matter of decades, VHS records and analogue cameras came, saw, conquered and went into obscurity when digital technology flushed them out. Henceforth, a computer could be given up as obsolete after a couple of years, suggesting a speed-up of the 'downward revaluations'. Asset stranding would, on this view, appear as natural to technological development as abscission to plant growth. Then there are more upsetting sequences in individual branches – English canals displaced by railways, or Malayan rubber by the synthetic variety – and, finally, those epochal disruptions that encompass the world economy as a whole and shed its leaves, break its branches, knock over the trunks like a once-in-a-century storm, although the common metaphor for these events is rather aquatic: in Schumpeterian and other theory, the long waves of capitalist development.[22] One prominent scholar of asset stranding has indeed pointed to the Schumpeterian version as proof that this thing 'occurs regularly as part and parcel of economic development'.[23] The main exponent of it, Carlota Perez, conceives of innovation as 'a bulldozer' crashing into the old way of doing things, pulverising fixed capital, clearing the way for the next surge of growth: what happened when steam knocked out water, when electricity revolutionised everything, when cars and mass production made whole strata of technologies outmoded, when IT mercilessly shook out face-to-face services.[24] The stranding of fossil fuel assets would be but another episode of such 'creative destruction'.[25]

A similar but more generative conclusion might be derived from Marx. It flows from the category of fixed capital. Capital as such is self-expanding value, value in motion, money always swimming through the world and swallowing it so as to make more of itself, as unable to stop as certain species of shark, who no longer take in oxygen when they cease moving and so suffocate and die: perpetual forward motion is their condition of existence. But for capital to stay in motion, some of it must freeze. To produce commodities that can be sold for a profit – to produce surplus value, in other words: the value to be subsequently reinvested for further valorisation – some means of production must be deployed along the way. A favourite example of Marx's is the steam engine. A capitalist in mid-nineteenth-century England who invested in the production of cotton thread would first buy a steam engine to power his factory. He would also buy bales of raw cotton, all of which would make it into this first, inaugural batch of thread leaving the premises. The material contained in the bales would then have passed over into the threads, baked into them, so to speak, and dispatched to consumers in this new guise. But the steam engine would remain behind after the completion of this first cycle of production, to animate a second and third and hopefully a thousandth run. It might take a few hours for the raw cotton to become yarn, but the engine 'is renewed only after some twenty years, say'.[26] The cotton would become the possession of the consumer, while the engine remained the asset of the capitalist. On these accounts, the former classifies as 'circulating' or 'fluid' capital, the latter as 'fixed'.[27]

One should not here confuse fixity with immobility in space. Ships and locomotives, Marx points out, belong to the category of fixed capital; to take a contemporary example, an LNG tanker that ferries fossil gas from Qatar to Germany travels a long way, but it has to make many rounds to pay off.[28] The fixity is a matter of function. The steam engine or LNG tanker is used in cycle after cycle, contributing to the production of so many units of yarn or gas, slowly wearing down and losing value – value transferred to the commodities sold on the market. At the end of its days, 'eventually', a piece of fixed capital 'expires and its entire value has separated off from its dead body and been transformed into money'.[29] While the functionality is the defining quality, it cannot be distinguished from materiality: an engine could impel spinning-machines only by dint of its physical properties. It functioned and endured because it was made up in a certain way. Fixed capital does not have to be immobile, but it must possess a durability – consist of, say, metal not soap – but since no material escapes the laws of thermodynamics, it too will meet its maker. An engine would someday break down; a tanker will rust and corrode. This is the ultimate entropic horizon of fixed capital, the physically determined expiry date, all the way up to which the capitalist will seek to keep it going.[30]

Most of this capital, granted, is also fixed in space. It spreads out across the world as an infrastructure or landscape of heavy pieces linked to one another, each 'localized by being incorporated into the earth'.[31] This imparts inertia to capital. The determinate physicality, the piecemeal, extended transfer of value – or, the long 'turnover time' – ties capital down to its fixed pieces. It becomes defensive about them. It will not lightly accept any interruption to commodity production, as that will mean loss of time that ought to be used, first, to amortise the pieces and then to let them do their work already paid for. To realise and prolong their value, capital will then restlessly guard the fixed portions of itself and strive for their full utilisation. Ultimately, the value of this fixed capital 'is measured by its contribution to surplus-value production', in the words of David Harvey; but due to its heft and rootedness, it is also the case that 'fixed capital slows everything down'.[32] Capital loses some of its proverbial agility. It might become loyal to old technological schools. The extant volume of fixed capital forms, says Marx, 'an obstacle to the rapid general introduction of improved means of labour', normally the best way to maximise surplus value.[33] Capital will not leave the side of its 'dead labour', another term for fixed capital – things once produced by living labour; now lifeless objects.

The flip side to this inertia, however, is that it also *exposes capital to its own development*. The longer the life of an engine, tanker or some other piece, the greater the probability that disturbing things happen while it lasts. A competitor could, for example, lay his hands on a more efficient steam engine. If it powered a double row of spinning-machines, cranking out twice the amount of thread in the same time, the erstwhile pioneer would have found himself burdened with a superannuated model, now rather a liability: Marx calls this 'moral depreciation'.[34] The risk inheres in any investment in fixed capital. The theory of this category is a theory of *inertia and exposure* as two sides of the same coin.

In fact, the risk is never far away: moral depreciation happens all the time, insofar as capital remains on the prowl for technologies of higher productivity. 'If the development of fixed capital extends this life, on the one hand, it is cut short on the other by the constant revolutionising of the means of production.'[35] In the game of competition, passé pieces will have to be jettisoned by owners who want to stay in the race. Devaluation is synonymous with the very development of the capitalist mode of production: 'a large part of the existing capital is *always* being more or less devalued' in the process.[36] Such bitter rejuvenation accounts for the perpetually unsettled character of this mode. It reaches a pitch of intensity in moments of crisis. 'Competition forces the replacement of old means of labour by new ones before their natural demise, particularly *when decisive revolutions have taken place. Catastrophes, crises, etc.* are the principal causes that compel such premature renewals of equipment on a broad social scale.'[37] The scale is then so broad that devaluation turns into wholesale '*destruction of capital*'.[38] This too is to be expected. Precisely because of the longevity of fixed capital, 'great catastrophes must occur' at some point or other, and then production might come to a full stop, just as feared. Commodities rot in warehouses. Raw materials lie unused. Machines stand idle, buildings remain unfinished – 'all this is destruction of capital'. Under such pressure, only the very best, least costly fixed capital can survive; called home to their lord far too early, a plethora of means of production 'go to the devil'.[39]

Crisis is the memento mori of the mode. But it is also a restorative force.[40] It is the autumnal storm that shakes the leaves off old vegetation so there can be a new season of growth: 'a crisis is always the starting point of a large volume of new investment.'[41] In the Marxian theory – on this score, close to the Schumpeterian theory of long waves – self-inflicted destruction is not only normal but cathartic for capital.[42] Marx speaks of a 'violent destruction of capital not by relations external to it, but rather as a condition of its self-preservation'.[43] It purges it of ancient dross. The obstacle to the rapid introduction of the latest tech is removed. Untrammelled, the oldest, least productive means having gone to the devil, capitalist development can leap into the next upswing.[44]

Lincoln, Lenin and Beyond

Where does asset stranding fit into this theory? It would seem to be something out of the ordinary. Fossil fuel assets will hardly be subjected to spontaneous redundancy. They will not be stranded either through everyday competition or epochal crisis: there will be nothing normal about this. The process will deviate from the standard definition, which is more apt for the 'write-offs' described by Marxian and Schumpeterian theory, whether on a micro- or macro-scale. Perceptive scholars have noted this first qualitative difference: asset stranding would occur through *a transition away from fossil fuels enforced by political actors*, be they governments or movements or some combination thereof.[45] These properties will not crash because other kinds of fixed capital delivering the same product become more profitable, or in one of those periodic shakeouts of industries, or as a cleaning of the slate for the next growth spurt; they will – if at all – be *crashed on purpose*, by someone who has decided that we cannot have any more of this warming. For it to encompass an entire economy, this someone will have to be a state.

The only way stranding could conform to typically Schumpeterian or Marxian processes of destruction would be for renewable energies to outcompete fossil fuels and their technologies, but we shall soon find strong reasons to believe that this will not happen.[46] The very superior cheapness of the former excludes, we shall see, such a non-political transition. In any case, the discourse of stranded assets rather formed around the notion of purposeful crashing. If the political impetus came from place-based and other resistance, the scientific foundation was provided by the concept of the carbon budget, giving rise to a more concrete version of a mathematical equation we have already encountered: a superabundance of fossil fuels + a limit to global warming + a subject bent on implementing that limit = stranded assets.

Two bourgeois economists expressed the equation well when they envisioned a scenario of the climate crisis cooking to a point where the patience of the public snaps and the state must turn on the culpable companies. 'Stranded assets result from the government strategically deviating from the previously announced policy' – or, in short, 'policy shocks affect capital accumulation'.[47] Mark Campanale, founder of the Carbon Tracker Initiative, the financial think-tank that did most to popularise the idea on Wall Street and in the City of London, imagined stranding this way: 'if we get extreme weather events at 1.5, the policymakers could turn around and say, "right, everything closes. Nobody drives a car today. Coal-fired power stations are off. Oil production stops."'[48]

Or, in the words of two other scholars, stranding will take place 'when the government suddenly wakes up' – a sort of climate-political emergency, akin, perhaps, to the revolutionary situation as defined above, but assumed on purely logical grounds: if the alternative is 'a global extinction event', surely someone will step in to suppress fossil fuels.[49] Exactly by what measures would be immaterial. Citigroup, in a report on the risk, reckoned that states might close down existing installations, or forbid new ones, or slap such costs on them that they would become unviable; or demand could simply be choked – either way, an enforced stranding highly abnormal for capitalist development.[50]

There was thus a mismatch between the standard definition of stranded assets and the object of the discourse, for which we can now propose a correction. We may depart from the definition of fossil capital as self-expanding value passing through the metamorphosis of fossil fuels into CO_2.[51] Asset stranding would then be *the politically induced destruction of value awaiting or undergoing valorisation within the circuits of fossil capital* – abbreviated, the political destruction of fossil capital. This was the spectre raised in the second decade. Note here that the 'assets' up for stranding are not exclusively assets in the narrow, financialised sense of recent writing on the topic: assets as things owned with a view to a revenue stream in the future.[52] The future is not all that counts in stranding. In fact, the past often weighs heavier. If a company has finally, after five years of construction, opened the gates to a coal plant, the owners' interest in the future is first of all retroactive: it is about recovering the advances of capital; or, allowing the piece to transfer its value to commodities, rather than seeing that value go to waste. The assets in question, the objects coming up for destruction are *any units of property with value* within the circuits. It was a worry for them that emerged in the second decade.

Apart from the small-scale prefigurations when the resistance won a battle or two, were there any precedents for such destruction? Some scholars have pointed to the abolition of slavery. In that case too, an entire category of assets – namely, enslaved human beings – had their value cut to zero, not because of write-offs that occur 'regularly as part and parcel of economic development', but because of an 'enforced prohibition', which, in the US, required victory in a civil war.[53] Other bad memories might have stirred as well. Assets have been sweepingly destroyed in more episodes than this one:

> Nothing attests more convincingly to the long-range Communist goals of the policies which the Bolsheviks pursued during the Civil War than the systematic assault on the institution of private property . . . The so-called Land Decree of October 26, 1917, deprived non-peasant owners of landed property . . . In January 1918 all state debts were repudiated . . . Each [measure] was intended to deprive persons and associations of title to productive wealth and other assets . . . The decree of June 28 ordered the nationalization, without recompense, of all industrial enterprises and railroads with capital of one million rubles or more owned by corporations or partnerships . . . The equipment and other assets of the nationalized businesses were taken over by the state,

whimpers Richard Pipes in *The Russian Revolution*. Logically, again, asset stranding would seem closer to Lincoln and Lenin than to Schumpeter and Marx.[54]

At a closer look, however, neither case exactly corresponds to it. Slaves were not fixed capital.[55] Cotton gins and steamships were, but they escaped abolition unscathed. Properties seized by the Bolsheviks, on the other hand, included the heaviest fixed capital (railroads) as well as the lightest fictitious capital (bonds); their hammer could strike pretty much anything in between. They were not out to attack any particular embodiment of capital – say, slaves or tobacco or fossil fuels – but driven by an inclusive aversion to the phenomenon.[56] With asset stranding, the destruction would have extra-economic origins – coming from 'policy', not market – as in both Lincoln and Lenin. It would aim straight at one type of material resource, as in Lincoln. But it would ricochet across the fields of fixed and, as we shall see, fictitious capital and therefore affect a sweep more similar to Lenin in its broadness.

More Lenin than Lincoln, asset stranding would brutally

activate precisely the exposure Marx theorised. There would be premature devaluation en masse because a 'decisive revolution' has taken place – 'catastrophes, crises, etc.' curving back upon the capital that caused them. We need not think of this as actual blood and gore coming to Wall Street. Civil war is not a necessary part of the definition. But the obliteration of value is a necessary part of the definition of climate action, and if it is initiated by some political actors, it might – unlike both the American and the Russian civil wars – run its course through intra-economic processes.[57] The spark would come from the outside, but it would burn through the fossil economy by mechanisms internal to the way capital operates. Asset stranding, we might say, would be Leninist-Marxist politics, in that order.[58]

This Exploration Will Cost the Earth

What value is up for destruction? It comes in layer upon layer. Let us proceed methodically: at the bottom are the geological deposits of fossil fuels that would have to be 'kept in the ground'; we have seen how 1.5°C requires that this fate befalls the generality. But 2°C would be more charitable. In early 2020, *Financial Times* informed its readers of the terrifying math as seen from the other side: 1.5°C would mean 84 per cent of reserves forfeited; 2°C implies 59 per cent surrendered; but 3°C – the 'best case for energy producers' – reduces the figure to 4, or, in other words, allows for almost all to be brought to the surface.[59] These were stocks already under capitalist ownership. Two years later, the Carbon Tracker Initiative estimated that oil and gas and coal companies together possessed reserves ten times larger than what could possibly be dug up without blowing the 1.5°C budget: a humongous amount to be stranded, already claimed and bundled up as valuable property – more precisely, in Marxian categories, landed property.[60] Such property rests on 'the legal fiction by virtue of which various individuals have exclusive possession of particular parts of the globe': a fiction entirely contingent on a particular set of social property relations, and one with very real consequences.[61] Landed property involves nothing less than 'the right of the proprietors to exploit the earth's surface, the bowels of the earth, the air and thereby the maintenance and development of life'.[62]

Fossil fuel reserves booked by a company are bowels of the Earth ripe for exploitation, but in the contemporary business of oil and gas, they are not money found on the ground. They are products of hard preparatory work. Before any reserves can be identified and claimed, there must be exploration. Once upon a time, such work did, perhaps, conform to the Tintinesque stereotype of poking the Earth with a few sticks and 'the inside of the earth seemed to burst out through that hole; a roaring and rushing, as Niagara, and a black column shot up into the air' and 'came thundering down to earth as a mass of thick, black, slimy, slippery fluid'.[63] In Upton Sinclair's *Oil!*, the most detailed account of petroleum exploration in Anglophone literature, narrating the boom in southern California in the 1920s, Dad, the obsessive entrepreneur, literally stumbles upon the black gold.[64] While out hunting quail with his son, the latter steps into a crevice filled with oozy bubbling oil.[65]

Yet even here, in the early days of petroleum, exploration was laden with fixed capital. Sinclair describes in detail the machines rolled out over the hills in search for oil. A steam engine labours day and night to drive the drill into the Earth, abetted by cement and cylinders and chains and mixing and riveting machines: 'it took money to drill an oil well out here in California.'[66] 'People kicked at the price of gasoline, but they never thought about the price of drill-stem and casings!'[67] It's the same story in *Cities of Salt*, the first fifteen chapters of which undoubtedly form the most powerful account of oil exploration written in any language. The arrival of oil in the Arabian Peninsula is heralded by an army of machines. The Americans burst into the oasis of Wadi al-Uyoun with crates and 'large pieces of black iron' and all manner of alien, roaring, flashing machines for locating the liquid.[68] Abdelrahman Munif had a PhD in petroleum economics and a career in the Iraqi oilfields behind him when he wrote this nonpareil masterpiece of fossil fuel fiction.[69] His hero, Miteb al-Hathal, 'the troublemaker', prophesises the death of the oasis and the end of the world as soon as the American explorationists show up. 'Be assured of this, people of the wadi – if they find what they're after, none of us will be left alive.'[70] When they set their drills and other mechanical accomplices to work, he can barely contain himself. 'In the wink of an eye they unleashed hundreds of demons and devils. These devils catch fire and roar night and day like a flour mill that turns and turns without tiring out and without anyone turning it. What will happen in this world? How can we kill them before they kill us?'[71] The catastrophe begins with fixed capital.

If oil exploration required the mobilisation of fixed capital already in interwar California and Arabia, however, that was nothing compared to twenty-first-century business practices. They were distinctly more 'capital-intensive', to use mainstream jargon. The tracking down of supplies involved increasingly advanced technologies, such as the latest models of 'airborne magnetometers', instruments attached to aeroplanes or helicopters – or, increasingly, drones – flying over the Earth and scanning its bowels.[72] With drones, companies deepened and widened the ambit of the search. They 'could be deployed to provide access to hard-to-reach areas, such as vertical or overhanging rock outcrops' and send back three-dimensional, 'hyperspectral' images, maps of the subsurface drawn in the stark colours of an oil spill.[73] Such nimble little pieces of fixed capital marshalled and condensed the ever-growing 'general intellect' of geophysical science in the hands of the explorationists, who, of course, had always relied on it.[74] Both *Oil!* and *Cities of Salt* depict them as carriers of mechanised knowledge.[75] But as the industry had to go farther afield to find reserves, the work became heavier with that load – particularly when it went offshore. Detecting oil in deep or ultra-deep waters demanded, among many other things, 'drillships', special-purpose vessels sailing between sites to plumb the depths, sending equipment down through the opening or 'moon pool' in the hull.[76]

We can here discern a historical tendency of great import. It has been forthrightly formulated by Equinor, the leader of the Norwegian pack of hunters, scouring through the waters of the

North Sea, the Arctic, the Gulf of Mexico, Brazil, Angola, Tanzania: 'most of the easy-to-find oil and gas in the world has already been discovered, forcing our explorationists to continuously come up with innovative ideas and utilising the latest technologies.'[77] Or, there is *a tendency of fixed capital in hydrocarbon exploration to rise over time*, as absolute volumes and in relation to any quantity of oil discovered. The longer fossil capital continues to operate, the more of the easily accessible oil and gas supplies are consumed, the greater the lengths companies must go to find new ones: and this is possible only by means of larger quantities of dead labour.[78]

The tendency received its first sustained literary expression in 2023, with the appearance of *The Black Eden* by Richard T. Kelly.[79] Set in the North Sea, it follows the two friends Aaron and Robbie, initially bound by the shared taste for diving, as they are pulled down the vortex of oil. In the late 1950s, companies have turned their searchlights into the waters off Scotland. A top advisor to the British minister of petroleum lays out the challenge:

> Where oil is drilled for, sir, it is won only by great expenditure and risk. In the case of the North Sea this truism would apply many times over. I admit, I struggle myself to imagine what sort of technologies could withstand that environment – certainly not the ones at hand. However, any party that sets itself to such a titanic endeavour should, I think, enjoy a measure of our encouragement.[80]

The American company leading the charge, Paxton Oil, deploys dynamite to blast through the seabed. Hired as a freshly graduated geologist, Aaron is told by Mister Paxton, the swashbuckling owner, that 'we ain't hunting for gushers no more. We gotta go deeper – however hard that is,' the sign of a new era: 'it ain't like there's gold under your feet no more. This oil is hiding some.'[81] Aaron becomes the chief explorationist on a rig, while down in Westminster, the government decides to back the charge, in all its expensive, hazardous, technologically daring character: 'this exploration will cost the earth. It will be highly dangerous – it will cost lives, of that you can be sure. It will require extraordinary skill, in conditions where the old tools simply will not cut, so that will call for innovation, too,' the advisor presses the case.[82] And indeed the specially devised installations turn out to be gargantuan. When Robbie first sets eyes on the platform where Aaron will provide the scientific expertise and he the muscles of manual labour, he has the impression of a mythological beast, 'some ancient skeletal sea-creature protruding from the waters as though shackled there'. On this island of metal, there is an 'overpowering stench of diesel' accompanied by a 'monstrous roar and vibration', something so large as to reduce humans to pygmies.[83]

It takes nearly 300 pages and more than one decade before any oil is discovered. On the way to the hidden gold, Aaron oversees the drilling of one dry hole after another, fears for his reputation, fancies himself an alchemist and feels himself going mad. His geology teacher, who once advised him to go into this business, has an epiphany: news of oil spills convinces him that the race leads into the abyss. He now counsels his former student to return to the surface before it is too late. But in the character of Aaron, investments in fixed capital are mirrored by those of a psychic nature: 'I've gone so far down a road now, made commitments. The further you go with something . . . it gets a lot harder to turn around.'[84] When production finally gets underway, the instruments are so massive that Kelly again turns to similes of the mythic and the ancient: there lies the newly constructed platform,

> that great skeletal beast that has dominated the view from above for so long – now complete and ready to serve. However brutal in aspect, it has a majesty of sorts, like some ancient pyramid, resting in its purpose-made concrete basin, waiting to be towed in the Firth for its great reckoning with the seabed a hundred miles out.[85]

Like any novel of fossil fuel fiction, *The Black Eden* is pervaded by forebodings of disaster, the lives of Aaron and Robbie ruined by oil, more literally so in the case of the latter. The plot is punctuated by two lethal accidents: in both, recently acquired pieces of fixed capital are accorded greater value than life. Investors would rather sacrifice workers than lose time for recuperating outlays. As they die deep under water, the divers are almost literally pressed down by the millstones of dead labour; although set in the 1960s and '70s, the imagery and narrative logic of this story are entirely in the spirit of the 2020s. So is the reaction to a proposal to revoke rights to drill. 'I reckon those companies who've sunk a load of money into buying the licenses might not be best pleased to see them cancelled.'[86]

Fossil Terre–Capital Goes Home to the Devil

What does this tendency imply for the prospect of closing up already booked reserves? Note: not reserves necessarily deep into development but reserves only just discovered and claimed. These form a type of property of a peculiar nature. It might seem as if they slip through the net of the category of fixed capital and end up on the other side; a steam engine in the hills of southern California or a drillship shuttling between the North Sea and Angola would seem a better catch. The reserves would then be commodities. But this would be an erroneous conclusion, for two simple reasons. First, a piece of fixed capital deployed in exploration does not *produce* the reserves, in the way, say, a spinning-machine produces yarn: rather it *uncovers* them, or makes them accessible for extraction. Second, the reserves are not normally put on the market, but rather jealously guarded as the most treasured possessions of the companies in question. ('"True?" echoed Dad. "Why, boy, we got an ocean of oil down underneath there; and it's all ours – not a soul can get near it but us!"' [Sinclair].[87] '"Under our feet, Ibn Rashed, there are oceans of oil, oceans of gold," replied the emir' [Munif].[88] 'You got rich – *you* got rich, not some other bastard' [Kelly].[89] Although it does, of course, happen that hydrocarbon companies trade reserves: in its sad annual report for 2020, to take one random example, Equinor reported that it had sold off its assets in the Bakken Formation to a smaller Houston-based company.[90])

Should we instead conclude that reserves are simply like land in the landed property of a farmer? When outlining his theory of landed property, Marx focuses on the case of wheat cultivation, but he tells us that 'instead of agriculture, we might equally well have taken mining, since the laws are the same.'[91] Here he probably has coal foremost in mind. And coal is indeed much like farmland: fairly easy to find; a lot must be done to prepare it for production, but the discovery as such is a reasonably straightforward affair, in nineteenth-century Britain as well as in, say, twenty-first-century China.[92] But with oil, things are different. This is laid out with clarity in a state-of-the-art handbook for the third decade, written by Hussein Abdel-Aal, an Egyptian professor of petroleum engineering with experience ranging from Texas to Dhahran:

> Oil searchers, like farmers and fishermen, are actually in a contest with nature to provide the products to meet human needs. They are all trying to harvest a crop. But the oil searcher has one problem the farmer does not have. Before the oil man can harvest his crop, he has to find it. Even the fisherman's problem is not as difficult, since locating a school of fish is simple compared to finding an oil field. The oil searcher is really a kind of detective. His hunt for new fields is a search that never ends; the needle in the haystack could not be harder to find than oil in previously untested territories.[93]

The same applies, naturally, to other minerals – think gold or rare earth metals – but there is a specificity to fossil fuels, in that they have for two centuries been the general energetic lever for surplus-value production and drawn capital of unparalleled magnitude into exploration.[94] Their place in the metabolic processes of the capitalist mode of production has become rather like that of grain for the early agricultural civilisations. The reserves serve as the substratum for all circuits of fossil capital; for the companies that own them, as the immediate basis for surplus value to come.[95] Equinor can pump oil and gas to sell at a profit only insofar as it has managed to unearth fields. They and they alone guarantee that the commodity – the hydrocarbons, in this case – will materialise: the capacity to produce surplus value is a function of discoveries made.[96]

And those discoveries presuppose, as we have seen, the sinking of fixed capital into the Earth. Something loosely similar happens in agriculture, when a farmer brings into cultivation fields of low fertility by, for example, using 'certain liquid fertilizers' and 'special ploughs' to master heavy, clayey soil, or building drainage ditches or levelling hills.[97] Marx here speaks of 'capital fixed in the earth' and comes up with the Gallicism '*la terre-capital*'. One of the least developed categories of Marxian political economy, 'terre-capital' signifies patches of the globe into which fixed capital has been incorporated, whether transiently (fertilisers, remote sensing) or permanently (ditches, rigs). Through such terrestrial extension, there emerges 'one of the categories of fixed capital'.[98] Unless we want to coin a new term – say, 'substratum capital' – this one appears the most apt for hydrocarbon reserves. To distinguish it from agriculture and other mining, we might want to speak of 'fossil terre–capital'. Unlike in farming, in a narrow but significant distinction, fossil terre–capital would here mean capital fixed into the Earth *in the process of making the desired subterranean land appear as such*.[99]

If some state or other entity were to declare such land undesirable, it would mean a terrific disaster for its owners, even before a single barrel of oil or gas has been pumped, for reasons that should be obvious: those owners must 'recover the capital spent in the pre-oil-production phase'.[100] The more capital spent, the greater the terror of leaving reserves unused. The tendency of fixed capital in hydrocarbon exploration to rise over time – or, the build-up of fossil terre–capital on the frontiers of oil and gas – makes for enhanced inertia and exposure: more to be lost from mitigation; all the more reason to resist it. A company might have spent a decade and 'many millions of dollars' to get a pool of oil and gas ready to go onstream; 'and also, all this time, the process is constantly repeating itself as more oil is being discovered, more oil is being developed, and more oil is being produced.'[101] The inertia and exposure enhances itself. As of the early third decade, exploration continued at full tilt – 'we will drill between 20–30 exploration wells each year moving forward,' Equinor announced.[102]

Along some frontiers, the work had only just begun. 'Harsh deepwater and ultra-deepwater areas, such as the Gulf of Mexico, North Sea, Africa coast [sic], South China Sea, and the Coast of Brazil' – the Arctic here unmentioned – 'have tremendous natural resources and billions of potential barrels that are yet to be explored,' in the estimation of one survey.[103] The profit bonanza of 2021 and 2022 sent drillships flying out on the seas, particularly off the northern coasts of Latin America. 'The untapped potential of deep-water reserves offers new appeal since shallower fields have been more thoroughly exploited,' commented *Wall Street Journal*.[104] If these efforts were successful – a question more of politics and profits than technical feasibility – the figures for fossil fuel reserves to be renounced for any given temperature limit would rise: if 2°C required 59 per cent in 2020, this could be, say, 69 per cent in 2030 solely due to a growth in the reserve base.[105] While the stranding would then have to be more comprehensive, it would also exact a higher tribute per unit of supply. More fossil terre–capital would have to go to the devil for every quantum of discovered hydrocarbons kept in the ground.[106] As for coal, the situation would be rather the reverse: because of their enormous superabundance, proportionately more of the coal reserves must be foregone for any given temperature limit; and for the same reason, less fixed capital is at stake.[107] This is one reason why the hydrocarbon sector looks likely to dominate the terrain of the overshoot conjuncture.

At the most basal level, then, asset stranding would mean *the political destruction of fossil terre–capital*, divided between the company owning the reserves and the agent, typically a state, owning the land and sea in which these have been found. The former usually acquires a license from the latter: the company a producer of surplus value, the state a collector of ground rent.[108] Needless to say, the state would lose out too, if this source of ground rent – or, government revenue – would be corked. States are usually not entities whose sole raison d'être is to produce fossil fuels for the market (think the state of Guyana vis-à-vis Exxon). They would retain a larger potential for reinvention after the stranding. But for the companies, the loss of fossil terre–capital would be the beginning of a process of complete demolition.

ExxonMobil Was Right

At this point, some might ask: but was not oil supposed to come to an end? In the years between 9/11 and the global financial crisis, a minor hysteria spread about the imminent peak and decline of oil production. No more oil was to be found. The depletion of the last reserves, acute shortages, incurable price shocks were just around the corner; in preparation, shelves filled up with books like *The End of Oil, The Final Energy Crisis, Half Gone, Out of Gas.*[109] The theory was that output from any given oil reservoir follows a bell-shaped curve. It rises and rises, until half of the oil is gone, and then falls and falls to zero. After the peak, volumes can only shrink; unlike previous oil crunches, such as in the 1970s – 'emotional and political reactions' – this one will last forever, as it is caused by absolute physical limitations.[110] Petroleum-dependent society hits a wall of vertical price rises. Conceived as one single oil reservoir, the world – so the proponents of the theory claimed – was rapidly nearing 'peak oil': one seminal text predicted that it would happen in the mid-2000s.[111] More precisely, another ventured that the global 'peak will occur in late 2005 or in the first few months of 2006. I nominate Thanksgiving Day, November 24, 2005, as World Oil Peak Day,' beyond which there would be only declension.[112]

No such thing transpired. There was no peak oil, just an unremitting rise in output – with the exception of the crisis blips of 2009 and 2020 – such that in 2019, global production volumes were 14 per cent higher than in 2005, when, according to the theory, they should have dropped by at least 10.[113] The US did experience a peak in 1970. This was the main empirical prop of the theory. But confounding it, output from this original palace of petroleum started growing again in 2009, until by 2018 the previous peak had been eclipsed and the US returned to pride of place as largest producer in the world, with no signs of let-up.[114] Norway hit a first peak in 2004: but on the back of the bonanza and frenzy, it moved towards the same heights again.[115] The curves did not have the shape of a bell. In the second decade of the twenty-first century, the notion of 'peak oil' underwent extreme moral depreciation, remembered, if at all, with cheeks flushing; rarely has a prediction been so rapidly disproven, and rarely have critical scholars – those who fell for the theory – followed such a flagrant red herring.[116] The problem was never too little oil.

The theory erred because it underestimated, most obviously, the productive forces of primitive fossil capital. Tar sands, shale, deepwater – all unconventional resources and exotic frontiers were written off as so many mirages, the dispelling of which would leave the world ever more dependent on a few sources of crude in the Middle East.[117] By the early 2020s, the situation was exactly the reverse: a greater diversity of hydrocarbons than ever, the US resurgent. Fracking, of course, was a game changer.[118] By pumping chemically slickened water and sand into shale rocks, producers could open thousands of fractures, passageways through which molecules of oil and gas would flow into a well: a technology for wringing hydrocarbons out of the most recalcitrant grounds. At the height of the peak oil hysteria, ExxonMobil intervened with an advertisement dismissing the theory as ignorant of how technology works. Not only did new fields keep coming online, but innovations such as 'multidimensional mapping tools and advanced drilling techniques have improved our ability to recover oil from previously discovered fields' – secondary fossil terre–capital, as it were, making more of the subterranean deposits appear in desired form.[119]

The theory erred, furthermore, because it underestimated the political power of capital: in the early twenty-first century, peripheries were prised open for owners of the most advanced machinery, from Iraq to the Arctic, Guyana to Ghana.[120] But the main mistake concerned technology. 'We're limited not by the amount of oil in the ground, but by how inventive we are about reaching new sources of fuel,' *Wall Street Journal* relayed the opinion of industry experts.[121] The *Journal* was quick to note, however, that technology comes with a fee: for every innovation wresting more oil out of stone or seabed, the costs rise. This was the rational kernel of peak oil theory, lost in the errors of a linear Malthusianism. When the most pliable fuels closest to the surface have been used up, capital moves on, arming itself with fixed pieces that allow volumes to keep growing – at the price of more value sunk into the ground. Here was that rare case when bourgeois optimists got it right. And that was the disaster. With the passing away of peak oil theory, there was much braggadocio and schadenfreude, but a more fitting emotional response would have been grief.[122] Had the peak happened on Thanksgiving Day 2005, the carbon budget for 1.5°C would not have been so close to depletion.[123] Instead, fresh forms of fixed capital rendered the fears null and void *and by the same measure expanded the value that would have to be destroyed*, a spiral continuing to push oil upwards *and* fortify the obstacles to mitigation.

Over the course of the 2010s, the discourse of stranded assets turned that of peak oil inside out. The world was drowning in supplies. The task was not to prepare for their drying up, but to get out of them as from an inundated city: an artificially induced 'peak demand'.[124] 'Reserves are already too high for global warming and are growing faster than ever before,' read one assessment from 2020; or, 'producers are now sitting on more carbon than ever'.[125] The outlook was for business as usual to continue *until the year 2200*, the default trajectory perpetuating itself through profits, growing reserves, more profits, and so on.[126] Another two centuries of fossil capital: there were no theoretical or empirical grounds for counting on less.

Layer upon Layer of Value to Be Destroyed

But pieces of fixed capital are not, of course, constructed solely for the moment of exploration. Rather they form nested dolls, with fossil terre–capital the innermost doll enveloped by fixed capital for the actual extraction of oil and gas, and further on. The tendency of fixed capital to rise pertains to *all* these moments, which are often blurred, so that, for example, fixed capital for production is also secondary fossil terre–capital; and as the dolls expand and fuse, they become ever more 'monstrous', to borrow a key adjective from *The Black Eden.*

Once hydrocarbons have entered the phase of actual production, they can be brought from fields to fireplaces only via links of fixed capital starting at the wells themselves. If these are located offshore, the oil will be pumped through a platform. Cities of steel, offshore platforms contain entire districts of fixed capital: towers, gangways, cranes, control rooms, helicopter pads, boat landings, not to mention the facilities for workers, ranging from bunk cabins to gyms and game rooms – all built around the central rig for round-the-clock drilling into the seabed. Here too, the second and third decades saw trends of digitalisation and automation, with robots, sensors and lasers coming on board.[127] Many platforms in the more mature parts of the North Sea and the Gulf of Mexico were then in their forties. Their expected lifetime had been twenty-five, but owners were constantly 'looking at how to extend asset life – working out options to get more from their existing assets'; flying drones around the platforms to inspect their structural integrity and identify needs for patch-up came into vogue.[128] So did new technologies for keeping corrosion at bay.[129] On the deep and ultra-deep frontiers, intense waves and winds and freezing or tropical temperatures nibbled away at the pieces, but owners strove to push back this entropic horizon by means of subsidiary machinery.[130] Engineers were coming up with designs that would keep the platforms going for one hundred years.[131] Were they to succeed, a structure launched in, say, the 2030s would have a preliminary expiry date in the 2130s, underwritten by an initial outlay of perhaps 1 billion dollars.[132]

Oil pumped from the ground must be stored on the way to combustion. Hence the storage tanks, the white or grey silos ubiquitous in the petroleum landscape.[133] Then there are the refineries, with lifetimes of up to sixty years.[134] 'There's going to be an awful lot of oil refinery capacity the world doesn't need in a 1.5 degrees world, that's going to have to be written down,' said Campanale in early 2023.[135] At that point, the largest refinery in Mexico, built to the tune of some 20 billion dollars, came closer to inauguration day, and Aramco was readying to bless not only Pakistan but also China with fresh new mega-refineries – soon in year zero out of an expected fifty or sixty.[136] Older generations kept being patched up and refurbished. ExxonMobil undertook the first major expansion of US refining capacity in a decade in Texas.[137] As anyone who has had the misfortune of visiting the Houston Ship Channel – possibly still the largest petrochemical complex in the world – will know, the sheer mass of fixed capital here in line for retirement is breathtaking.

It is no less so in the phase of transportation. 'Oil and gas pipelines act as veritable arteries inside the Earth. Using extensive steel and plastic pipes, they transport gas and oil throughout the planet,' Abdel-Aal waxes lyrical.[138] Then there are the super-tankers crisscrossing the high seas, newer models tending to be 'stronger, more maneuverable, and more durable than their pre-decessors'.[139] Not to be forgotten is the equivalent infrastructure for coal, stretching from mines via railroads and storage facilities to ports and ships. And then we still have not mentioned the plants where those fuels are burnt for electricity.[140] All in all, in the layers above the fields and seams, there resides 'the largest network of infrastructure ever built, reflecting tens of trillions of dollars of assets and two centuries of technological evolution', in the balance because of climate policy – not existing policy, to be sure, but the possibility of one.[141]

So far, we have stayed within the circuit of primitive accumu-lation of fossil capital, where fossil fuels are the commodities holding the promise of profit: the molten core of the totality.[142] But the fuels are then used as accessories in every other form of production. In the former circuit, fossil fuels are the output; in the latter, an input in the process of producing profit – what we might call fossil capital in general. Consider the steel industry, responsible for nearly one tenth of total CO_2 emissions in the early third decade. Conventional blast furnaces were by this time still melting most iron ore. With a turnover time spanning decades, they were built to generate heat above 1,000°C by the burning of coke, their cauldrons of fire and lava streams fed with this and no other hyperpotent kind of coal. A 'policy shock' would cause 'cascades' of asset stranding rippling into steel plants, as well as any number of other establishments predicated on fossil fuels.[143] If it were to shut down the production of fuel (the primitive circuit) it would also slake the derivative fires (the general circuit). Both circuits must come to a complete end: this is the meaning of mitigation in its original form. But whereas companies producing oil and gas and coal must cease to exist as such, a company manufacturing steel can in principle survive by exiting the circuit of fossil capital in general – by using other inputs than coke, that is.[144] Still the transition might be traumatic.

Whether a piece of fixed capital in the general circuit would come in for destruction depends on its physical properties.[145] An electric arc furnace running on electricity from a gas-fired power plant to melt scrap metal into steel could just as well run on electricity from a wind farm; a welding robot has no umbilical cord to fossil fuels specifically. But a blast furnace can have no second life after coke. Insofar as investment in irredeemably fossil fuel–dependent fixed capital within the general circuit proceeds apace, stranding would mean destruction of massive, indeed growing amounts of value here too. And on the other side of the primitive circuit, there are the suppliers of its fixed capital – the Halliburtons of the world, the manufacturers of drillships and offshore platforms and conveyor belts for mines and all the rest, who would lose not only a market, but possibly also tools of use only in such construction.[146]

Then there are the airports of the world. While non-fossil aircraft propulsion may be on the horizon – we shall return to some possibilities below – a policy for 1.5°C would have to close much aviation down, because it would be decades before any such replacements could be delivered at scale; likewise for 1.7°C, probably 2°C too. Built to last for decades if not centuries, airports embody capital as deeply sunk and widely sprawling as can be.[147] And what of the hotels built around them?[148] And what of the financial investors bankrolling all of this? Consider a bank that has furnished a series of newly opened hotels with hefty loans, and now the hotels must close because traffic at the airport has been choked . . . and what then of the bank that has, in turn, invested in this hotel-investing bank, while also waiting for steel companies to repay their loans for blast furnaces?

Liquidating Lundin

But before the loans, there are the stocks. They take us back to the reserves. To own a share in a publicly traded company is to have a legal claim to a slice of the surplus value it is undertaking to produce. The price of the share depends on the value investors expect to see.[149] 'The market value of these securities', Marx recognises, 'is partly speculative, since it is determined not just by the actual revenue but rather by the *anticipated* revenue as reckoned in advance'; and in the case of fossil fuel companies, the anticipations are founded on the reserves, the bedrock for surplus-value production in this line.[150] It follows that the market value of the shares of these companies is determined by the reserves on their books.[151] Most highly appreciated are 'proved' reserves, or those considered a safe bet for extraction; some way behind stand the 'probable' backups, trailed by the 'possible' cache, deemed to have the highest probability of being overestimated or even unusable.[152] When a hydrocarbon company announces a credible discovery, the stock market pays its compliments.[153] To take but one example, in 2011 the curtain was lifted on a particularly huge oil field off the southern tip of Norway, by the proud Swedish discoverer Lundin Petroleum.

(A company with a noteworthy history, to be told here only in brief. There once was a family of German aristocrats who owned a shipyard in Odessa. It specialised in steamers for the river Dnieper. During the revolution of 1905, the young boy of the family, Willy von Wagner, concocted a 'stink bomb' to throw at the revolutionaries, an admixture of his own urine and dog poop. When the First World War broke out, the family decamped to Vienna, where it would later move in White émigrés circles; attempts to return to the shipyard and estate ran aground on Bolshevik power. Willy's sister Maria ended up in Sweden. She married a chemist, surname Lundin, with whom she sired a son in December 1932: they gave him the name Adolf H. (H. for Henrik). Tragedy hit the family in 1940, when Willy, a fully committed Nazi, died during military action in Norway.[154] After the war, Adolf H. Lundin built one of the most successful enterprises of the golden age of Swedish capitalism; a rabid anti-communist, aficionado of the apartheid regime, sponsor of Ronald Reagan's election campaign in 1980, Adolf H., who never modified his name, amassed a fortune from minerals, focusing increasingly on oil and gas.[155] With his wife, a scion of Swedish cement capitalists, who had likewise been staunchly supportive of Nazi Germany during the war, he sired the sons Ian and Lukas.[156] He discovered parts of the North Field in Qatar, later recognised as the world's largest trove of fossil gas. In the late 1990s, Lundin Oil, as his enterprise was then known, entered Sudan and egged on its army to burn down villages and commit massacres so as to open up southern oil fields for extraction; as of 2023, a trial against the company for complicity in war crimes was still ongoing.[157] But in the early 2000s, the Lundin enterprise, now run by Ian and Lukas, shifted its focus to Norway. Seventy years after Willy fell, the sons of Adolf H. found a particularly huge oil field.[158])

On the day of the announcement, the stock markets rewarded Lundin Petroleum with a 30 per cent rise in its share.[159] The single largest discovery of oil in the world in 2011 – 'it will have a huge positive impact on our future production', Lundin exulted – it drove a doubling of the price over the course of that year.[160] (On one count the following year, the Lundin stock had risen more than that of any other European company over the past decade.[161]) But only in the next few years did it become clear just how huge this particularly huge oil field was: as of the early third decade, by far the largest in Europe, measured in proved reserves remaining.[162] Known by then as Johan Sverdrup, the treasure had been divided between Equinor, main operator of the field, Total and Aker BP.[163] A new page was turned in the history of the Lundin enterprise when it merged with Aker BP in the summer of 2022, under a deal that transferred its Norwegian assets to that company. To mark the occasion, Ian Lundin gloated that the family firm had 'grown into something none of us dared to dream of, with the per share value having grown around 150 times [in two decades], providing a compound annual average return to shareholders of 28 per cent for over 20 years' and – no mere boast – having 'flourished into one of the leading exploration and production companies globally'.[164] Johan Sverdrup was its biggest gift to the world.

The field was officially inaugurated on 7 January 2020, around the time when, on the other side of the world, the Australian bushfires culminated: 'some people are saying we should stop producing oil altogether, for the sake of the climate. But we believe Johan Sverdrup is a prime example of exactly why we shouldn't do that,' Equinor submitted.[165] The first phase of the project included four platforms; a second phase, completed in late 2022, added a fifth plus twenty-eight new wells to this 'highly profitable project' (Equinor again).[166] A reporter from *Bloomberg* then paid a visit to the site, which looked like five yellow-and-grey Meccano blocks linked to each other over the deep blue sea. She witnessed employees gathering for a speech from the Norwegian oil minister 'in a room resembling a hotel lobby, with soft leather chairs, a chess set and glass jars filled with candy. Down a flight of stars in the canteen, a grand piano stood off to one side': she marvelled at the 'surprising level of comfort' for an establishment this far out into the sea. 'The platform is so large that workers get from one end to the other using three-wheeled push bikes, traversing the 1 kilometer distance through enclosed walkways that link its five individual parts.'[167] Johan Sverdrup was then on the way to cover 7 per cent of oil demand in Europe. A third phase was on the anvil, expected lifetime *at least* fifty years, further exploration in Norwegian waters on the uptick.[168] ('Will you leave profitable barrels behind in your energy transition journey?' one representative of the largest Swedish bank asked the CEO of Equinor. 'No,' he responded. 'We don't plan to do that. We plan to develop the oil.'[169])

As of the early third decade, mitigation in Europe had thus come to mean, by absolute definition, the strangling of Johan Sverdrup a few years into its life in valorisation. So, imagine the Norwegian state has a change of heart. Imagine it proclaims that all oil production in its territorial waters will be terminated by 2034, so as to retain a chance of staying within the carbon budget for 1.5°C: this would be an out-and-out disaster for some, not only because of the fossil terre–capital and other fixed capital liquidated, but also because the share prices of every company involved – Equinor, Aker BP, the various later iterations of Lundin – would collapse. The loss of Johan Sverdrup alone would send their valuation on stock markets plummeting. Indeed, an 'off-limits' sign on even a minor set of proved reserves would hang a question mark over their future as whole, in the eyes of those who buy and sell shares: might their business model be nearing its end? How can we trust that other reserves will not also share the fate of limitation? If this deposit is put out of bounds, why not also deposit X and Y, including those yet to be discovered? Shares of companies engaged in the primitive accumulation of fossil capital would be sensitive to *any* restriction on the tapping of booked reserves – one reason among many that mitigation in the third decade, if not before, took on a quality of all-or-nothing.

A Tangle of Fossilised Circuits

Shares melting away towards zero value would be disastrous, first of all, for the issuing companies, because they rely on them for capital to accumulate. This has been the case since the childhood years of fossil capital. The rise of steam power in the second quarter of the nineteenth century proceeded in tandem with the ascension of the joint-stock company form, the two so tightly interlinked in British economic history as to merit a separate study of the connection; the classic example would be railways.[170] These required money beyond the reach of any single person. 'The world would still be without railways if it had

had to wait until accumulation had got a few individual capitals far enough to be adequate for the construction of a railway,' Marx commented in 1867: but railways were built 'in the twinkling of an eye, by means of joint-stock companies'.[171] Only by pooling money together in corporate entities with shared ownership could investors fund the requisite fixed capital. Because fossil capital advanced by spreading out heavy pieces of machinery, it had to boost the joint-stock company form, the scales of both growing in a relation of mutual dependence. 'The shares in railway, mining, shipping companies' – the three ventures characteristic of the steam era – 'represent real capital', money injected into metabolic processes so as to come out larger on the other side, not only as inflated wealth but also in the physical form of more 'railways, mines, steamships, etc.'.[172] There was, of course, in Marx's reading, something imaginary and fantastical about all this wealth traded on paper, but it rested on real goings-on in pits and boiler rooms and spurred on their proliferation.[173]

Fossil fuels and financial flows were tied together from the beginning in the knot of fixed capital, and so they have remained. In the event of stranding at some point in the twenty-first century, victimhood would be dual and joint: companies active in the primitive circuit would be asphyxiated of capital for further accumulation *and chunks of the stock markets would go down with them*. If the stock market is the place where legal claims to future surplus value are traded, based on the forecasts of such value; if one segment of companies have the basis for their value contracting or even outlawed in one go; if these companies have hitherto been highly valued and much traded, then the markets themselves might implode.[174]

So much for the market for shares or stocks or 'equities' – titles of actual ownership. Alongside it, there is the market for credit: money lent to a company on the condition of repayment with interest. Credit, likewise, has the function of taking the entrepreneur beyond the confines of his own pocket. Here we encounter the circuit of interest-bearing capital, or money lent out by its owner to course through the process of production, where the borrower puts it to work and engages it in the creation of surplus value, some of which returns to the lender, at the agreed-upon date, in the form of interest.[175] Money is hired out to finance sundry purchases, not the least of fixed capital. It enables projects unaffordable for even very rich individuals – say, a coal mine in the Cape Colony, or a pipeline running from the Niger Delta to Europe, or five oil platforms linked as one.[176] Indeed, the rise to prominence of interest-bearing capital in advanced capitalist countries is rooted in the need to fund lumpy, long-term investments in fixed capital.[177]

'So much financing we need that's just fresh air for now,' whines one entrepreneur trying to enter the emerging offshore business in *The Black Eden*.[178] The more expensive the means of production, the more dependent the individual capitalist becomes on borrowing from other capitalists; the longer the time it takes before they throw off actual commodities, the greater the need to arrange for repayment in the future.[179] It follows that *increased fixed capital formation necessitates increased integration* into equity as well as credit markets – or, to use a pregnant Marxian phrase, into '*the common capital of the class*'.[180] The more the primitive and general circuits pass through fixed pieces – stretching from ever-larger platforms to ever-larger airports – the more integrated they have to be into the circuit of interest-bearing capital (and, we might add, dividend-expecting capital). Only by dipping into the widest and deepest pools of the class can capital make the pieces tower high.

If integration of this kind had been going on since the nineteenth century, it had gone so far as to raise the stakes to stupendous proportions by the early twenty-first. In an illuminating paper on asset stranding commissioned by the European Parliament in 2022, Winta Beyene and her colleagues stress the fact that 'banks have traditionally been large lenders to the fossil industries. This, combined with the fact that the fossil fuel industry is very capital-intensive and requires financial resources for its operations, substantiates a big impact that the banking sector may have on future fossil fuel procurement' – and, of course, vice versa.[181] Flipping the script, *any limitations on fossil fuel infrastructure would endanger the common capital of the class by which it has been financed.* The dangers have grown pari passu with the size of fixed capital. As Beyene and her colleagues observe, total borrowing to oil and gas increased globally by an average of *15 per cent per year* between 2006 and 2014, and the trend then continued after the signing of the Paris Agreement; but this was not a matter of hydrocarbons exclusively.[182] In 2020, some eight tenths of investment in coal plants in the People's Republic of China came from its banks.[183] If those plants were to be shuttered early, the banks might never see the loans repaid. They could go into the red and default.[184]

Beside the banks, there are the asset managers that pool money from rich clients or pension funds to fill up their portfolios: they too have their destinies intertwined with fossil fuels. In a revelatory examination of ownership structures in the US, conducted in late 2020, Adam Hanieh demonstrated that the so-called Big Three – Vanguard, BlackRock, State Street – had commanding positions in the primitive circuit, holding the top three shareholder positions in ExxonMobil, Chevron, ConocoPhilips, as well as in the largest producer of shale oil, the three largest independent oil refiners, the largest gas network and the top five electric utilities in the country. Other types of financial conglomerates had eggs in the same baskets. Goldman Sachs and Morgan Stanley owned several percentages of companies with names like Plains All American Pipelines. It was banks and institutional investors like these – JPMorgan not to be omitted: largest, most unstinting of all – that provided the funds for the late efflorescence of oil and gas in the US. In short, there was no telling where finance ended and reserves, platforms, pipelines, refineries and the rest of the physical infrastructure began.[185]

But the integration still did not end there, because there was also the circuit of commercial capital, the oldest form of money-making – buying cheap and selling dear. A capitalist in this circuit does not himself produce any commodities. He is just a merchant. He parts with his money to acquire the goods others have made, brings them to a distant location on the market and sells them at a higher price, so that his money returns to him with an increment.[186] In the early twenty-first century, the titans of commercial capital were the 'commodity traders' – Glencore, Vitol, Trafigura, Gunvor, Mercuria – whose *Geschäft* it was to purchase primary commodities at one point in space and time and sell them at others. In 2021, these invisible hands were finally and revealingly profiled in *The World for Sale: Money, Power, and the Traders Who Barter the Earth's Resources* by Javier Blas and Jack Farchy, a masterpiece of business journalism, which left no doubt about their origins.[187] It was oil that gave rise to late commercial capital. When countries in the Middle East and elsewhere in the global South nationalised their oil in the 1970s, they were in dire need of middlemen to pick it up and ferry it to northern markets: a niche that soon became a network of highways to extreme fortune.[188] For Glencore, coal was more important; for all the others, oil and subsequently gas were the chief substances.[189] We have already seen how the former shared in the bonanza of 2022. Vitol more than tripled its profits during that year, shuttling hydrocarbons around the globe, ranking as the fifth largest company in the world by revenue – a tiny bit smaller than Amazon, far larger than Apple.[190] Trafigura more than doubled its profits and ranked twelfth, above Volkswagen.[191]

What would climate action mean for this circuit of capital? 'Vitol, Mercuria, Gunvor and Trafigura rely on oil trading for the bulk of their profits,' Blas and Farchy observe, before considering whether the materials needed for renewable energy – cobalt, lithium, nickel – might plug the gap and concluding, wisely, as we shall see, that they will not. 'It's tough to see how those markets will replicate the billions of dollars the commodity traders currently make each year from trading oil.'[192] Indeed, taking out fossil fuels would open a chasm in world trade itself. In 2021, the most traded product in the world by value was crude petroleum; adding refined varieties, a total of 8 per cent consisted in shifting around oil.[193] By volume, this one fuel accounted for 22 per cent of cargo exported and imported over the seas; adding coal and gas, a total of 45 per cent of seaborne trade carried fossil fuels.[194] Many of the vessels would have no life after mitigation. By carrying capacity, more than one quarter of the trade fleet was made up of oil tankers: in a defossilised world economy, there would be nothing to fill them up with.[195] Any chance of staying below 1.5°C would thus require the abolition of the bulk of not only the circuit of commercial capital, but a good deal of the entire system of world trade.[196]

If the circuit of commercial capital is utterly integrated in world trade, it is also entangled with that of interest-bearing capital: commodity traders rely on borrowed money to buy the goods they will sell.[197] As of the early 2020s, all major banks and institutional investors had fortunes wedded to them. Further blurring the circuits, oil companies such as BP and Shell had their own commodity trading departments, while commodity traders ventured into production: in a move taken out of the pages of the third volume of *Capital*, Glencore hit the ceiling of profit-making solely from buying cheap and selling dear and resolved to establish direct control over coal extraction. It would need to run its own mines. To purchase such productive assets – in Colombia, South Africa, Australia – and build the fixed capital, Glencore concentrated financial firepower from multiple sources and turned into a publicly traded company, the flotation in 2011 the largest ever on the London stock market. Trafigura, Mercuria, Vitol likewise invested in pipelines, ports, refineries and facilities for storing oil and gas.[198] Commercial capital became less and less distinguishable from primitive fossil capital, both circuits infused with credits and equities: a tangle exceedingly difficult to unwind. If primitive fossil capital could be analytically isolated, it was, by the early 2020s, thoroughly enmeshed with fossil capital in general, interest-bearing (and dividend-expecting) as well as commercial capital, a series of circuits together constituting *fossil capital as a totality*. Now imagine all this coming crashing down.

And with It All Transactions Based on It

Dressed in black tuxedo and bowtie, addressing similarly dressed insurers dining around candle lights in the heart of the City, Mark Carney, the governor of the Bank of England, gave a speech on stranded assets in September 2015, just weeks before

Paris. Tone sober, he began by reviewing the evidence of the climate catastrophe in motion, and then he warned that it risked catching up with the time horizon of investors, leading to what he called 'a climate "Minsky moment"'.[199] By this, he meant a sudden crash in the whole financial system. A decisive event in the formation of the discourse, Carney's speech was followed by widespread concern that the markets of the world were in for a totalising collapse, often referred to as the bursting of 'the carbon bubble'. On the day mitigation begins, the shares in fossil fuels pop; debts can no longer be repaid; investors switch to 'fire sales' of associated assets, as if they were burning their hands; no one can trust banks and asset managers filled with deals impossible to honour; because every room is connected to every other, the contagion is uncontainable; the dinner guests run from their house on fire, until they reach some assembly point panting and naked, or something to that effect.[200]

Why Minsky? A post-Keynesian economist of Menshevik descent, Hyman Minsky developed a theory of 'financial crises as systemic, rather than accidental, events', bound to recur because investors base their decisions on expectations of future profits, and these are by nature subjective.[201] Investors might become extremely fired up about something. Then they all buy that one thing, incur debts to buy more of it, raise the price of the thing in question so it becomes even more attractive to own, build a Ponzi scheme of investment until something shakes their confidence and panic sets in and they all try to dump their vastly overvalued assets. For Minsky, 'a capitalist economy *endogenously generates* a financial structure which is susceptible to financial crisis': as in the Schumpeterian case, a theory with a distant affinity to Marxism.[202]

One strand of Marxist crisis theory has likewise zoomed in on the tendency of fictitious capital – claims to value yet to be produced – to run ahead of itself and everyone else, until it starts levitating. The aggregate expectations become so excessive that the material reality of value production cannot possibly live up to them.[203] That is when the flyer realises that he is flapping his bare arms in the air and crashes: what happened in the tulip mania, the railway mania, the dot-com bubble, the subprime mortgage bubble and the other orgiastic flights of speculation that riddle capitalist history. All have ended with fictitious capital being brought down to earth. When such capital starts spiralling 'onwards and upwards into the stratosphere', writes Harvey, 'the quantitative limits of real surplus-value production are quickly left behind, only to assert their limiting power in the course of a crisis.'[204] The function of the crash is to re-establish 'a measure of concordance' between actual value and its fictitious representations, after which accumulation may resume in a somewhat more reasonable manner.[205] Minsky and Harvey here converge on a fly-away model of crisis.

While that model resembles asset stranding and the bursting of the 'carbon bubble' in some respects, there are subtle differences. For Minsky, regulatory intervention rescues the economy from total meltdown; to prevent crashes from happening again, there ought to be more of it, along familiar (post-)Keynesian lines.[206] From the perspective of the financial system, in the case of asset stranding, however, *regulatory intervention is the problem*. Without it, the revenue streams would simply continue to flow. For Harvey, such intervention is, as in Minsky, what capital needs and even demands. He offers an analogy with psychoanalytical undertones: investors active on financial markets are like teenagers, who crave the satisfaction of every lust and demand autonomy, while in fact they still fill their bellies in the household. 'When things go wrong they come running home to mommy and daddy.' The state, 'being an indulgent and loving parent', then invariably opens the purse, the kitchen: time for the bailout, the stimulus package, in accordance with an all too well-known script.[207] While this is the story of early twenty-first-century pre-climatic crises, it does not, again, quite correspond to asset stranding. Here the parent would rather chase the teenagers down as they deal in hard drugs: the crisis and correction *begin* with the state intervening, sovereign and superego-like.[208] There is no rebalancing between fictitious and real capital. Rather, both go to the devil; or, an entire built environment of value production is deliberately imploded, when the use value of fossil fuels is finally classified as negative. The crash is not induced by speculators flying into hallucinatory expectations about how rich fossil fuels can make them, but by society abruptly deciding that no one can get rich from fossil fuels any longer. This is not surplus-value production asserting its 'limiting power' on the layers above, but limits drawn to surplus-value production itself, in one material form.

Nor would asset stranding be preceded by a discordance between fictitious and fixed capital, to be rebalanced by the crash. The problem is instead defined by their utter conjunction. United they fall, when the substratum is pulled from under them – a type of crisis that activates the self-destructive mechanisms endemic to capital itself, but only because of an exogenous intervention, of a kind that Marx, perhaps, intuited on one page in the third volume of *Capital*, where he likens property in the Earth to the property in other human beings. Those who cash in on land they own are a bit like those who enslaved Africans. Both are inclined to believe that the money they make comes rightfully to them, as owners of the assets in question.

> It appears to the slaveowner who has bought a Negro slave that his property in the Negro is created not by the institution of slavery as such but rather by the purchase and sale of this commodity. But the purchase does not produce the title; it simply transfers it. The title must be there before it can be bought, and neither one sale nor a series of such sales, their constant repetition, can create this title. It was entirely created by the relations of production. *Once these have reached the point where they have to be sloughed off*, then the material source, the economically and historically justified source of the title that arises from the process of life's social production, disappears, and *with it all transactions based on it*.[209]

Suspended somewhere between Lincoln and Lenin, this is the Marxian crisis theory that comes closest to asset stranding. It posits not a contradiction within the relations themselves, but rather a 'point where they have to be sloughed off' – the point where slavery was abolished, or, on this page of *Capital*, where landed property itself will be done away with for good. That point is the moment of political intervention. Here, it happens to have a tantalisingly ecological substance. 'From the standpoint of a higher socio-economic formation, the private property of particular individuals in the earth will appear just as absurd as the private property of one man in other men' – an analogy rather more narrowly reserved, in the discourse of asset stranding, for private property in the *fossil deposits* of the Earth. Then follows one of the sentences most beloved by ecological Marxists. 'Even an entire society, a nation, or all simultaneously existing societies taken together, are not the owners of the earth. They are simply its possessors, its beneficiaries, and have to bequeath it in an improved state to succeeding generations, as *boni patres familias*,' or good heads of the household.[210]

What Marx here envisages, if we allow ourselves some interpretive licence, is a contradiction between capitalist relations and basic sustainability, reaching a point where some 'material source' of property titles must be placed at a remove from said relations, and then 'all transactions based on it' come to an end. This is an exegesis of mitigation. Cutting emissions down to zero would mean this precise thing: all transactions based on fossil fuels disappear. We can then begin to see more clearly the referent of 'revolution' and the similar expressions abounding in the discussions of climate politics in the early 2020s. Capping global warming at 1.5°C or indeed any approaching limit would mandate a blow to the capitalist mode of production of a breadth and depth hitherto unseen. It is to be expected that any anti-revolution born out of these conditions will be corresponding in scale and determination.

Counting the Reasons for Intransigence

What sums do all these transactions contain? What money are we talking about here? With the layers of interconnections – vertical, horizontal, transversal – and the profusion of scenarios that could play out in whatever way, plus the multiplicity of variables, in addition to differences in methods and concepts and data, it is no surprise that estimates of the potential losses have varied widely.[211] Most have focused on the primitive circuit. One study found that the stranding of reserves would directly rob the producers of fossil fuels of between 13 and 17 trillion dollars, if temperatures were to halt at 1.8°C or 1.5°C.[212] By way of comparison, the great recession of 2009 reduced world GDP by a little more than 3 trillion.[213] Another study rather put the sum at a maximum of 4 trillion for 2°C, while yet another reached a whopping 185 trillion in a scenario with a fifty-fifty chance of staying below that limit.[214] In 2022, total world GDP was 103 trillion.

The first capital destroyed would belong to the private giants. *Financial Times* estimated that 'big oil and gas companies' would lose one third of their value (as of 2020) if the carbon budget for 1.5°C were to be respected, which might have been a serious underestimation, possibly derived from the questionable assumption that there would still be time to exploit most proved reserves before mitigation kicks in.[215] More realistically, 'under no scenario is there a lasting place for oil and gas companies in their current form', least of all in one of 1.5°C without overshoot: *all* their value would sound more like it.[216] Irrespective of their linkages to other types of businesses, these companies sit at the pinnacle of global capital. Of the ten largest companies in the world measured by revenue in 2020, six sold oil and gas (ExxonMobil did not enter the top tier that year, ending in place eleven; but two car companies did).[217] In the year of pandemic doldrums, only two remained; as of this writing, the update from the profit bonanza has yet to come. Nothing currently indicates that the primitive circuit is about to fade into the margins of this mode of production – if anything, the opposite.

Blast furnaces would contribute about half a trillion US dollars, according to one report from 2022, revising previous estimates seven times upwards while still using relaxed assumptions for stranding.[218] Finance proved trickier to calculate. Estimates of the portion of the main stock exchanges tied straight to fossil fuels varied from the trifling to the preponderant.[219] For Moscow, on the eve of the invasion of Ukraine, it was 48 per cent; for Riyadh, more than 70; for London, 15 – but in absolute terms, the shares were concentrated in London and New York.[220] One 'stress test' of the European financial system in 2017 found that its fifty biggest banks had about one tenth of their total port-folio of equity holdings in fossil fuels, but if indirect links were accounted for – holdings in 'climate-policy relevant sectors' also including utilities, transport, energy-intensive industry – the share rose to four tenths.[221] As of the early third decade, 900 companies responsible for 95 per cent of hydrocarbon production had bonds on the market worth a total of 4 trillion dollars.[222] Moody's floated the figure of 22 trillion in the financial firms of advanced capitalist countries exposed to a transition away from fossil fuels; a fine-grained study of Dutch firms came up with 160 billion or 10 per cent of all assets likely to be wiped out in this single country; a case study of Mexico yielded results ranging from a couple of per cents of the financial system to one third of the bank capital – all guesstimates, bedevilled by the difficulties of drawing lines around the risk.[223]

For world capitalism as a whole, then, the sum total of the losses would be anybody's guess. 'When it comes to the fossil fuel system, the write-downs would be four or five times bigger than the size of the financial crisis' of 2008, in the educated guess of Campanale – but where would such self-amplifying mega-write-downs come to a stop?[224] Efforts to prevent the feedback loops of a hothouse planet might send such loops into the capitalist world economy instead. As one asset manager from London put it in 2023, there could be little doubt that 'there are large, large balance sheets, which in a net-zero world will be entirely useless' – and, mind you, even larger such sheets in a *zero* world – or, 'the effects of writing off stranded assets would be felt across the business world. It would be one of the biggest ever shifts in the allocation of capital,' in the estimation of *Financial Times*.[225] It would go beyond both Lincoln and Lenin, because it would, by definition, not stop at the borders of one nation.[226]

In which nations would capital be dealt the hardest blows?

The obvious candidates would be those where the largest reserves are located: oil in the Middle East, coal in Asia.[227] The former presents a special case, due to the utter soaking of the dominant classes of the Gulf in oil (and gas). With countries like Kuwait, Bahrain and Saudi Arabia relying on hydrocarbons for 60 to 90 per cent of their government revenues, the foundations of royalist-capitalist class power would be blasted apart.[228] Fossil fuel reserves already under development in the early third decade and in danger of stranding were, in absolute numbers, concentrated to four countries: China, Russia, Saudi Arabia and the US.[229] While China held the most coal, the costliest stranding of mines would rather happen in Russia and the US.[230] But coal assets did not necessarily correspond to their physical location. Few social formations in the early twenty-first century were as thoroughly based on the black rock as South Africa: but investment in this sector still, as of 2022, came primarily from the global North. If South African mines were to close, the largest losses would be sustained in places like London and Frankfurt.[231]

The same separation in space between reserves and owners extended, of course, to capitalism as a whole. You can sit in Stockholm and own part of a field in Guyana. In the most comprehensive and detailed study to date, Gregor Semieniuk and his colleagues traced some 44,000 assets in the production of oil and gas via nearly 2 million companies to their ultimate owners, and nearly all roads led to the North. They compared a business-as-usual scenario yielding 3.5°C over the twenty-first century with a policy scenario keeping to 2°C and found that 40 per cent of the physical stranding would sit in the OECD – but in terms of *value* destroyed, the share would be 60 per cent, rising to 88 in the financial sector. Countries like Nigeria and Kazakhstan would 'export' their stranded assets. France would 'import' losses as big as those incurred inside Saudi Arabia. The largest net transfer, from place of physical origin to place of ownership, would go to the US; but the British Virgin Islands and Switzerland would also come under heavy blows. This threat geography reflected, of course, a still imperialist world economy, in which northern capital held resources in the South through countless strings: and the hands were predominantly private: over half the losses in the 2°C scenario would fall on 'private persons', or the very rich. In the US, 82 per cent would hit the wealthiest tenth of households. This was a world in which propertied people – in the North above all – had, as Semieniuk and his colleagues observed, 'a potentially perverse incentive' to 'accept inertia' and 'earn dividends from the continued operation of fossil-fuel production'.[232] Counting assets liable to stranding was an exercise in counting the reasons for intransigence. For every dollar, a threat of loss; for every such threat, a reflex to protect one's capital from it. These were some mathematical numbers of an imagined climate revolution and thereby, as the other side of the coin, the urge to overshoot.

5

How to Kill a Spectre

'A new spectre is haunting the fossil-fuel dependent world: asset stranding,' began yet another peer-reviewed article on the topic, published in 2020.[1] The allusion to *The Communist Manifesto* was here less of an empty cliché than usual. But the thing about spectres and ghosts, of course, is that they do not quite exist. Asset stranding, around this time, remained an entirely hypothetical event. And yet it exerted causal power in the world. To make sense of this, we might turn to Mark Fisher's musings on 'hauntology', a concept he borrows from Jacques Derrida's *Specters of Marx* and defines as '*the agency of the virtual*, with the spectre understood not as anything supernatural, but as that which acts without (physically) existing' – although 'agency' and 'act' are here misnomers, since the spectre does not have such subjective properties. Its virtuality is, rather, objective, its quasi-existence inferred. But it comes to possess an efficacious 'spectral causality' through the subjective perceptions of those acting inside the circuits of accumulation, rather like the ghost in the corner becomes a causal force through the child's interpretation of the shadows. It presses in on them from two sides: as that which is '*no longer*, but which remains effective as a virtuality' – the apparition of a revolution once experienced as real – and that which has '*not yet* happened, but which is *already* effective in the virtual (an attractor, an anticipation shaping current behaviour).' Fisher here invokes not only the *Manifesto*, but also Freud: psychoanalysis as the 'study of how reverberant events in the psyche become revenants'. And finance, he notes, with its swings between mania and anxiety about the future, is the pre-eminent province of spectral causality.[2] How does one handle such creatures?

An Arsenal of Lukewarm Weapons

The haunting began, as we have seen, in the late Kyoto era, when the nerves of the capitalist classes of the world were badly jangled, after the brush with catastrophe around 2008. The fear of a carbon bubble was informed by fresh memories: Mark Carney gave his speech in a moment of lingering stress disorder.[3] But fossil capital had, of course, seen figures moving in the dark long before that, and conceived of strategies to dispel them. Carney invoked one of these when, alluding to the remaining carbon budget for 2°C, he said: 'if that estimate is even approximately correct it would render the vast majority of reserves "stranded" – oil, gas and coal that will literally be unburnable, *without expensive carbon capture technology, which itself alters fossil fuel economics.*'[4] The reference here was not to technologies for vacuuming CO_2 out of thin air, but to large contraptions that could be physically attached to coal and gas plants to prevent the carbon from escaping. Think of them as filters on chimneys that catch the CO_2 molecules in the process of emission; not after it, as in removal technologies. Once caught, the gas could then be locked back underground. Some components of such devices had been in operation since the 1970s, when companies started using them for something called 'enhanced oil recovery' – squeezing more oil out of dwindling wells by pounding CO_2 into them.[5] Apply the necessary pressure from underneath the oil and more barrels, more value would pour forth. Such deployment of concentrated CO_2 represented just the kind of technologies ExxonMobil bragged about – not incorrectly – as a weapon against peak oil.

In more ways than one, 'carbon capture', or 'carbon capture and storage', or CCS was a precursor of overshoot ideology. Long before the discourse of stranded assets, primitive fossil capital had, of course, caught scent of the threat to its business: the very reason it organised the denial of climate change in the early 1990s.[6] When parts of this class fraction broke off from the denial during that decade and, instead, nominally recognised the existence of the problem, they had to come up with some ingenious technology for making their assets reconcilable with the solution – or, rather, the *promise* of one such tech. During the 1990s, some oil and gas majors began to see great potential for carbon capture as a charm against the menacing spectre.[7] They talked up its promise whenever necessary, and they were not alone: in 2004, the International Energy Agency threw its weight behind it in a lengthy report on the subject. Advancements in energy efficiency and the rollout of renewables, the Agency claimed, would 'at best, only partly solve the problem'.[8] A full solution would demand the use of carbon capture, which could serve as 'an essential "transition technology"', not just in the near-term but 'for the next 50 to 100 years'.[9] Stick that filter onto the chimneys, and the chimneys would no longer be part of the problem.

Carbon capture was an expensive technology and so would need to be combined with other mitigation options as well, the Agency argued, but the benefits compared to the alternatives on the table were indisputable. What would the most economically rational distribution be? The Agency asked its own model that question and was served the following result: carbon capture could take care of almost half of all emissions cuts by 2050.[10] It would 'allow for the continued use of fossil fuels while at the same time achieving significant reductions in CO_2 emissions'.[11] Indeed, the model prophesied that the total use of fossil fuels *could almost double* in the period between 2000 and 2050, as long as power plants and refineries built during this time had a CO_2-capturing device somewhere sucking on the chimney.[12] Coal would benefit most of all. Carbon capture would 'result in a significant increase in the use of coal' and 'a lower use of renewables and nuclear'; and because coal was 'an established fuel' that could help ensure the security of energy supply, this was actually a good thing. Carbon capture, the IEA found, would make coal 'a more sustainable option'. It could henceforth be called 'clean coal'.[13]

If this was not clear enough, the IPCC released its own report on the matter the year after, reiterating some of the same messages and laying out the state of knowledge on the tech. In an unusual rendering of the superstructure explicitly serving the base, it had come about, at least in part, to rope fossil fuel–producing countries onto the climate train and banish their nightmares of phase-out. Daiju Narita studied the process leading up to this report and found that 'some countries including the US, Australia, and Canada, in addition to Saudi Arabia' had long been the main standard-bearers of carbon capture.[14] They wanted the IPCC to 'promote' it because it was 'a way to sustain their way of economy'.[15] Because most of these countries had long obstructed all types of climate action, giving in to their request could be a way to break the stalemate and get some negotiations going. In the words of one of the report's lead authors, carbon capture had the 'potential of finding a way out of that problem, by adding something to fossil fuel use, coal, even more so than oil. That would enable to continue [using] fossil fuels while solving climate problems.'[16] The report itself backed this up with reference to the usual ideological tools: 'one aspect of the cost competitiveness of CCS systems is that CCS technologies are compatible with most current energy infrastructures.'[17] Making carbon capture a central part of the package would, in terms soon about to come in wide usage, stave off asset stranding.

Immediately following these reports, carbon capture enjoyed a moment in the limelight. In 2005, the G8, as the group of heavyweight nations was then called, met in the upmarket Scottish village of Gleneagles and swore to have twenty demonstration projects up and running by 2010.[18] Australia, Canada, the US, Norway and others with similar interests announced generous subsidy schemes to launch the technology on a trajectory to the stars. The EU alone emptied its pockets of 1 billion euros for large-scale demonstration projects in the energy sector; combined with yet another pot of money, that became 3.7 billion in total.[19] In the media, carbon capture was hyped as 'groundbreaking' and 'game-changing' for climate politics.[20] Yet by the time the Paris summit loomed on the horizon, the tables had turned. Most of the schemes had precious little to show for them.[21] One after the other, projects that gained funding were delayed or cancelled.[22] Just days before Paris, the UK scrapped its own £1 billion 'competition' for carbon capture projects.[23] Auditors in 2018 observed that 'neither of the [two] programmes succeeded with the deployment of CCS in the EU'.[24] In one of them, no projects saw the light of day at all. In the other, the Union had paid €424 million to six projects: one ended early, four went nowhere after money ran out, one stayed at the scale of a small pilot project. Europe's subsidy programme, the auditors concluded, 'has not contributed to the construction and entry-into-operation of any CCS demonstration project'.[25] In the analysis of one of the authors of the original IPCC report, the death spiral commenced after the flop in Copenhagen, where the air simply went out of the bubble: 'without a global signal that climate change mitigation must be taken seriously in investment decisions, industry finds little reason to invest in deploying CCS on a large scale.'[26] There was, it turned out, little reason to fear the spectre when it appeared sickly and alone.

Never Let a Promise Go to Waste

But the spectre was back just a few years later, re-energised by climate movements and financial crisis in combination. To kill it again, carbon capture was no longer a favoured weapon. When Carney gave his speech, he did not actually advocate that technology: he merely took note of it, perhaps keenly aware of its deflated promise. Instead, his preferred solution was 'disclosure'. If only companies would be transparent about how much of their assets were liable to stranding, stakeholders could assess the risk and assign it a proper price. The market would then rid itself of such assets, reallocate capital by its own accord and pre-empt the feared political destruction. We can here recognise all the usual precepts of bourgeois ideology, including its unfailing rationalism and optimism. It is never too late for another neo-liberal solution: 'private industry can improve disclosure and build market discipline without the need for detailed or costly regulatory interventions.'[27]

The immediate effect of the Carney speech was thus a raft of initiatives from the Bank of England, other central banks, the G20 and similar bodies for inspiring voluntary disclosure and 'raising awareness' – words from the IPCC – in the belief that coming clean about the problem would make it go away.[28] This did not happen. Disclosure did not happen. As one investigation from Lord Stern's research institute confirmed in 2022, seven years after Carney, a miniscule fraction – some 2 per cent of nearly 500 companies in 'climate policy relevant sectors' worldwide – had gone public about their risk exposure.[29] Any disclosure that did happen was haphazard or half-hearted or outright deceptive.[30] Reallocation most evidently did not happen. The impact on investment flows of these initiatives was nil.[31] If they had anything to show for them, it was a continuous circulation of the *notion* of an impending 'climate Minsky moment' – more talking, but no cure.[32]

This kind of talk kept the hauntology going, and it quickly spread from central to private banks. Citigroup, as we have seen, wrote a report on asset stranding in 2015 – only to continue to pour money into the primitive circuit, including some of the most egregious pipeline projects in North America (such as Enbridge Line 3 and Line 5); outside China, it remained the world's biggest investor in coal.[33] A remarkable document crying out for psycho-analytical interpretation was produced in 2020 by JPMorgan. 'We cannot rule out catastrophic outcomes where human life as we know it is threatened,' it readily acknowledged.[34] Even more candidly, 'the earth is on an unsustainable trajectory. Something will have to change at some point if the human race is going to survive.'[35] What could possibly change? 'A sizeable shift from fossil fuels to renewables is technically feasible,' analysts at the world's largest bank duly recognised. But it would come at a cost – namely, 'the premature scrapping' of fossil fuel assets, the emphasis here being, somewhat surprisingly, on all the coal plants that would have to die. So, 'it isn't going to happen.'[36] 'Most likely, business as usual will be the path that policy-makers follow in the years ahead.'[37] In a sign of the shift already underway, JPMorgan ended by noticing two new, if risky, spectre-defeating strategies on offer: carbon dioxide removal and geoengineering. Then it continued to pour money into the primitive circuit, including some of the most egregious pipeline projects in North America (such as Enbridge Line 3 and the Coastal GasLink); in all the world, no other entity invested more money in fossil fuels.[38]

BlackRock listened to the pulse on the street in 2019. Citing the mass mobilisations of that year and the accusations heaped on his company, CEO Larry Fink, in a letter to investors published just days before Covid-19 reached the US, judged that 'climate policy' was coming closer. 'In the near future – and sooner than most anticipate – there will be a significant reallocation of capital,' threatening a crash that would make all previous crashes look like dinner receptions, because it would be systemic and terminal. Hence BlackRock would now be so provident as to exit coal.[39] Coming from one of the Big Three, the announcement made waves at the time; but then for some reason BlackRock kept its assets in Glencore, BHP, Adani, RWE and, two years after the Fink letter, ranked number one among asset managers investing in coal.[40] It also secured a lucrative deal with Aramco to own and fund about half of its gas pipeline network.[41] In the years after his divestment pledge, Fink pulled back from the idea, the company now explicitly refusing to desist from funding new projects in any of the three fossil fuels.[42] But in his annual letter of 2023, the CEO struck upon another kind of exit: technologies for capturing carbon. 'We are creating opportunities for clients to participate in infrastructure and technology projects, including the building of carbon capture storage pipelines and technology that turns waste into clean burning natural gas.'[43] This looked like a comeback for carbon capture onto the stage, and that indeed it was; but it was also something more. For the content of that term had changed subtly and the idea moved from the filter-on-the-chimney stage to the *post factum* removal stage, or the astrochicken stage, if you will.

But just as oil supervened on coal in the middle of the twentieth century without displacing it, removal developed alongside attempts to resurrect point-source capture, or CCS. In the spring of 2022, apprehensive representatives of Saudi Arabia intervened in the negotiations within the IPCC about the exact formulations in the next 'summary for policymakers' with the following demand. The Kingdom 'insisted that the estimated value of stranded assets only reflect the unabated part of fossil fuels, saying new technologies will make fossil fuels low carbon' – a reference to both categories, the distinction between them, as so often, muddled. Following this protest, 'delegates agreed to indicate that "Depending on its availability, CCS could allow fossil fuels to be used longer, reducing stranded assets."'[44] It could not get much clearer than that: the use value of these non-existing, non-proven means of production lay in the shielding they afforded against the political destruction of fossil capital. CCS could continue to have that use value, despite its dismal performances in the real world, because it was precisely the *promise* of it that was the point; and the same logic was now extended to technologies for removal.

In the formal UN negotiations, the same rearguard battles were underway. When there was talk of phasing out coal at COP26 in 2021, Australia declined to consider 'wiping out industries'.[45] China, India, South Africa and the US all joined in its defence, and the final text from the summit called for the need to address 'unabated' coal – code for 'coal is just fine, if complemented with carbon capture'.[46] The following year, in Sharm el-Sheikh, Saudi Arabia made a fuss about the recurrence of talk about phasing things out, arguing that the focus should be on emissions, not fuels, and that references to renewable energy 'should be complemented by abatement and removal technologies'.[47] Sometimes the superstructure really sits smack on top of the base.

I'm Not Worried about the Stranded Assets

In the interventions from Saudi Arabia, in texts from JPMorgan and BlackRock, the logic of overshoot was taken from computer models and into the calculus of accumulation. Fresh material efforts to fight the spectre were underway, and the fear of it kept being aired, sometimes with exquisite honesty, as by Benjamin Zycher of the American Enterprise Institute, writing in *Financial Times*. A transition from fossil fuels would 'destroy some substantial part of the economic value of the pre-existing energy-using and producing stock of physical and human capital. Earthquakes cannot yield economic benefits; the same is true for policies that wipe out the value of significant parts of the economy.'[48] ('There is slave property of the value of $100.000.000 in the State of Virginia, &c., and it matters but little how you destroy it, whether by the slow process of the cautious practitioner, or with the frightful dispatch of the self-confident quack; when it is gone, no matter how, the deed will be done, and Virginia will be a desert' – a forerunner from 1832.[49]) In 2021, the *Wall Street Journal* had moved the worry about asset stranding from op-ed to extensive feature, suggesting objective grounds for considering this the next source of crisis.[50] The banks of New York received a letter of alarm from their superintendent.[51] During the early Paris era, it seems safe to say, the spectre of asset stranding decisively entered the collective consciousness of the capitalist classes of the world.

What of their primitive core? In the years following McKibben's *Rolling Stone* article, before the Carney speech, a large enough segment of investors, still jittery from 2008, began to raise questions. In annual shareholder meetings and other fora for interacting with their money-makers, they asked: how risky is my investment in your company? Do you know to what extent your assets are exposed to stranding? Can you tell me? What are you going to do to ensure the value of my capital?[52] Inconvenienced, the fossil fuel giants had to engage with the spectre head-on and produced a spate of responses, concentrated around the year 2014.[53] In its annual report, BHP admitted that governments 'are contemplating the introduction of regulatory responses to greenhouse gas emissions', which 'may adversely impact the productivity and financial performance of our operations'.[54] The 'carbon bubble' might come to exist. But then the company decided that its 'overall asset valuation is not at material risk' and that the vastness of its coal and oil assets 'uniquely positions us to manage and respond to changes and capture opportunities to grow shareholder value over time'.[55] Stranding, perhaps, but not for us: we will stay the course.

Of greater weight, ExxonMobil sought to soothe shareholders in 2014 with a special report, dismissing the fear as overblown. 'We are confident that none of our hydrocarbon reserves are now or will become "stranded".'[56] The reasons for the confidence were threefold. First, demand for the goods will continue to grow over the twenty-first century, guaranteeing that they remain eminently saleable. Second, 'although there is always the possibility that government action may impact the company, the scenario where governments restrict hydrocarbon production' – say, by cutting emissions by 80 per cent until 2050 – 'is highly unlikely.'[57] Third, the company itself was taking pre-emptive measures to minimise any future harm to business. Under the headline 'managing the risk', it called attention to novel technologies for doing so: 'for example, ExxonMobil operates one of the world's largest carbon capture and sequestration facilities' in Wyoming and another in Australia. Picking a favourite verb in the financial argot, Exxon-Mobil said it was 'leveraging' this experience 'in developing new methods for capturing CO_2 which can reduce costs and increase the application of carbon capture for society'.[58] A few years later, it had graduated into fully-fledged overshoot ideology and was explicitly referencing the IPCC to clarify that carbon capture technologies would also be 'critical to [sic] enabling removal of CO_2 from the atmosphere'.[59] The message, in one line: you can entrust your capital to us, because in the unlikely event of policy, we have technologies up our sleeves. But the confidence was never fully stable. Small shareholders with activist inclinations continued to pester the board, and in November 2021, Exxon-Mobil made its glummest admission to date. 'If the world shifts away from fossil fuels more quickly than anticipated,' it warned, according to a report from *Houston Chronicle*, 'the value of its oil and gas assets could fall, although the company said it could not now estimate if or how much it could fall.'[60] Not going away, the spectre would have to be fought off on a continuous basis.

The competitors of ExxonMobil lined up to send the identical message: fossil fuels will remain indispensable, governments will not put their words into action, and in any eventuality, we have emerging technologies for neutralising the climatic effects of carbon.[61] Shell said as much in a very dismissive letter to shareholders in 2014.[62] In another incantatory ritual of the early Paris era, strangely symmetric with the COPs in its insistence on all being fine and well, Shell repeated the same thing in 2018 and 2021 – but this did not free the company from the need to address the threat and post all manner of scenarios and sensitivity tests and strategies improving the 'resilience' of its property in the face of it.[63] In the years after the Special Report, this corporation released a steady stream of scenarios overflowing with removals and overshoot and validating the continued use of oil and gas all the way to the end of the century.[64] These were then served up as arguments for why there was 'a low risk of Shell having stranded assets, or reserves that we cannot produce economically, in the medium term'.[65] As for BP, 'I'm not worried about the stranded assets,' said the CEO in an address to the International Petroleum Week in London in 2018. According to the report in *Bloomberg*, his next two sentences were: 'we try to make investments today which we know will have economic *long* lives for a *long* time. If you're in the low part of the cost curve, whether it's natural gas or oil, it's decades.'[66] The latter two sentences explained why the first *had* to be true: because the investments were so long-long, stranding would be too ghastly to contemplate. Which was, of course, why it constituted such an irrefutable worry and had to be commented on all the time.

ConocoPhillips took a more pugnacious stance. 'This is a movement to frighten away investors and capital, and God help us all if they're successful because there won't be enough supply,' Marianne Kah, the corporation's chief economist, declared, laying into the divestment campaigns and the discourse they had instigated – which, again, did not save ConocoPhillips from the headache of contingency planning in a world where these enemies might one day have the upper hand.[67] Chevron, in a report from 2017, saw them taking on a multitude of guises. There could be 'interruption of the company's operations due to war, accidents, political events, civil unrest, severe weather, cyber threats and terrorist acts', alongside or mixed up with the usual bugbear of 'environmental statues and regulations'. At this point, Chevron, for the three typical reasons, still held any risk of stranding to be 'minimal and certainly manageable'.[68] But in a subsequent report from 2021, the emphasis was on manageable. Now the corporation was alert to the existential danger, which, fortunately, however, could be dealt with, as explained by the IPCC in its scenarios for managing 'temporary overshoot' with removal.[69] Technologies of removal would henceforth be 'essential' and 'central' and 'key pillars' to Chevron itself as it ploughed on.[70] In this document too, overshoot ideology had fully come home to the base.

The commodity traders were just as alive to the threat. 'Our business will probably die over the next ten years,' the CEO of Vitol admitted in 2019, the year before he actually died, perhaps now resigned to mortality; but others were determined to keep fighting it off.[71] 'A number of divestment campaigns advocate a halt in coal investment, on the basis that future climate change policies will render coal resources and infrastructure "stranded assets." We do not believe that this is a material risk to our business,' Glencore professed.[72] Shareholders need not worry about being 'prevented from realising the full value of our fossil fuel assets', for coal is the cheapest, most reliable source of electricity, without which developing countries cannot develop out of poverty.[73]

And then there was Aramco. Standing before the men (and a few women) of the oil companies assembled in their capital city Houston, CEO Amin H. Nasser in 2018 declared that 'I am not losing any sleep over "peak oil demand" or "stranded resources"' – which was, of course, an acknowledgement that he did in fact lose at least some sleep over it, or else he would not have had to deny it. (Compare a child that keeps saying: I'm not losing any sleep over the ghost in the corner!) In fact, Nasser was agitated. He called for more exploration, more investment in supply by at least 20 trillion dollars over the next couple of decades. But 'this staggering amount will only come if investors are convinced that oil will be allowed to compete on a level playing field, that oil is worth so much more, and that oil is here for the foreseeable future.' The buzz around and 'misplaced notions' of stranding constituted 'direct threats' to this expansion.[74] In an interview with Reuters four years later, Nasser reconfirmed his loss of sleep. 'The pressure and the rhetoric is – don't invest, you will have stranded assets. It makes [it] difficult for CEOs to make investments': hauntology at full blast.[75]

Before we turn to the salient psychic dimensions of asset stranding, we might rephrase the sentences following the opening line of the *Manifesto*. All the powers of fossil capital have entered into a holy alliance to exorcise this spectre: JPMorgan and Black-Rock, ExxonMobil and Shell, Chevron and Aramco. Two things result from this fact. 1. Asset stranding is already acknowledged by all the powers of fossil capital to be itself a power. 2. It is high time to meet this nursery tale of the spectre with support for every revolutionary movement against the existing order of things and bring to the front the property question, no matter what its degree of development at the time. Meanwhile, the force that shapes the future is that of exorcism: the holy alliance calling forth the overshoot conjuncture.[76]

The Psychology of Asset Stranding (or, A Different Kind of Climate Anxiety)

When the divestment movement and its more respectable off-shoots, such as the Carbon Tracker Initiative, got going in the early second decade, their stratagem was to play on the edgy nerves of investors, whisper in their ears or dress up like ghosts in the corridors of the business world and scream, 'Boo! This asset is contaminated with risk!' – a sly interpellation of the risk-averse side of the financial personality, meant to bring about the desired reallocation of capital. The idea was to leverage the famous herd instinct of speculators to get them to flee fossil fuels. The propagators of the discourse hoped that it would bring about its own object.[77] Advice could be given on strictly capitalist grounds: 'smart investors can already see that most fossil fuel reserves are essentially unburnable', making any further investment in them 'very risky', wrote Lord Stern in 2013 – clearly a plea for a particular smartness rather than a description of what investors did.[78] If only the chatter was sustained insistently enough, the threat would begin to take shape and be avoided: a gamble not all that different from disclosure. Was it any more effective? Divestment, it should first be noted, had to fail on its own direct terms, since selling off fossil fuel assets, by definition, meant that they merely ended up in the ownership of someone else, likely with fewer scruples.[79] A slightly different purpose was to surround the assets with so much uncertainty as to provoke a run on them and affect a general devaluation. Evidently this did not happen, any more than disclosure made the slightest dent in the curves.[80] But then there was the achievement purely at the level of discourse: the construction of the spectre, even if it did not scare anyone away from fossil fuels.[81] What should the scorecard be here?

We have seen that actors in the primitive circuit often recognised the danger of asset stranding by negation – I am not worried, I am not losing any sleep – recalling an attitude Freud reported from some of his clients. "'You ask who this person in the dream can be. It's *not* my mother.' We emend this to: "So it *is* his mother."'[82] To negate something in this high-pitched fashion is, 'at bottom, to say: "This is something which I should prefer to repress." A negative judgement is the intellectual substitute for repression; its "no" is the hall-mark of repression, a certificate of origin.'[83] The preference here would have been for the problem to just go away, or at least for a situation in which it could be left uncommented. But with every new endeavour to rebut the danger, representatives of fossil capital produced proof of its seriousness, if only in potentia.

'Most international oil companies understand the threats ahead,' the *Financial Times* estimated in 2020.[84] Now that they were forced to negate asset stranding, they had to come up with arguments stabilising the future of fossil fuels, renormalising investment in them and renewing confidence.[85] Some were phrased in a revealingly imperious tone. 'The idea of coal, oil, and natural gas reserves becoming "stranded assets" from unanticipated or premature write-downs is bogus,' began a piece by Jude Clemente in *Forbes* in 2018. Because fossil fuels have grown in lockstep with GDP since the mid-nineteenth century, the future will also mean 'more coal, more oil, and more gas' – 'hardly stranded, new investment in energy exploration and development is mandatory.'[86] One cannot live a good modern life without fossil fuels and, more particularly, the poor of the world want their share of it too. ExxonMobil argued that the real stranding in any transition would afflict them.[87]

Glencore, BHP, BP, Total, Equinor all said the same thing, but none said it with greater verve than Chevron. 'There are still 1.2 billion people in the world without electricity and more than 2.7 billion people who burn solid fuels, such as wood, crop residue and dung, to cook their food,' and it is for *their* benefit, for the dung-burning wretched of the Earth and for the international 'stability' that their satisfaction will secure, that the charity Chevron chooses a future of more hydrocarbons.[88] The past five millennia of class struggle may predispose some to reach for their Kalashnikovs when the rich say that they have to get even richer for the sake of the poor, but one cannot rule out that these companies were empirically correct in attributing growth in fossil fuel consumption to the lower classes. If so, any exceeding of 1.5 °C (or 1.7 °C, and so on . . .) would constitute not so much paupericide as an exercise in collective suicide by the poor. We shall shortly examine this case.

As the discourse morphed into an ideological struggle, the two camps fought to make their preferred realities come true: one by saying the assets are doomed, in so doing dooming them; the other by saying they are safe, in so doing saving them – a psychological cat-and-mouse-game, producing no resolution but evidence enough that a degree of anxiety had been implanted into the breast of fossil capital.[89] It was a 'realistic anxiety', in the Freudian sense. It was not free-floating or unprovoked. Nor was it phobic in nature, nor paralysing. 'Anxiety' of this kind 'is a reaction to a situation of danger' – 'that is, to an expected injury from the outside'.[90] The prototypical injury is, of course, castration; and in the extended sense of the term, we can indeed consider the anxiety in question a species of castration anxiety: 'a fear of being separated from a highly valued object'.[91] Anxiety of this sort can be useful for the ego, as a signal due to which it 'can adapt itself to the new situation of danger and can proceed to flight or defence'.[92] These are some basic psychological coordinates of overshoot.

The anxiety here was, it is important to note, separate from what we normally refer to as 'climate anxiety', because it didn't concern the impacts of global warming per se.[93] In the discourse on asset stranding, a distinction was early on established between 'physical risks' and 'transition risks', the former referring to something like what happened in May 2023, when hundreds of wildfires in the Canadian province of Alberta shut down its oil and gas production.[94] Intensifying hurricanes in the Gulf of Mexico, Chevron noted, could ravage platforms.[95] Storms may inundate refineries. (There is a theoretical precedent for this in the second volume of *Capital*, where Marx discusses how bad weather accelerates the wear and tear of fixed capital.[96]) But, as of the early third decade, it was abundantly clear that transition risks were uppermost in the minds of investors. Brett Christophers interviewed individuals working at twenty-one investment institutions and found these to be typical statements: '"right now, the regulatory risks are much more material to us than the physical risks."' '"For now, at any rate, it is not climate change itself that matters to company value, but the response to climate change in terms of political action. Of course the risk will eventually be a mix of the two (physical and political). But not yet."'[97] We may hypothesise that, for the time being, the most feared injury will remain deliberate and not accidental castration. Insofar as climate suffering predominantly afflicts propertyless people at a distance from the circuits of fossil capital, and insofar as that suffering *may* prompt political action that *may* target the source, the physical risks will be overshadowed by the transition risks.

Pyrrhic Capital

On the other hand, there is evidence that investors did not care much at all. Their actions, on the whole, spoke loudly about their unshaken faith in fossil fuels. If they were genuinely anxious about stranding, they should have been less gung ho about them.[98] The hauntology would, from this perspective, rather appear as impotent as any other 'agency', and the reason is not hard to locate, for the idea of stranded assets had one very manifest weakness. 'We take it as a given that climate science mandates a severe climate policy response,' stated one paper on stranding from 2022 – but how could that be taken as a given?[99] There was a streak of pious rationalism at the heart of the concept. Because humanity is governed by reason, it seemed to say, the carbon budget will translate into stranding. Shrewd and smart investors could pour scorn on the notion, as indeed they did. When Blood and Gore published their advice in 2013, the pages of *Wall Street Journal* filled up with letters mocking them for their unremunerative sentimentality: 'I propose that someone place a bet', wrote one, and wager 1,000 dollars

> that fossil-fuel companies will perform better than renewable companies over the next decade. I put up as my companies Chevron, Exxon Mobil, Royal Dutch Shell, Schlumberger [the world's largest provider of offshore drilling services] and Conoco Phillips. Mr. Gore and Mr. Blood are welcome to put up the solar and wind companies of their choice, selecting from the New York Stock Exchange and Nasdaq. Let's compare stock prices in 2023.[100]

It is not clear if Blood and Gore took the bet, but it is clear who the winner would have been.

Jude Clemente of *Forbes* had a comforting message to the readers of that magazine. Look at how toothless the Kyoto Protocol turned out to be, and now we have the Paris Agreement which is even 'non-binding: i.e., it has no legal standing', so why fret.[101] And, of course, he was right. Another study based on interviews with dozens of investors from North America and Europe found this to be a common attitude: three decades of blah blah blah have produced absolutely nothing in the way of 'serious action', so why believe that any frontal assault on private property in fossil fuels would ever come to pass?[102] Every vacuously incantatory COP summit then raised the stocks of business as usual further, in a counter-loop with its own self-reinforcing power.[103] The 'transition risk' seemed about as real as the risk that Palestinians would march triumphantly into Jerusalem, with several mechanisms heightening the improbability: the more investors invested in fossil fuels, the more 'rational' it seemed for investors to invest in fossil fuels – the herd instinct in the opposite direction.[104] The more mitigation was pushed beyond the short time horizon of fictitious capital, the less reason to select other assets.[105] The longer the talk of it remained just talk, the more vindicated the conviction that all of these banks and asset managers and corporations were too big to strand – bleeding into forms of denial, including literal: 'there is so much money in the market that couldn't give a flying fuck about this stuff,' said one of Christophers's informants. 'You still get investors in the US who don't believe in climate change. These guys honestly couldn't give a stuff.'[106]

The aggregate outcome of these trends was, of course, the *production of even more value that would have to be destroyed* should anyone ever do anything about global warming – assets building on assets, the carbon bubble blown to ever greater circumference. Everyone from JPMorgan to the IPCC understood this logic.[107] The more surplus value 'capital has realized by rooting in the soil like a pig in potatoes', the larger the slaughter of value will eventually have to become.[108] In 2023, Campanale could thus look back on a decade of failure, which a fortiori made his mission in life even more called for: 'the amount of fossil fuels that have been financed has grown over the decade, not decreased. Hundreds of billions of dollars are being raised through the banking system, through the equity markets and private markets to fund more fossil fuel expansion, when we can't burn even what these companies already own.'[109] It has often been said that if warming were to be capped at 2°C rather than 1.5°C, the stranding would be less bloody.[110] But this is to look at it from a static viewpoint. In 2022, say, the budget for 2°C was still filled with enough air to save capital from the worst carnage, because the limit remained distant; but *once temperatures actually near* 2°C, so much more capital will have been sunk into fossil fuels in the meantime – the very reason for the nearing – that it would far surpass what 1.5°C would have required. The destruction would then have to be even more remorseless to avoid 3°C, and so on.

Another distinction established early on in the discourse was that between an 'orderly' and a 'disorderly' transition, the former signifying a smooth, gradual, IAM-like process in which policy-makers inform investors that a change is coming and then do exactly what they have promised: perfect information, time to prepare and close shop without tears.[111] One survey published in 2019 presciently added that in this scenario, 'the wider socio-economic context is also free of major disruptions (no wars, no pandemics, no disruptive climate-related physical impacts).'[112] The first three years of the third decade killed it. Now remained solely a 'disorderly' transition, the result of an ever-widening discrepancy between business as usual and budget constraints reaching a point of criticality where some event, in the articula-tion of climate and politics, makes the whole order come crashing down; and for this scenario to be plausible, no pious rationalism is needed.[113] It is the absence of reason, the denial of reality that pushes the world towards it.[114] The minimal rationalism that, perhaps, remains here is the expectation that a human survival instinct will at some point kick in and force a readjustment to reality. Because fossil capital will have won so much until then, it will stand to lose all the more. 'Climate change', one stranding study put it, 'is a "ruin" problem – i.e. it will result in a system exposed to irreversible harm that can eventually lead to total failure'.[115] The apocalyptic magnitude of climate breakdown is mirrored by the apocalyptic magnitude of capitalist collapse, if any constraint is ever put on the former.

But no such collapse will be induced without politics. This is the concrete meaning of the revolution being born out of the historical delay of mitigation: asset stranding will take on ever more frighteningly Leninist-Marxist proportions. The more the spectre is ignored, or the more knives are plunged into it, the more powerfully high it rises. The counter-loop trips over itself and reverses into its opposite: 'the problem with climate risk is the longer you wait, the bigger the problem becomes,' readers of *Bloomberg* could learn in January 2023.[116] Not even the profit bonanza could extinguish the fear, but rather stoked it from behind; nervous shareholders were back with questions to Exxon-Mobil.[117] There was a sense that the arrogance of fossil capital in this period masked, and worsened, a structural and psychic fragility. By never ceasing to be victorious, it painted itself into a corner filled with its worst nightmares – a dynamic so contradictory that it would have to break out in some other direction.

The Contradiction of the Last Moment

Climate politics becomes more existential over time.[118] Mitigation after a long delay comes to imply swift decapitation of fossil capital, whose interest in defending itself thereby hardens pro rata.[119] The more it has expanded, however, the greater its political clout will also be: a capital incorporated into one thousand new coal plants will be more capable of dictating policy than one with only ten left.[120] A profit bonanza makes it, at the same time, more exposed to destruction and better placed to fend it off.[121] More specifically, if a fresh round of investment occurs close to a temperature deadline, the interest in missing it will be overwhelmingly strong: if the fixed capital is in its forties or fifties, retirement will be far less disastrous then if it has just sprung up. From the moment 'when fixed capital has developed', every 'interruption of the production process acts as a direct reduction of capital itself, of its initial value'. It follows that the fresher the fixed capital and the greater its total scale, 'the more does the *continuity of the production process* or the constant flow of reproduction become an externally compelling condition for the mode of production founded on capital', as observed by Marx – or, 'the more the system expands, the more it hardens into what it has always been', by Adorno.[122] We might call this the contradiction of the last moment. Every time the last moment for climate protection is approaching, the systemic pressure for barrelling past it will be harder to tame.

These are the primary drivers, first for passing a temperature limit, then for developing other projects for managing the fallout, which, of course, grows in proportion. But if we recognise this, we must also realise that the pressure will remain constant or rather build up throughout the overshoot conjuncture and *work to ensure that these other projects are substitutes for and not supplements to mitigation.* If the original interest is in deflecting value destruction, it will stiffen insofar as it succeeds in pushing into overshoot. It will operate relentlessly to make adaptation and removal and geoengineering stand in for the mitigation that must not happen, because it would be so existential in scale. This is the fundamental reason for regarding rationalism-optimism as a delusion, and it marks out the multiplying fronts of climate politics: for each category of measures, there will have to be a struggle against it substituting for the liquidation of fossil capital; in each, defeat in this struggle will set off its own secondary and tertiary spirals of disaster. We shall return to those in the sequel to this book.

Passive and Active Capital Protection

If the three other projects work perfectly fine, asset stranding may be forgotten and never heard of again. Its role in history, its spectral causality will then have consisted in impelling developments in a different direction, where the climate problem is brought under control by means other than the mitigation spoken of in earlier phases. Capital will have succeeded in protecting both the climate and itself. The property question will be buried under layers of new technology – but if the outsourcing of climate management to these three affiliated projects is ineffectual to some degree, it will rear its head again. Or, if they are launched with the firm belief that mitigation must also happen, a return to the mission of asset stranding is inevitable. Which will it be? The only thing we can be certain of is that dominant classes enter the overshoot conjuncture with an interest in *protecting value awaiting or undergoing valorisation within the circuits of fossil capital from destruction* – abbreviated, capital protection.

Taking a leaf from Antonio Gramsci, we may further distinguish between *passive* and *active* forms of such protection.[123] We shall consider active capital protection to be efforts directly undertaken by agents within the various circuits of fossil capital for said purpose – say, Aramco doing this or that to ward off the threat. Passive capital protection, on the other hand, is performed *on behalf* of the bearers of value, by others. A good example would be the IMAGE team slowing down emissions reduction rates in its IAM, because 'fast reduction rates would require the early retirement of existing fossil-fuel-based capital stock', and instead coming up with BECCS. The problematic of base and superstructure here resurfaces. In the case of overshoot ideology, the correspondence is of such a nature as to suggest direct determination, as in a compass. The structural interest in protecting capital functions like a magnetic north pulling towards it the needle of ruling ideas about how to deal with climate. It may be invisible, far from any particular paper or report or model or summit, but it exerts such a deep attraction as to make them pivot into alignment. Passive protection is passive also in the sense of unresistingly coordinating activities with the operations of fossil capital. On the other hand, it might well provide the initiative, as when IAM scenarios are first developed by modellers and later used by corporations: we are dealing with a dialectic.

The capital protected, however, is more than a heap of assets. These are accretions, the weight of things generated, not the process itself. We have seen Kelly use the metaphor of pyramids for fixed capital in offshore extraction, and the pyramids of the Old Kingdom did indeed incarnate a physically tangible inertia. But to understand their function in pharaonic Egypt, one would have to study the logic of its kingship and tributary state – as in, why were they built? The drive to form more fixed capital in fossil fuels is not self-explanatory. Why, for instance, could not investors build up assets in renewable energy instead? Or did they? What were the underlying drivers that took on the material form of *this* type of fixed capital and all its appurtenances?

With or without a Fight

When governments and IAM wizards expounded on programmatic overshoot, they did not, of course, deny that some form of endgame for fossil fuels was part of the long-term – in some cases, very long-term indeed – vision for the future. It was just that they saw the phase-out stretched out over time, taking place in a controlled fashion that avoided sudden, large-scale asset stranding. Overshoot would play out in parallel with conventional mitigation, and carbon dioxide removal in particular would be rolled out in addition to emission cuts, not as a substitute for them (the rationalist-optimist doxa). Only some part of emissions – those commonly categorised as 'unavoidable' – would need to be maintained; the rest was slated for eventual reduction to zero. In this vision, overshoot was but a temporary condition, born out of the necessity to buy 'us' just a few more years while mitigation got up to speed and the transition ran its natural course, no revolution needed. In crossing the first threshold, the ideological architects of the conjuncture denied that the gates were thereby opened to barge past future ones as well: this exceedance was to be the one exception, a mere hiccup on the circuitous path to Paris.

The processes that were supposed to make this happen would be propelled by durably capitalist dynamics. With some minimal prodding by states, perhaps some well-placed subsidies here and there, powerful new industries would ascend to dominance. By the sheer gravitational force of their competitive advantage, they would suck innovations and money into their orbits. Perhaps not all at once, and perhaps not very rapidly either, but nonetheless inevitably, 'green capital' would take over from the brown.[124]

The latter would be eroded out of existence, a process surely marked by stresses and strains, but – because the transition was left so much to capital's own devices and thus spread out over a prolonged period – far less destructive than the asset-stranding alternative: not so much a revolution as a tender retirement, a slow deactivation of one set of energy technologies as another comes to fruition. The haunting here ends simply because more valuable assets are discovered elsewhere, and all the attention in the room gradually shifts to them instead.

But how plausible is such a scenario? Does it chime with hitherto observable trends in the real world? How likely is overshoot to be temporary, rather than dragged out towards infinity – in other words, how inclined will capital be to relinquish fossil fuels without a fight? Answering these questions necessitates a deep dive into the technological and political-economic components of conventional mitigation. Did obstacles in them contribute to the overshoot conjuncture, and if so, what are the odds for their unbidden dissolution before the next limit is transcended as well? It is to these issues we now turn.

6

We Are Going to Be Driven by Value

When the overshoot conjuncture emerged in the early third decade, its greatest paradox was this: the closer the world edged to 1.5°C, the easier it became to avoid that fate by ditching oil and gas and coal for the renewable alternatives. For every year of continued and rising emissions, the technical feasibility of ridding the metabolism of the human species of fossil fuels stood clearer, more firmly substantiated, more widely publicised. Never before had it been so easy to live without them.

A Flow of Opportunities

Visions of an economy run entirely on the power streaming in from the sun are nearly as old as the fossil economy itself. One notable case is that of Svante Arrhenius. Best known as the first scientist to calculate how a rising atmospheric concentration of CO_2 would drive up temperatures, in a paper from 1896, soon after he also sketched a programme for turning the world towards the 'power lavished on us by the sun in amounts that never seem to ebb'.[1] Coal and petroleum ought to be left in the crust. Humanity would be wiser to make use of the flow of energy, a fraction of whose supplies would cover any conceivable needs. Basking in figures of how much sunlight hits the Earth's surface, Arrhenius singled out 'sun machines' as 'indubitably the most essential aid for humanity in the future' and pointed to the prototypes just invented; of special promise were desert countries and the tropics, where the sun 'almost always shines', waiting for humanity to install the contraptions for harnessing it. Wind power presented nearly as great opportunities, particularly, and luckily, in northern locales poorer in sunshine.[2] Combined with some water, these two sources could power the present and every imaginable future world economy several times over – but not to save the planet from the rising temperatures attendant on fossil fuel combustion, of which Arrhenius took a positive view. His identified tendency to global warming would make the Earth a more pleasant place: the vegetation more luxurious, the harvests more plentiful, the north freed from harsh cold and the fear of new ice ages.[3] Instead, the pressing problem that mandated a switch was the scarcity of fossil fuels, soon to be felt in the form of high prices. Peak coal was coming, because capitalists thought nothing of squandering the subterranean wealth; but thankfully, the alternative was there for the taking.[4]

Arrhenius put forth these arguments in the 1910s, and in the following two decades, there was at least one country where they had a degree of influence: the Soviet Union. As Daniela Russ has showed in a remarkable piece of historical excavation, there was a craze for solar power among early Soviet scientists, engineers, poets and others, who took a fancy to the vision of a world electrified by the sun – not in the next few decades, but centuries or even millennia into the communist future. Utopian fantasies accompanied pioneering research into both solar and wind: rudimentary devices for generating heat as well as motion; installations designed to irrigate infertile plains. Again, this spring season for renewables was not motivated by any concerns about global warming; rather, they were admired for their potential to overcome the material limitations of coal and oil and allow humanity to extend its metabolism to every nook and cranny of the Soviet Union and, by extension, the planet.[5] Worries about the limitations faded, in the Soviet Union and elsewhere, until they returned with a vengeance in the 1970s. A first paper to outline a full transition to solar and wind appeared in *Science*, focusing on Denmark, which, the author claimed, could have it completed by 2050; now worries over dwindling oil framed the vision.[6] But 'climate disruptions' had come into view as an additional reason for the switch.[7]

The vision, then, was for a 100 per cent flow economy – one where all energy use would be derived from the sources renewed on a daily basis by the sun, powering every instrument, heating and cooling every object, moving every vehicle, 24/7, tutti quanti.[8] After a long hiatus, during which Arrhenius, the early Soviets, the *Science* paper and other harbingers were ignored, the vision crystallised just as the Kyoto era came to an end. Actual developments in the technologies of solar and wind had, of course, progressed during the late twentieth century, if only in fits and starts; but it was only in the early twenty-first, on the back of advances made, that a 100 per cent flow economy came to be rigorously outlined.[9] Between 2009 and 2011, a team of scientists around Mark Jacobson at Stanford University published a series of eye-catching articles demonstrating just how close at hand and workable the vision was.[10] Fossil fuels could be eliminated from the world economy without any needs sacrificed. Substitutes from the sun were stomping in the stalls. Research into 100 per cent renewables then rushed out across the fields in the 2010s, but it was not without detractors.[11] Ideologically, they fell into two camps. There were the ecomodernists, who faulted Jacobson and others embracing the vision for counting on a decrease in total energy consumption – when it ought to increase – and for discounting nuclear power.[12] There were the eco-anarchists and/ or -primitivists, who complained, conversely, that these people upheld an unfounded faith in modern technology and failed to realise that renewables are every bit as bad as fossil fuels.[13] Here the favoured solution was not nuclear, but a reversion to self-sufficient, extremely localised communities with 'a frugal lifestyle' and/or a reduction of the human population to 'one billion or so'.[14] Modernist, anarchist, primitivist, the critics advanced a number of technical objections to the propositions from Jacobson et al., who rebutted them patiently and painstakingly.[15]

By the early 2020s, the critical choir had mostly fallen silent. The visionaries had come out on top of the scientific debates.[16] Ploughing ahead, they presented detailed roadmaps for switching first 130 and then 145 countries – accounting for 99.7 per cent of emissions from fossil fuels – to 100 per cent renewables, 'ideally by 2035, but by no later than 2050'.[17] The efforts culminated in Jacobson's aptly titled monograph *No Miracles Needed: How Today's Technology Can Save Our Climate and Clean Our Air*, the most comprehensive manual published to date.[18] It could be done: the evidence was overwhelming, just as humanity put its toes on the threshold to 1.5°C.

The Time Problem Solved

The contours of the transition were envisioned as follows. First, and most effortlessly, all electricity generation would have to be sourced from renewables. No more coal or gas combusted in power plants: every watt drawn from the flow. Second, and in parallel, but with some more effort, all use of energy that can be electrified must be so: no more internal combustion engines when electric vehicles take over; no more blast furnaces when electric arcs melt the ore. Energy previously released by the burning of fossil fuels – oil in cars, coke in steel plants – would be provided by the flowing currents of electricity. Every little gadget would be caught in this net, down to the suburban man's leaf blower, 'possibly the most annoying fossil-fuel machine today'.[19] Third, all thermal energy – heat in factories, heat in homes – would need to be generated by renewables for them to reign supreme. In every proposed scheme, solar and wind would together make up the lion's share, the former usually projected to dominate by the middle of this century.[20] Jacobson and colleagues ruled out building new dams. Biomass for burning did not feature. But slices of tidal, wave and geothermal power were included, the latter an outlier by emanating from hot springs and steam inside the crust; in all other respects, the energy would be supplied from the flow, or the practically immediate results of solar

radiation, neither snared in vegetation nor sequestered underground but streaming, blowing, rising and falling throughout
the lands and seas of the surface.[21]

Could such a system be relied on? Would the energy be available on demand? The wind famously does not always blow,
and the sun sets at dusk. The integration of the flow in weather
swings and diurnal cycles makes for intermittency, a drawback
highlighted ad nauseam by devotees of the status quo. It vexed
the visions from the beginning: Arrhenius recognised that nature
builds no storehouses for wind power, like lakes and rivers for
hydro. The flow offers exuberant supplies, but 'the great art is
to collect this energy and store it in times of plenty for times of
dearth. Without a doubt, the spirit of innovation will celebrate
great triumphs in this field, which will become epoch-making
for the future welfare of humanity.'[22] The visionaries of the early
twenty-first century would consider him proven right: there was
now a wealth of options for ensuring reliability.

First, and perhaps most importantly, grids spanning a spectrum
of landscapes could be integrated into one. The wind always
blows somewhere, and if it does not, then the sun shines or the
waves bob or the old dams are filled to capacity: the trick is to
tie the cables together (a trick already tried by the Soviet com-
missions for wind power in the 1920s).[23] Shortfalls in one corner
can be smoothed out with surfeits from another, particularly if
the grid encompasses several regions and even continents. Some
places are blessed with more of both solar and wind than anyone
can dream to ever use – Cairo, say, having the permanent intensity
of the desert sun all around it, plus the winds of the Red Sea a
stone's throw away – while others sit at one pole: the North Sea
countries swept by winds de trop. At the opposite pole, southern
Europe has all the sunshine. By the early twenty-first century,
unlike in Arrhenius' time, there was no technical obstacle for
plugging one into the other.[24] High-voltage cables overhead or
undersea could unite a continent or two in supergrids that would
make blackouts exceedingly improbable, but one would not necessarily have to go so far to combine solar and wind. There were
emerging designs for putting panels *on* the blades of turbines,

capturing both types of flow from the same installations.[25] Less avant-gardist: wind and solar are known to be complementary, simply by dint of seasonal variations in weather, particularly in the Northern hemisphere. There is a lot of sun in the summer months, a lot of wind in winter; the art would be simply to exploit the complementarity.[26]

And then there was also – second – a growing assortment of technologies for storing the flow and releasing it later: batteries, tanks, flywheels, fuel cells, reservoirs into which water could be pumped up in times of electrical plenty to be discharged into turbines below in moments of dearth.[27] The latter alone would guarantee energy always at hand. Small tanks tucked away from rivers, recycling water in a loop, had a potential storage capacity several times larger than what a 100 per cent flow economy would need.[28] Third, the infrastructure could be oversized on purpose. Because the potentials of wind and, in particular, solar are practically limitless, one could build more farms than needed on average, as a buffer for when the flow sags.[29] Fourth, if none of these solutions were sufficient, there would always be the possibility of adjusting demand to supply: producing ice for refrigeration and running waste water plants when the grid is full, but not when it is half-full; feeding electricity into parked cars during nightly hours of surplus.[30] In some mix, these four solutions – interconnection, storage, oversizing, demand management – obviated intermittency and thereby defeated a central argument from the detractors, be they modernist or anarcho-primitivist.[31] A flow economy would not need fossil fuels (nor, for that matter, nuclear) as a baseload.

The Space Problem Solved

But would there be any land left for humans and other species to live on? Would they not be squeezed out by all the spinning, shining new energy parks and ancillary structures? The flow of energy is defined by its incorporation into landscapes, the terrestrial spheres through which humans live and move. The stock of oil and gas and coal, on the other hand, is buried in chambers below. These opposite profiles have bred a prejudice against the former: an economy fully reliant on the flow would have to turn over immense areas to power generation. Fossil fuels are said to be 'compact' and 'dense', their exploitation requiring trivial amounts of land in relation to the profuse work they ignite.[32] Moving to renewables would force humanity to gather fuels thinly dispersed on the surface, in a shift from a 'vertical' to a 'horizontal' energy regime: all the tools and utensils hitherto stowed away in the cellar spread out in the living room.[33] This view of the matter is grounded in Ricardian-Malthusian theory. It overlooks some basic circumstances. When the stock is brought into the landscapes, as barrels of oil or barges of coal, room must be made for these commodities to circulate. Precisely because they are exterior, injected into the landscapes from the outside, an entire infrastructure must be devoted to their entry and further transmission to points of combustion: all the derricks, mines, pipelines, terminals, together constituting, as we have seen, 'the largest network of infrastructure ever built'. A 100 per cent flow economy would lift this boot off the neck of the planet. All the land taken up by the production and distribution of the stock in its commodity form could, once the facilities have been safely discarded – the fixed capital destroyed – be freed up for other use. The question must then be rephrased. On balance, which of the two regimes would demand the most space?

Such an accounting exercise is fraught with uncertainties. No estimates of the aggregate acres of the globe allocated to fossil fuels could be found in the early 2020s; but Jacobson gave the figure 1.3 per cent of the US land area, while counting on the construction of all farms and plants needed for 100 per cent energy to swallow 0.31 – an emancipation of space, not a constriction.[34] One study likewise found that swapping coal for solar over a couple of decades would release American land.[35] But then the US was uncommonly encumbered by legacy infrastructure. For the world as a whole, Jacobson and his team estimated a total flow footprint of between 0.17 and 0.22 per cent of land: on average, no more would have to be set aside.[36] Everyone agreed that the type of renewable most consumptive of land is biofuels, since vast fields must be planted with crops for burning; large dams have likewise tended to drown whole villages and vistas, biodiversity ravaged by monocultures and mega-lakes alike – some of the reasons for omitting both from the schemes of Jacobson and most others.[37] The helio-aeolian core of the flow has a different spatial logic. Solar and wind melt into landscapes, and for that very reason, unlike the stock, *their capture can be combined with other activities.*

No farmer can sleep well at night with a pipeline crossing her fields. The devastation experienced by farmers and fisherfolk in the Niger Delta and the Ecuadorian rainforests testifies to the unviability of the combination. Indeed, the very emergence of 'place-based resistance' in such sacrifice zones belies the notion that 'fossil fuel extraction is not particularly space intensive'.[38] But wind turbines must have space in between them, so that the blades of one do not collide with the other; and in that space, it is perfectly safe to grow crops, tend animals, maintain a forest or do virtually anything else: no substance foreign to the landscape will suddenly leak into it, sticky and stinking. The base of the tower itself takes up minimal land.[39]

Any early twenty-first-century traveller to Germany, a country that had built up a degree of flow infrastructure alongside its fossil patrimony, could spot the difference. The open-pit lignite mines looked like craters from some monstrous meteorites, in which nothing could live, whereas wind turbines were planted like flags among all manner of other vital land use; and every time the former expanded, they had to raze the forests and villages standing in the way (turbines included, as we have seen). There was placed-based resistance against wind farms too. The far right specialised in it, in Germany and the rest of the global North.[40] The same people who cherished vast expanses of coal mines and oil fields tended to rail against the aesthetic despoliation inflicted by the poles and rotors, but opposition could also stem from any other preference – Sami herders, for instance, fearing that their reindeer would avoid winter pastures impaled by farms.[41] Clearly it is not a question of the flow occupying a lot of surface area and the stock little to none.[42] Rather, each has its own spatial appearances and affordances and connotations, to be weighed against other interests in the land and combined with them, or not. As for wind (but never for coal), there is often, if not always, an exit from terrestrial constraints: moving offshore.[43]

As for solar, much like for wind in Sápmi, land can constitute a real problem. This power source comes in two forms: photo-voltaics, PV or 'solar panels' in the vernacular, their function now common knowledge, and concentrated solar power, an instal-lation that places mirrors in a circle or square and turns them towards a tower; the arrays collect the heat from the sun and shoot it into the centre, where it is converted into electricity. If part of a savannah inside a designated national park is given over to a concentrated solar power plant, the loss of land is absolute. The same if a wheat field or timber plantation is cleared and a PV farm instead mounted on the ground. In scenarios where solar grows in this fashion, the land requirements rise to up to 5 per cent of a country's area if this single source generates 80 per cent of the electricity by 2050.[44] But PV has a promise its bigger brother lacks.[45] Solar panels are modular, scalable, conjugable in ways that come closest to realising the potential of the flow to mesh with rest of the landscapes.

No one wants to sleep under an oil derrick. But panels can be favourably affixed to the roofs of huts and villas and any other kinds of buildings – say, the concrete boxes of the Palestinian refugee camps in Gaza and the West Bank, some of the most over-crowded places on Earth, defined by their lack of access to land (and deprived of electricity to boot).[46] No coal mine could fit into their alleys. PV does not have to drive people from their homes, but can be fitted onto pre-existing structures, redefining the built environment as a potential canvas for power generation.[47] Guests in the attic, not forces of expulsion; by the second and third decade, the cutting-edge modes of coexistence were 'agrivoltaics' and 'floatovoltaics'. In the former, panels are elevated on stilts and combined with agriculture flowering around and beneath them. The panels will need to be positioned and tilted so that plants below receive enough sunlight; but the provision of shade may also be a service, increasing the yields of crops ranging from soybeans and sweet potatoes, lettuce and broccoli, eggplants and cherry tomatoes to cassava and even corn (while those of wheat might rather go unaffected). Under the shade, less soil moisture is lost to evaporation, less photosynthetic productivity to heat stress: advantages higher prized in a warming world. Panels do not spoliate the fields below. They uplift them.[48]

Animal husbandry goes well with PV too. Sheep have been found to love the shade and huddle in it during hot days – agrivoltaics providing adaptive benefits over and above its spatial thrift in mitigation.[49] Rainfall collectors have been attached to panels bestriding fields in India.[50] (No one has yet proposed gutters on pipelines for irrigating nearby plots in need). Panels can be installed on top of greenhouses; under and between olive trees; in unused corners of fields; on degraded agricultural lands, of which there might be no shortage.[51] By the early 2020s, some 12 per cent of the surface area of the globe was devoted to crop cultivation, not to mention the three times larger area devoted to livestock, hinting at the potentials of agrivoltaics to shrink the footprint of solar towards some vanishing point – but perhaps floatovoltaics put it in the shade.[52]

Panels can be built on buoys and anchored with mooring lines. The photovoltaics will then float in the water, such as in already existing dams, reservoirs, ponds, canals, lakes in abandoned quarries or pretty much any other body of water nestled into human environs.[53] Lignite mines in Germany would make for excellent bases, once extraction is ended.[54] Not a square meter of new land would have to be claimed. No lack of suitable water would hamper the arrangement: covering one tenth of the world's dams would generate as much electricity as *all its fossil fuel–powered plants*.[55] Countries that could easily meet their entire electricity demand by putting solar rafts on a fraction of their hydropower basins included Canada and Brazil, Zimbabwe and Sudan, Myanmar and Laos.[56] If placed like a lid on a pumped storage reservoir, the PV would possess an inbuilt battery. On dams providing drinking water as well as electricity, it would preserve the precious resource by cutting evaporation, otherwise rising with every tenth of a centigrade: imagine the gains to Lake Nasser or Kariba.[57]

In the early third decade, agrivoltaics and floatovoltaics were just beginning to get off the ground. Both had their learning curves ahead of them. The next step, presented in a first scientific paper in 2022, was silvovoltaics: interleaving forests with solar trees. If panels extend from a trunk like layered leaves, they can generate two orders of magnitude more electricity per square meter than flat, single-storey panels; below and alongside them, bamboo and any other timber trees can grow undisturbed.[58] All major forms of human land use were thereby opened for synthesis with PV. More combinations might lie in store (offshore floatovoltaics, floating concentrated solar power . . .). We can then venture the following hypothesis: continued technological development in hydrocarbon production will *expand its land footprint per unit of energy*, whereas the same process in solar power generation *will have the opposite effect*. The former tendency is the spatial aspect of the growth in fixed capital formation, producers having to go to greater lengths to find and send back oil and gas, constructing ever more props and passages along the way.[59] The latter will work towards a situation where the sun hits spots and objects that exist not only to be hit by the sun – the state of nature of the flow, as it were.

With a bit of Marxism and common sense, we can then also turn the Ricardian-Malthusian view on its head. As regards life at the surface, a transition from stock to flow would lead away from a horizontal energy regime, in which fuels circulate as commodities through a million coiling channels, towards a vertical successor, oriented towards the sun and the wind that appear *above* people, organising their metabolism in storeys and levels.[60] There will be less shuffling around, more direct downloading. In the last instance, the difference is a function of the stock materialising as fuel only through labour and assuming the form of a commodity, two conditions that can never apply to sunlight or wind. Fossil fuels impose a zero-sum game on the land. The strange commodity exacts zones and corridors exclusively for itself. In the helio-aeolian flow, the land rather unfolds its leaves like a sunflower.

Woes of the Transitional Period

But would there be materials enough to build all these blades and bases, panels and pontoons, barrages and pumps and everything else required for a 100 per cent flow economy? Or would the planet be pillaged for a second time? One version of this worry is the contention that materials for solar and wind would come from a prolongation of the original pillage, since they are parasitic on fossil fuels. In a text from 2021, a prominent anarchist critic of renewables, Alexander Dunlap, pointed to wind turbines made of steel and stated that 'industrial steel manufacturing is impossible without burning coal, as metallurgical coal – or coking coal – is a vital ingredient in the process': hence no difference between wind and any fossil source of power.[61] The factual statement was falsified in that same year, when a Swedish plant delivered the first shipments of steel not impregnated by coal.[62] Not long after, one survey picked a bouquet of eight-six technological options for defossilising the industry, spanning everything from renewable sources of electricity to recycling (to which few materials are as amenable as steel).[63] But Dunlap did have a point of sorts.

Suppose the manufacturing of wind turbines commences in year 0, when steel manufacturing is still based on coke. Suppose the transition in the steel industry gets going in year 10 and is completed in year 20: new turbines will then carry traces of coal for the first two decades, washed away only at the beginning of the third. The lingering dirt is in the nature of the process. In a transition from one type of energy economy – or 'regime' – to another, the tools of the latter will have to be assembled with those of the former.[64] The only escape from this tautology would be a 100 per cent flow economy arriving ready-made from another planet where it is already fully established. Dunlap and similar-minded critics contend that, because renewables have not yet had their life cycles purified of fossil fuels, they are equally bad: the energetic version of the classical anarchist demand that every step towards the desired end-state must be a perfect picture of it – which is to guarantee that no journey out of the status quo ever takes off. A transitional period is rich in intermediates and impurities, by definition.

As for the mining of materials in a mature fossil economy and a mature flow economy, the two can scarcely be compared. In the former, extraction is inflicted on the Earth without cease, because that is how the fuels come to the surface: the oil and gas and coal going up in smoke must be succeeded by new quanta brought up from below, day in, day out. In the latter, extraction withers away. The flow requires no more mining than wild apples. Once the ladder or stick for reaching them has been built, the picking can unfold with minimal material throughput.[65] This qualitative difference between the essences of stock and flow reduces the question of materials to, again, the transitional period: will there be enough to *construct* the means of production for the post-fossil era? Or would the extraction, however temporary, leave indelible scars on the planet? As of the early third decade, the world produced roughly 350 times more coal than materials required for building turbines and panels and their auxiliaries – lithium, cobalt, copper, rare earth minerals – and 190 times more oil.[66] If the rate of extraction of these transitional materials were to increase twelvefold, their output would still reach merely 3 per cent of all that coal. Such numbers suggested that the planet would be spared further wounds by a swift transition: it could never cause anything like the volumes extracted and destroyed by the late fossil economy.[67]

This was scant solace, however, for the people and other species whose land had suffered that fate at the hands of, for instance, companies mining lithium for the batteries of electrical vehicles.[68] Even if only a fraction of the larger problem, destruction of such kind was a reality already in the early third decade. Could it be overcome, as handily as the time and space problems? Because the technologies of wind and, in particular, solar power were, by this time, in states of extreme flux, no one could say with any certainty what materials would be indispensable the next decade or even year.[69] Lithium had a reinvigorated contender in sodium, easily obtained from salt.[70] Silver, pasted in a thin layer on panels, the one material whose limitations might put a real damper on their production, could be replaced with copper, and so on.[71] A radical programme of continuous recycling – including from scrapped wind mills and solar panels – could help reduce materials consumption to some degree.[72] One would have to be somewhere on the anarcho-primitivist spectrum to deny this sort of technological progress. One would have to be an ecomodernist in the clouds, however, to believe that the transition could begin to float above the ground and put zero pressure on resources. In fact, precisely its historical delay means that it will – if it ever begins – need to be so rushed and executed within a timeframe so short as to call forth the problem of bottlenecks. A mine is not opened in a minute. It takes time before it yields it goods. How, then, can extraction of the materials required for the reconstruction deliver on schedule? There will be stress on the land (not to mention the labour).[73] After intermittency and area, judicious visionaries acknowledged that materials remained the one 'formidable challenge', the more so the longer it was postponed; it might even place a question mark over their whole approach.[74]

Consider lithium. Electrifying the entire car fleet of a country like the US would be possible, in the sense that there is enough lithium to assemble the batteries of the current design (as of the early 2020s); but it would risk ruining the ecosystems and – often indigenous – peoples of the mining areas, in Chile, Argentina, Bolivia. The alternative would be a modal transition. Instead of shifting the existing, and growing, fleet to electricity, one could shift travellers onto buses, subways, bikes, walking lanes: modes of transportation breaking the car dependency and minimising the demand for lithium (or any other material). The battery of an electric Hummer is several hundred times larger than that of an e-bike. Why then even contemplate keeping and supposedly greening the former?[75] The visionaries turned out to suffer from a kind of black-box conservatism: they took the present constellation of productive forces as given; all they ever wished for was to put it on another energetic footing. In *Electrify: An Optimist's Playbook for Our Clean Energy Future*, one of their number, Saul Griffith, went out of his way to allay any fears of negotiating the American lifestyle. 'We don't need to switch to mass rail and public transit, nor mandate changing the settings on consumers' thermostats, nor ask all the red meat-loving Americans to turn vegetarian' – phew. The renewable future will not tinker with 'the major objects in our lives – our cars, homes, offices, furnaces, and refrigerators. All of these objects will just be electric. There is no need to fear this future.'[76] Mark Jacobson was impeccably patriotic about also proving that the tanks and fighter jets and armoured personnel carriers of the US army could be powered by renewables.[77] (It would save the lives of American soldiers, as there would no longer be oil supply lines vulnerable to ambushes.)[78] The question of whether these vehicles should be carried over into the flow economy appears not to have struck him, and it could be extended further. Do people in the global North really need their leaf blowers and livestock? Would the transition have to show such deference to the things that be? Should the skin of property be changed, or the most invasive snakes also culled in the process?

A Trio of Immature Tech

Such conservatism, however, was also part of the signature cogency of this research programme. It took the sum total of the existing productive forces, subtracted the stock and showed that the remainder could be run on the flow, by means of a subset of productive forces that likewise existed. The basic technologies of wind and solar power have been around since before Arrhenius and the early Soviets. In the 2010s and early 2020s, they had decades and years of evolutionary leaps behind them. The difference from a removal technology like BECCS could hardly have been more fundamental: these forces existed in the physical world. No one had, by this time, seen a BECCS plant;

but the majority of humanity had probably set eyes on a wind turbine or solar panel at some point. Nor were these like nuclear fusion, sometimes touted as the gateway to post-fossil cornucopia, always three decades down the road; as of the early 2020s, the general assessment held that it might become a reality around 2050, a working alternative to fossil fuels perhaps towards the end of the century.[79] But if time is short – especially if it is too late, in some senses of the term – one should not entrust mitigation to phantom or embryonic technologies. They ought to be present in the here and now, tried and tested, ready to be scaled up: and such was indeed the status of the cornerstones of the 100 per cent vision.[80] The issue with them was not one of innovation so much as adoption. Neither the solutions to intermittency nor the arrangements of coexistence on the land – agrivoltaics, floatovoltaics, silvovoltaics – depended on the breakthrough of some fancy new tech, but rather on the conjoining of familiar tools, such as panels and buoys, or panels and poles.[81]

The visionaries of a conservative overhaul thus demonstrated that the incumbent technomass could be maintained if fossil fuels were to be given up for renewables, on the basis of technologies ready to go. They succeeded, with three exceptions – all significant, all ambiguous. First, there was aviation. Aircraft flying on electricity from the flow appeared to be if not a castle in the air, then inconveniently far into the future. Although the first solar-powered plane with a person in it flew already in 1979, flying hundreds of people between continents would require marvellous improvements in battery technologies, packing at least eight times more energy into every kilogram than what the most powerful exemplars achieved in the early 2020s; anything less would fail to keep the machines in the air long enough, or take up too much space.[82] Lithium was the presumed material. All aircraft, in other words, are worse than Hummers in the sky. But here too, there were undeniable advances: in those same years, the first contracts were signed for delivery of planes called 'Alice', a model carrying nine passengers almost 500 kilometres – not enough to fly from Berlin to Vienna, but not so far off.[83] All-electric aircraft on short-haul flights seemed to be within reach.[84] NASA announced the invention of a new type of battery far exceeding the performance of the lithium types, and conservative visionaries already drew maps of the airports of the future, looking exactly as before, except for solar panels producing electricity in situ and charging stations where the planes would dock.[85] Leaving aside the question of its desirability, electrification of the world's aircraft was a prospect somewhere between the speculative and the practical, contingent on breakthroughs still in the future.

Long-distance shipping was stuck in roughly the same spot. Ferries propelled by batteries managed only short voyages. But container ships and bulk carriers could already be switched to substitutes for heavy fuel oil: hydrogen or ammonia produced with electricity from the flow; various designs for reviving sails; a modicum of PV onboard; all of these and more in combination.[86] By the early twenty-first century, nearly 100 per cent of commodities transported on sea were ferried through the burning of fossil fuels.[87] Around 96 per cent of manufactured goods had also been infiltrated by such fuels as feedstock – as raw materials, that is; as in plastics fabricated out of oil or fertilisers out of gas, finding their way into everything from tampons to tofu. A behemoth with a finger in all kinds of commodities, the petrochemical industry used fossil fuels for both burning and building.[88] The former could be turned over to the flow right away, much as in the steel industry; the latter was a harder nut to crack. Some feedstocks were lazily taken from oil and gas and their by-products lying around in refineries – notably hydrogen, which could instead, just as for ships, be acquired by means of electricity from solar and wind, splitting the two H atoms from the one O in the molecule of water. Other feedstocks centred on C, or carbon, as the irreplaceable element for further processing. Ethylene and propylene – the main precursors to plastics, polyester, rubber, going into apparel, adhesives, refrigerants and so on – derived from C. Fossil fuels naturally and conveniently supplied this element to the industry: could it be derived from some other source? For now, we may simply observe that by the early 2020s, substitutes for fossil fuel feedstocks in the petrochemical industry were under development, but in laboratories rather than in plants.[89]

In this condition, the trio of aviation, shipping and petro-chemical feedstocks formed an exception to the rule of existing productive forces immediately transferrable to the flow.[90] By the early 2020s, there was a difference between, say, the electricity in a socket and a ship in the harbour: mature technology allowed for the former to be sourced from the flow, but not the latter. The difference was like that between an adult and perhaps not an embryo, but a toddler or adolescent. Technological development, however, it is important to remember, does not occur in a political vacuum. What if the authority regulating international shipping announced a total ban on heavy fuel oil, to come into effect by 2050? It would be a mighty spur on shipping companies and the manufacturers supplying them to speed up the maturation of alternatives.[91] The sum of viable mitigation should not be equated to that of the existing productive forces. The latter would rather grow with the former.[92] Anything else would presume that technological development is immune to political pressure, rendering everything from the pyramids via the atomic bomb to the Covid-19 vaccines inexplicable. In fact, given the laissez-faire situation of the early twenty-first century – the *absence* of something like bans on fossil fuels – the remarkable element was rather the maturity of so very many alternatives.

Just How Unnecessary the Transgression Is

If we apply a parsimonious yardstick, we would have to conclude that the visionaries had failed to cover a little less than one tenth of the total fossil-fuelled pie, the share taken by the immature trio.[93] In the early 2020s, the blueprint, then, was for a 90 rather than 100 per cent flow economy (on the premise that none of the existing productive forces be left behind: no planes grounded, no ships scrapped, no plastics avoided). Jacobson and the other researchers working in this vein envisioned a transition extending from this moment in time until mid-century, preferably 2040. Eighty per cent of it would be completed by 2030.[94] As of that year, there would then be two tenths of the original pie left to dispose of. We have seen that the IPCC called for emissions to be halved by 2030: the proposal here was to cut them significantly deeper, with means of production already in existence, so as to keep a safe distance from that guardrail. It follows that, when the 2020s dawned, *humanity had been given a task that it was able to solve, since close examination shows that the problem arose when the material conditions for its solution were already present* (and as for the last tenth, those conditions might have already been forming). Had the transition been initiated by the time of the Covid-19 pandemic, there would have been *no technical obstacles standing in the way.* The same applies to earlier dates – initiation at the time of the Special Report, or Paris, or Copenhagen, or even the Villach and Bellagio workshops – but the closer the world came to 1.5°C, the more fully developed were the material conditions for averting it.

We have seen JPMorgan observe that a shift to renewables 'is technically feasible'; and the scientific literature could hardly have been in more resounding agreement, harping on the theme – 'no significant technical or economic barriers could be identified' a typical conclusion.[95] What, then, was the nature of the obstacles? We know some things about it from the previous chapters, but here it is worth pointing out that the studies tended to end on a note of planning. Every available solution demanded it. Enveloping regions in a supergrid, siting pumped storage reservoirs, oversizing solar and wind capacity, managing and regulating demand, combining turbines and panels with other forms of land use – nothing would work without planning. Not least importantly, to keep the materials problem within bounds and minimise disturbances to wildlife, 'careful strategic planning is urgently required'.[96]

All this planning might have sounded unappealing to some. But it did not constitute a *technical* obstacle, the way, say, the lack of oxygen makes colonising Mars difficult. Imagine, instead, a couple in crisis setting themselves the task of cooking a dinner together and sharing a pleasant evening for once. All ingredients are in the fridge, a wine bottle has been opened, the chopping board laid out, but now the two start arguing again and forget the onion so it burns and, before long, the quality time has turned into an all-out row: here, the problem sits not in the technology, but in the dysfunctional relationship. Or, imagine instead, purely as a thought experiment, that humanity possessed all the instruments needed for colonising Mars, but that they could be utilised only if it submitted to the Khomeinist interpretation of sharia law and the government of *velayat-e faqih* hitherto practised only in Iran – possibly a complicated thing to do; quite probably off-putting to many. These situations corresponded to the one of the third decade, insofar as relations between people prevented them from accomplishing what was needed and necessary. This might be of some import to the interpretation of the trajectory of human history: *the overshoot conjuncture emerged for reasons unrelated to technological feasibility.* Humanity could continue to do the things it did and still stay below 1.5°C, if it would just let go of fossil fuels – the kernel of the conservative vision and all the evidence it marshalled.

Note that two questions are at stake here: one normative and one descriptive, historiographical. One might take the position that a transition from stock to flow would be ethically bad, and that it would be more desirable to have generalised puritanism/ privation or 7 billion people 'or so' removed from the planet, the two principal submissions from the anarcho-primitivist camp. This could and, we believe, should be disputed. Moving to an 80, 90, 100 per cent flow economy over one or more decades would prevent mass suffering on a scale that should make any reasonable moral standards sanction it – even if, as is likely, the full mobilisation of solar and wind in such short a time would generate its own negative consequences, to be minimised but perhaps not eliminated. An economy that provides decent lives to 9 billion people or more cannot exist without *some* mark on the planet. If they are both criminal, the flow – even in its worst possible renditions – and the stock here compare like shoplifting and genocide. Destruction first of the land and then of the climate is inherent in the production of fossil fuels; the latter is absent from solar and wind, the former avoidable through planning.[97]

Once fossil fuels have been excluded, as well as nuclear and new hydropower and biofuels, solar and wind are all that remain (plus a smattering of tidal, wave, geothermal). Only purists with airy-fairy notions of human life and/or reactionary cravings can oppose them too. The ethical problems rather concern how their power should be used and what for.

The more interesting question here, however, is the descriptive, historiographical: what explains the fact that no transition was initiated when 1.5°C came into view? Even if the critics of renewables were right in denouncing their destructiveness, this would not take us closer to an answer. Rather, such a trait should have spoken in their favour, considering the proclivities demonstrated by this mode of production. Capitalists have never cared for indigenous peoples or wilderness areas. Effects on them are, in this regard, beside the question. The inference must be that the overshoot conjuncture, the continuous emissions increase, the fossil fuel frenzy were all caused by factors other than the technological impossibility of a transition *and if that was true for the early 2020s, it will be so even more for the years and decades ahead* (unless the productive forces were to suddenly cease to develop). Something else must explain the refusal of capital to shift from fossil fuels to renewables. A key suspect would, of course, be cost and price.

Return of the *Gratisnaturkraft*

The capitalist classes did receive the news: in 2020, the International Energy Agency announced that solar power was now 'the cheapest source of new electricity generation in most parts of the world' – indeed, photovoltaics offered nothing less than 'the cheapest source of electricity in history'.[98] It should have been an auspicious start to the decade. The next year, the World Economic Forum informed its audiences that the cost of large-scale solar projects had plunged by 85 per cent in ten years; during the pandemic itself, most of the solar and wind that came online were cheaper than the cheapest fossil fuels.[99] The double whammy of pandemic and war then put strains on their supply chains too, but the prices of fossil fuels rose more, the divergence accelerating.[100]

These were the latest manifestations of secular trends. Solar panels first entered the market in 1958; six decades later, their price had decreased by a factor of more than 3,000, making photovoltaics not only the cheapest source of electricity in the annals of human history, but also the energy technology undergoing the most spectacular price collapse ever recorded: there simply was no precedent.[101] Used and commercialised around half a century earlier, the price of wind turbines did not drop as rapidly. But between the early 1980s and 2020, the cost of electricity from installations onshore fell by a not unimpressive 80 per cent, and the fall sped up in the 2010s.[102] This decade marked a bifurcation. At its beginning, solar and wind still tended to be costlier than fossil fuels; from its middle onwards, as the long degression gathered pace, their prices were decidedly lower.[103] The turning point came earlier than anticipated.[104] It should have been the beginning of the end for fossil fuels.

The drivers of the trends were manifold and fairly well documented: for photovoltaics, the move from more or less artisanal manufacturing to mass production was key. In any such move, the labour time necessary for producing a commodity contracts, and then so does the exchange value; if it takes twelve hours for a skilled artisan to assemble a solar panel, it will cost orders of magnitude more than if factory workers churn it out in a few minutes. Much of the early twenty-first-century productivity burst famously occurred in China.[105] In a textbook version of technological development under capitalism, a large number of companies there fought to cut back on labour and mechanise production so as to undersell each other.[106] Panels had the advantage of being modular, granular devices easily replicated and scaled, improved by learning from thousands of iterations and then standardised in new runs and onwards in virtuous cycles.[107] Wind turbines benefited from similar mechanisms. Towers became taller, capturing stronger winds farther above the ground; rotors grew in diameter; manufacturing was simplified into iterative series; windier sites could be accessed, not the least offshore. Nothing indicated that these trends would come to a halt anytime soon.[108] Rather, further improvements in the pipeline (to use an inappropriate cliché) pointed to a continued decline in the price for both sources of the flow; to take but one example, bifacial panels, capturing solar irradiance from two sides rather than one, were poised for another leap.[109]

Obviously, these trends resulted in and from a phenomenal *growth* in the installed capacity. The number of panels and turbines in operation rose as fast as their prices fell. Ever more of them could be seen with the naked eye. Does that also mean that the world had in fact embarked on the transition? Very emphatically not so. A transition would be like tearing down one house and moving into another. That is something else than acquiring multiple houses and spreading out furniture and goods in all of them: the distinction between an energy transition and an *addition*. A transition would mean the world *closing down* fossil fuels, moving out from them, breaking up with them, terminating their combustion and replacing it with the flow. An addition is the process clearly visible from the beginning of the twenty-first century: turbines and panels springing up alongside all the unending fossil infrastructure, *supplementing* but not displacing it.[110] For the climate, it mattered naught if China built a lot of solar panels, as long as it also maintained its coal-fired power plants – and built more of them. This, of course, is just what happened, in 2022 for instance: the People's Republic added 125 gigawatts of solar and wind capacity, while coal not only did not diminish but grew by some 27 gigawatts, plus another 106 greenlit.[111] It was true, of course, that electricity production in some places – Germany, the UK, the US – became less dominated by coal over time, causing emissions to creep down. But this was still a far cry from the eradication of fossil fuels in these economies. In the world economy as a whole the pattern was 'all-of-the-above', to quote the motto of Barack Obama.[112] Talk of *transition* would be merited only if there were 'an active suppression of fossil fuels' in synchrony with the buildout of alternatives; and from this, the world was very far away indeed in the early 2020s.[113] The frenzy plainly took it in the opposite direction.

This is not to gainsay the many strides made by solar and wind: PV capacity growing by a factor of twenty-four during the 2010s and wind quadrupling; 83 per cent of all new inputs to grids in 2020 coming from them; the share of renewables in total global energy use hitting a new 'high' of 5.5 per cent in 2022; a first tentative boom for floatovoltaics in Asia.[114] All of these, however, were compatible with business as usual. Not even a rising share in percentage points would denote a transition. It is theoretically possible to have a world with 99 per cent energy from the flow and 1 per cent from fossil fuels and still see the latter combusted in growing absolute quantities. Such quantities are the only that count for the climate.[115] Measured in them, the burning of fossil fuels in electricity generation – the department most easily transitioned – expanded surely and steadily in the three decades leading up to 2022, even as their share remained constant.[116] And constant their overall share really did remain: in 2022, fossil fuels accounted for 82 per cent of total energy consumption in the world, the same proportion they had claimed for decades.[117]

To proclaim the transition underway would, in other words, require more than the relative share of the flow creeping upwards: fossil fuel combustion would have to *decrease in absolute terms*, the infrastructure subjected to devolution, the expansion reversed. Bandying about the notion of a transition already happening in the early century was an exercise in obfuscation. It belonged to the hurrah-optimism of the period: 'the transition from fossil fuels is well underway. Each year sees an increase in the amount of electricity generated from renewable sources,' an editorial in *Nature* proclaimed in 2017, prematurely and illogically.[118] 'The energy system transition that would be required to limit global warming to 1.5°C is underway in many sectors and regions around the world,' the IPCC adjudicated in the Special Report – words objectively pulling the wool over the eyes of the public, as was indeed the general function of this optimism.[119] It could be found even in the critical literature, as sloppy references to the transition somehow already being a thing.[120] In the early 2020s, its status was rather the same as that of asset stranding: a hypothetical event, not yet even attempted.[121]

If a transition *were* to take place, it would snuff out air pollution. The second largest cause of death worldwide, beaten only by heart disease, it killed 7 of the 55.4 million people who lost their lives in 2019.[122] In an economy 100 per cent powered by the flow – the only worthy aspiration – no fuels would be set on fire, no gases rising, no fine particulate matter swirling.[123] It would trim down total energy consumption. The burning of fossil fuels is notoriously wasteful, much of the energy (or exergy, if you will) contained in them simply lost as heat. Of the energy pumped into the tank of a car, 70 per cent is dissipated in the internal combustion engine before the rest reaches the wheels; but when the energy is instead administered as electricity from a socket, only 20 per cent goes to waste. Producing steel from scraps in electric arcs instead of in blast furnaces cuts energy inputs by a factor of ten. To this must be added all the energy spent on seeking, finding, mining, transporting and refining fossil fuels, activities brought to a close as extraction withers away. In sum, a transition could by itself cut total energy consumption *by half or more* (and to this might be further added – as in Jacobson's vision, for all its conservatism – shifts to public transport, carpooling, improved insulation and other measures slimming energy use).[124] It would arrest global warming.

Aside from all these very major benefits, due to the secular price trends, a transition would also represent *a net saving in the narrowest monetary sense*, strictly counted in dollars and euro and yuan.[125] By how much? Researchers at Oxford University in 2022 estimated that a 'fast transition' – completed by 2050 – would yield a pay-off to the world economy by anywhere between 5 and 15 trillion US dollars. These would be the expenses saved when agents on the market no longer need to cover the bills of fossil fuels. (A slow transition, eliminating them by 2070, would be far less economical.)[126] The upfront cost of rolling out the flow infrastructure would be quickly recovered because the fuel is gratis.[127] The latter circumstance makes all the financial difference in the world. If this world were minimally rational, it would, for every conceivable reason, jump at the opportunity.

The Scandal of Overshoot

Many an observer was caught off guard by the price collapse. But one community was especially disgraced by it, namely that of the IAMs. No modellers had counted on anything like the trends observable IRL early in the century. All proceeded on the assumption of a price descent so modest and slow as to deviate from a reality racing ahead: the cost for solar panels modelled for the year 2050 was *higher than that actually observed in the late 2010s.*[128] The same held for concentrated solar power and offshore wind.[129] These particular productive forces had, in other words, overrun the plotted finishing line more than three decades ahead of schedule, and none as fast as PV. The IAMs projected an annual fall in the cost of PV of less than 3 per cent over the 2010s; in fact, it was 15.[130] They were nearly as far off the mark on capacity installed – an expected annual growth rate between 15 and 30 per cent for the period 1998–2015; an actual growth rate of 38.[131] The IAMs fed into the Special Report thought PV capable of generating some 12 petawatt of electricity per hour come 2050; scholars of PV put the potential between 41 and 96. The models informing the policy choices of the European Commission believed PV equal to the task of supplying less than 20 per cent of electricity by mid-century: empirically grounded estimates advised *at least half.*[132] The underestimation of the cheapness and capabilities of the flow was as thoroughly consistent as could be.[133]

What accounted for this historic miss? There was nothing gradual or linear about the collapse; but IAMs, as we have seen, were premised on change conforming to such a profile. They further presumed the existence of a 'floor' below which the price of solar and wind could never fall; but not only did the two violate this presumption in practise, there is, as we shall soon see, no theoretical reason to believe that such a floor can exist.[134] There appears to have been 'status-quo bias' in the models.[135] They harboured 'a preference for inefficient combustion, in particular by relying on coal and bioenergy'; the scenarios they spewed forth were 'optimistic on deployment of lumpy energy-systems technologies, such as carbon capture and storage, while insufficiently reflecting empirically observed innovation dynamics in more granular technologies such as solar photovoltaics'.[136] Where did this bias come from? Possibly from primitive fossil capital itself: companies like ExxonMobil and Shell generated their own energy scenarios, which might have leaked into the IAMs, directly or via the scenarios drawn up by the International Energy Agency.[137] Some critics have suggested that 'interests from industry' lurked in the underrating of renewables, but if so, their influence was obscure and uncertain, largely because of the opacity of the IAMs.[138] Rarely if ever did they disclose their parameters. The capacity for self-criticism and correction remained limited: despite the miscalculations being well-known from at least the late 2000s, the IAMs *persisted*, for reasons again murky, in predicting unreasonably high prices and low penetration of PV in particular.[139] Further research would have to determine if this capital protection was active or passive in nature, or a mix of the two.

But capital protection it was. The effect of the error was to *make mitigation look far more expensive than it needed to be*. This was particularly embarrassing for the IAMs, because price and cost were supposed to be their forte; but the political consequences were, for that same reason, fateful.[140] Ask an IAM what the optimal course of action would be, and it would not answer 'roll out the photovoltaics'. It would say 'go slow, take it easy, letting go of fossil fuels anytime soon would be adventurist, because they are financially the safest option. Overshoot and subsequent removal would be the more prudent choice.'[141] IAMs tended to accord a greater role to BECCS than either solar or wind as a source of energy by mid-century.[142] Indeed, they presumed that the former would generate up to *four times more* than the latter, and this at a time – 2023 – when still only one full-scale BECCS plant existed in the whole physical world.[143] They even leaned heavily on CCS as a filter to clean coal, despite the fact that, at this same date, after more than two decades of ballyhoo, exactly one full-scale power plant with such equipment existed in that world.[144]

What if actually existing technologies were factored in? In 2022, the first paper by an IAM team accounting for the real developments of the solar and wind ended up modelling almost no overshoot and no removal: their rationale had all but dissipated.[145] If the secular trends were accounted for, the IAMs could no longer recommend overshoot – in its programmatic form, a proper scientific scandal. The idea was as unnecessary as the thing itself. Imagine if, back in the early 2000s, the EU had stumbled upon a Dutch modelling team that harboured the same degree of optimism about the growth trajectory of actually existing renewables that it ended up bestowing upon BECCS. Perhaps it would have proven rather more challenging for fossil capital to keep the spectre at bay. Instead, and despite awareness of the flaws of the IAMs, the IPCC prodded climate governance to turn towards overshoot and removal and pick BECCS over solar and wind.[146] This was in keeping with the irreal turn. But the distortions and surrealities aligned, as ever, with the poles of social power.

Priorities of Profit

Who decides whether to bet on stock or flow, or if both, in what proportions? Who determines where investments go? Primitive fossil capital has more say than most, in particular the oil and gas companies, through whose immoderately deep pockets much of the new means of energy production perforce arise. The alternative is not unknown to them: by the early 2020s, they had been closely acquainted with the flow for about half a century. Consider BP, widely seen as the most environmentally enlightened – or, depending on perspective, unctuous – among the supermajors. It started manufacturing solar panels already in 1980.[147] By 1999, BP Solar had become the largest vertically integrated company in this line of business, producing cells, installing plants, distributing electricity, mastering the full chain.[148] In the following year, dawn of the millennium, BP thought itself deserving of a fresh logo called 'the Helios mark', aka 'the sunburst' – a sun with bright rays shading from yellow into green – and asked to be henceforth known as 'Beyond Petroleum'.[149] But eleven years later, the company liquidated its solar subsidiary. All panel factories were closed. The reason was not kept secret: solar failed to throw off a profit.[150] Hopes for the soul of the company revived when the new CEO Bernard Looney in February 2020 – on the day after the WHO had officially christened the disease 'Covid-19' – responded to the pressure built up on the climate front in previous years, by announcing that BP would cut its production of oil and gas by 40 per cent until 2030 and increase that of renewables by a factor of twenty, flabbergasting the business press. 'BP means business,' wrote *Forbes*.[151] It pledged to be 'a net zero company by 2050 or sooner'. Looney had understood that 'trillions of dollars will need to be invested in replumbing and rewiring the world's energy system' to prevent the carbon budget from running out; he wished to be part of the solution. There would have to be an extreme makeover of his company.[152]

In late 2022, Looney snuck into COP27 in Sharm el-Sheikh as a registered delegate for Mauritania, in which country BP had recently developed a major gas field.[153] During that annus mirabilis, he had begun to beat a retreat from his promises. 'One of the misconceptions about our strategy is that we're going from oil to renewables. That is not what we are doing,' he told the head of the sovereign wealth fund reinvesting profits from Norwegian oil in, among other companies, BP.[154] In earnings calls, his mantra was 'resilient hydrocarbons'.[155] After all this time, upwards of a century since the British first struck oil in Khuzestan, it was still the hydrocarbons that made the money, now more of it than ever. Barely had the year of the biggest bonanza ended before *Wall Street Journal* reported that 'Looney plans to dial back elements of the oil giant's high-profile push into renewable energy', since he was 'disappointed in the returns' from those diversions. 'He has told some people close to the company that BP needs to do more to convince shareholders of its strategy to maximize profits in areas where it has a competitive advantage, including [*sic*] its legacy oil-and-gas operations.' The CEO wished to 'clarify' that ecological virtues 'aren't distracting the company from its ability to deliver profits' – meaning, firstly, that the high-flown ambition to cut oil and gas production by 40 per cent, which had made such waves, because no other company had said anything like it, was lowered to 25 per cent (to begin with).[156] Secondly, investments in hydrocarbons would now go up, not down. Thirdly, the flow was out, again. BP would quit solar and wind for a second time, because they made the smallest profits of all kinds of energy in its portfolio – or, as the chief of the 'gas and low-carbon' department further clarified, 'we will not grow renewables for the sake of growing wind and solar.'[157] What, then, set the priorities? 'We're going to be driven by value,' explained Looney. 'That's what we're going to be driven by. And if we see value, we'll do it. And if we don't, we won't.'[158] We are going to be driven by value – an utterance with, as we shall shortly see, a great deal of Marxist truth content in it.

As embodied in the company of BP, capital here kept the flow somewhat like a mistress in a separate house, to whom it paid a visit when the mood at home soured, bringing a token gift and mumbling about coming commitments only to run back to the toxic partner, from whom it had no real intention to divorce.[159] But the treatment was arguably crueller still. One company quicker on the ball than BP was Exxon, which launched a solar energy research programme in the early 1970s in reaction to the oil crisis, set up two subsidiaries, found them unprofitable and sold off the remnant to BP Solar in 1999.[160] Exxon, however, saw enduring value in PV for one purpose: to power lights on oil platforms. This cut operational costs and improved profit margins. From the 1970s, PV for lighting spread as best practice to all majors engaged in offshore drilling, the first but not last instance of the flow chained in the service of the stock – bring out the gimp – while Exxon also took the lead in spurning it in any other role.[161] The windfalls of 2021 and 2022 were received as a vindication of this strategy. Having resisted any dalliance with the flow, ExxonMobil made greater profits than BP. 'We leaned in when others leaned out,' CEO Darren Woods congratulated himself.[162] Solar and wind were useless in the eyes of ExxonMobil, because they did not rake in any profits – but they were so cheap as to be irresistible *in the pursuit of profits from oil and gas extraction*. In 2020, the company signed contracts for drawing 70 per cent of the electricity to its Texas fields from these two sources.[163] Flow for moving oil rigs up and down: less than a promiscuous energy addition, an absolute subservience to the core business.

Chevron followed the same path, researching solar in the 1970s and setting up its own special subsidiary; but this one seems to have focused on internal service from the start.[164] In 2003, Chevron opened the world's largest array of photovoltaics of its kind in California, named with the telling oxymoron 'Solarmine'.[165] It powered the pumps on the nearby oil fields. Chevron knew as well as anyone else how to produce electricity from solar, and how advantageous it could be, and how fruitless it would be as a product in itself. 'Chevron's path to net zero', stated CEO Michael Wirth in 2022, 'has no place for renewable energy like wind and solar. Instead, Chevron will focus on maximizing fossil fuel profits' – an interminable quest, which by then had subsumed the flow as servant; in 2020, the company made another push to feed its California fields with solar.[166] 'Lost Hills' was the name of one (yet another Freudian slip). 'Electricity is one of Lost Hills field's largest operating expenses, so having solar will be an important factor to help keep those costs down and maintain the planned oil field life,' one spokeswoman from Chevron explained the logic.[167] Here, then, were the companies outwardly saying that the world cannot possibly live on solar and wind alone, because they are too unreliable and whatnot, inwardly relying on them for the most cost-efficient production of their commodities. Solar boomed across the Permian Basin, the panels built with enough space between them to make room for rigs – petrovoltaics, as it were – and benefit from their 'magnificent cost advantage over gas-fired power plants. The marginal cost of solar is zero.'[168] Occidental, the largest oil producer in the Basin, plugged solar into its fields 'to provide lower-cost electrical power'.[169] But it would not sell any of it.[170]

From Louisiana to Libya, PV was used as the optimal solution for protecting oil pipelines against corrosion, shielding the metal with a cheap electrical current.[171] Platforms basked in the light of the sun; but in Norway, wind was more munificent. Lundin rerouted electricity from turbines in northern Finland to Johan Sverdrup.[172] Equinor constructed the world's largest floating off-shore wind farm, a composition in which the turbines stand on buoys rather than on the seabed, allowing them to move farther from land, into the strongest winds: and this to power platforms in the North Sea.[173] Norway could not stop patting itself on the back for the efforts to electrify its continental shelf. Wind and water from dams dispatched through undersea cables provided renewable ignition to Johan Sverdrup and the other carbon bombs, but Equinor had crass reasons not to bother much about the flow *per se*.[174] For the rest of the 2020s, it counted on an internal rate of return – jargon for profitability – of 30 per cent from oil and gas. For offshore wind, it put the figure between 4 and 8.[175] Hence it invested 28 times more money in the former than in its renewables segment.[176] In recent decades, a pattern emerged in which Norwegian primitive fossil capital developed a bit of wind for sale in moments of weak profits from oil and gas, only to revert to its true love as soon as these normalised – 'green flings', researchers called these episodes.[177] It behaved, in short, no better than anyone else in this circuit.

The profitability differentials posted by Equinor were typical. ConocoPhillips disclosed internal rates of return above 20 per cent for oil and gas but expected maybe 7 for offshore wind, maybe 5 for solar, and so it did not trouble itself with the latter two (except for powering oil fields).[178] BP put the rates in the same range.[179] Total did build a renewables segment in the late 2010s, which failed to make a profit, before yielding a solid 10 per cent in 2021; but this compared to nearly 40 for oil and gas.[180] When asked about previous ambitions to become a leading pro-ducer of renewables, the CEO of this company responded that he did not want his employees to be driven by that 'story. I want them now to be driven by delivering the profitability.'[181] In the record year of 2022, Total increased investment in gas and oil by 50 per cent and cut back on renewables by 11.[182]

Shell was the supermajor closest to BP in character. In the late

1990s, it formed Shell Solar, soon the world's fourth largest vertically integrated solar company, punished for paucity of profit and sold off in 2009. Shell and BP thus moved in tandem and dumped solar *just as it reached grid parity* – becoming, that is, as cheap as fossil-fuelled electricity. That was the moment when solar lost its lustre, leading these firms to undertake a 'recarbonisation' as they pivoted towards the Canadian tar sands, then the latest frontier of oil.[183] After the 2018–19 wave of climate mobilisations, Shell too vowed to be 'net zero' by 2050 – only to reconfirm its fealty to fossil fuels after the bonanza, CEO Wael Sawan downplaying any future role for solar and wind, echoing Looney.[184] 'If we cannot achieve the double-digit returns in a business, we need to question very hard whether we should continue in that business. Absolutely, we want to continue to go for lower and lower and lower carbon, but it has to be profitable.'[185] Or, 'we will invest in the models that work – those with the highest returns', meaning that very recent vows be broken, wind projects scrapped, oil going steady, gas boosted: recarbonisation all over again.[186] (But solar and wind were good enough to power Shell platforms.[187])

The choice came as naturally as a coffer of gold over a loaf of bread. In the late 2010s, BP was the most magnanimous company, by virtue of allocating 2 per cent of its capital expenditure to 'clean energy'; supposed to be a lot, its total renewables capacity then equalled roughly two large gas-fired power plants. Shell dispensed 1.33 per cent. ExxonMobil and Chevron gave a miserly 0.22 and 0.23 respectively – but these shares included things like PV for platforms and pipelines, and most of the rest was biofuels.[188] Close to 100 per cent of investment, in other words, went to the stock *even before* the bonanza. There might have been an additional motive here: investing in the flow could set in motion a snowball of asset stranding.[189] At bottom, however, the preference was a direct reflection of the profitability differentials. As one business consultancy noted in early 2023, 'unless there is an official ban on such investment, investing in oil and gas will continue to be directed by the well-established and most potent indicator: the rate of return.'[190] An official ban would mark day one of the transition. The real trends, as this consultancy demonstrated, were for investment in oil and gas to *rise* in the twenty-first century, from low levels in the mid-1980s and 1990s; measured as a share of world GDP, it was nearly three times as large when the Paris Agreement was signed as when the UNFCCC was negotiated. And investment invariably rose more when the price of oil and gas did so.[191] Capital, it seemed, took the world deeper into fossil fuels *just as – and because – they became more expensive than solar and wind.* In *The Price is Wrong: Why Capitalism Won't Save the Planet*, Brett Christophers has shown in meticulous detail how this could be the case, driving home the reality that price is the wrong metric for understanding the transition that still, by the early 2020s, wasn't happening.[192] Profit is what matters for those who decide what means of energy production to invest in.

Does this mean that the preference belonged only to a clique of capitalists, a retrograde cabal of oil and gas companies that could not bring themselves to separate from their honey? Or did it express the predilections of a wider set? First, we should notice that the differentials those companies reported from their own activities and scenarios reappeared across the sectoral divide. One study calculated 'annual profit margins' for the period 2011–20 and found it to be, on average, 4.9 per cent for integrated oil and gas companies and as much for enterprises mining coal. But firms manufacturing panels and turbines made on average −0.2 *per cent* – bleeding losses, that is – while project developers came at a piddling 0.7.[193] The mean profit mass of an oil and gas major was nearly 6,000 times larger than that of the typical solar company.[194] Imagine, then, that you have a certain amount of money in your purse and enter the marketplace and face the choice of investing in stock or flow. What would you bet on?

All the companies whose behaviour we have examined here were publicly listed, joint-stock, shareholder-owned entities, which means that what they did, they did on behalf of those owners. BP did not come around to the resilience of hydrocarbons out of its own private perversions. Already in 2020, months after his net-zero pledge, Looney acknowledged that it was 'probably going to be in oil and gas for decades to come, because how else is that $8 billion dividend going to get serviced?' – a reference to the money shareholders could legitimately expect.[195] 'We must perform. Our shareholders expect and deserve nothing else,' he would quiver.[196] The problem BP faced in the years after his pledge was not that it made no money – to the contrary, as we have seen, it was 'getting more cash than we know what to do with' – but that *its main rivals made so much more.* Between February 2020 and February 2023, ExxonMobil registered a 'total return' of about 110 per cent, Chevron more than 80, Total almost 60, Shell almost 40 and BP 'only' slightly more than 20.[197] Now one could think that all-time profits would be a good moment to pour some of that cash into renewables. In fact, this had been the stock answer to the question of how such profits could be justified in times of a (so-called) transition – BP et al. would use them to fund the requisite equipment – but when the money truly came raining down, the result was to *withdraw* from renewables, under the pressure of shareholders who might otherwise bolt.[198] 'This isn't some form of altruism or some form of charity,' Looney expounded on the nature of his entity in an interview with *Time.* 'We can create value for our shareholders through this shift' – back to fossil fuels – and 'we will create more value through this shift than we would if we keep doing what we're doing.'[199] It was the impersonal voice of self-expanding value that spoke through Looney as through a dummy.

The priorities were those of the common capital of the class. Between 2016 and 2022, fossil fuel companies raised 3.6 trillion US dollars in various types of credits on global debt markets. Producers of renewable energy attracted 160 billion – a ludicrous fraction, entirely sound *sub specie capitalis.*[200] Banks had good reason to place their money in the former. 'A big driver of continued fossil fuel investment is that fossil fuel firms have remained significantly more profitable than renewable energy firms,' in the words of Beyene and her colleagues.[201] Behind primitive fossil capital stood the representatives of fictitious capital, who might have held even greater sway over the future of energy: in the flow business, companies needed their money to get projects going. Above all, countries in the global South were bathing in the most intense sunshine and waiting for finance to sponsor the realisation of the potentials.[202] But the stock promised several times more money in return.[203] 'We are not mandated to care about the planet,' one of Christophers's investors confirmed.[204] To the asset managers, the mandate was known as 'fiduciary duty' – another two words for maximising profit – which compelled them to keep banking on fossil fuels.[205] A provisional conclusion would be that the choices made by Looney, Woods, Wirth, Sawan and their like were the choices of *capital in general.*

A Scissors Crisis for the Twenty-First Century

But why could the flow not give a good profit? Was this an aberration of the moment, which just happened to coincide with the approach of 1.5°C? Or should we expect it to recur deeper into the conjuncture? First, we should take note of the tendential movements of the prices of fossil fuels over the longue durée: adjusted for inflation, the price of coal had, by the early 2020s, been fluctuating very mildly around a constant level for 140 years. It had no tendency to go either up or down. In real terms, offering an identical amount of useful energy, a lump of coal fetched the same price on the world market in 2020 as it did in 1880. As for oil and gas, real prices were similarly stable until the early 1970s, when they set off on a jagged, modest but perceptible rise.[206] The movements for solar and wind, on the other hand, as we have seen, formed rather a free fall, with the result that the prices of stock and flow assumed almost the shape of a scissor: one blade going up, one down.[207]

The downwards trajectory of the flow was sharper than the upwards of the stock. The latter did not exhibit the steep incline expected by peak oil theory. Instead, the prices of oil and gas moved slightly upwards, while coal did not budge. But how could this be the case over such a very longue durée, during which the productive forces did anything but stand still? Between 1880 and 2020, the technologies for taking coal and oil and gas out of the ground developed as fast as in any other business, labour productivity increasing in the typical manner of capitalist development: and yet their prices did not come down. This mystery is easily dispelled by the fact that greater efforts had to be made to extract the same amount of fuel, once the seams and fields within easiest reach were depleted.[208] More precisely, *greater quantities of labour* had to be deployed in the process of extraction; more precisely still, greater quantities of *dead labour*, or labour congealed in machinery, or, in other words, fixed capital. The rise in fixed capital formation was so sharp over so long a time that it counteracted the increasing productivity in the case of coal and more than cancelled it out in oil and gas. Peak oil theory would predict that prices of the latter would shoot up to the point of utter unaffordability; but as we have seen, it gravely underestimated the prowess of the productive forces, which rather unlocked fresh riches of hydrocarbons, at the cost of increased mobilisation of fixed capital. Hence a gently rising curve, not a crazy spike.[209]

These empirically observable tendencies for fossil fuels obeyed the law of value. Labour was the centre around which their exchange value revolved. The law of value specifies that commodities are exchanged against each other, through the medium of money – commanding a price, that is – in proportion to the varying amounts of labour time required to produce them.[210] This holds for any given moment as well as over time. Prices move as total necessary labour time does. 'The law of value governs their movement in so far as [a] reduction or increase in the labour-time needed for their production makes the price of production rise or fall': more labour – dead, above all – for getting fossil fuels out of the ground translates into higher prices (or, in the case of coal, negates what would otherwise be a decline).[211] And the very same law makes it possible to cash in a profit. The collective workforce producing a commodity must first do so in a quantity that covers its basic needs. The value of the goods from this initial phase will match that of the things the workers use to reproduce themselves – food, clothes, shelter; perhaps also car, smartphone, health insurance – but there is no reason why their labour must stop at that point. They might as well stay at the machines or in the mines for several hours more. They will then perform surplus labour, the value of which falls to the capitalist: and this is the *sine qua non* of profit, its absolute precondition and determinant frame. 'There is no profit on production if there is no surplus labour and an attendant surplus product,' Anwar Shaikh has confirmed these rudiments of Marxian political economy.[212] If it is old fare, however, the theory also possesses a singular explanatory power for the scissors of the early twenty-first century. By the same measure as the two diverge, profits will be more and more concentrated to the upper blade.

The stock can only be brought to the surface through labour. It is there constituted as a commodity, which commands value on the market. This marks out the *possibility* of a profit, just as for any other enterprise producing and putting up a commodity for sale. Workers in a coal mine can go on digging coal long after they have filled wagons sufficient to pay their wages, and this has, indeed, never been much of a problem: ever since the Elizabethan leap, coal has been a fount of surplus product. Oil and gas share the same baseline of profitability. Over and above it, they offer at least three potential sources of super-profit. First, oil and gas companies can cut their costs of production, chiefly by introducing all manner of technologies, be they drones for identifying cracks in platforms and saving on maintenance or solar panels for generating the cheapest electricity. If, say, ExxonMobil adopts some improved method that reduces its costs below those of its competitors, it can sell the oil at the common market price and earn a profit higher than theirs.[213] The labour required for hauling up one barrel is – if only briefly – lower than for the rest of the pack. Then the rest also adopt this best practice and catch up, evening out the profits, establishing a starting line for the next part of the race, and so on.

Second, companies may control fields from which resources are wrested with relative ease. Perhaps the oil is closer to the surface or less viscous than usual. Less labour, less advanced machinery is then required to fill one barrel. This amounts to the same situation as the first: favourable exceptions to the ruling quantum of necessary labour time, which allow the fortunate owners to produce oil and gas cheaper than their rivals, while selling it at the same going price. However, because this second source of super-profit derives from the properties of particular patches of land (or seabed), it has often – from Marx to Shaikh – been considered irreproducible, a unique endowment from which competitors are locked out and which they cannot copy, the way they can, for example, order the best PVs to their platforms too.[214] But if ExxonMobil has exceptionally bountiful fields in the seas off Guyana, other companies will in fact try to seize hold of equivalent – or even better – domains. This is the mission of exploration, the labour congealed in fossil terre–capital. If they succeed – if, say, Equinor finds its own bumper fields in the Arctic – they will run neck and neck with the winners from a moment ago; as with technological development, the playing field will be levelled for the next round. Whether based on exceptional instruments or reservoirs, super-profits may slip through the fingers of their owners, and the cumulative effect is merely the formation of even more fixed capital and the seizure and eventual depletion of even more of the best fields. But at any given moment, excess earnings remain possible: there can always be a new solution or deposit that cuts the necessary labour time below the average and revives the fortunes, and so the turbulent chase continues.

Third, fossil fuels – and, again, hydrocarbons in particular – are shipped and pumped and trucked across tens of thousands of kilometres, from one corner of the globe to the other. This too derives from their status as commodities begotten by labour. Pieces of the stock can be bundled up and ferried away to any point. But this makes them vulnerable to disruptions along the way: if a war breaks out; if one warring party refuses to sell its oil or gas until some demands are met; if one imposes a blockade on another; if a revolution topples an entire regime of oil production, supplies might be cut off and prices suddenly soar. Every time this happens, astonishment ensues. But since the oil crisis of the 1973, it has happened so many times – precipitated by the Iranian Revolution, the first American invasion of Iraq and the second, the first Russian invasion of Ukraine and the second, these being among the more memorable episodes; interspersed among them, lesser events like strikes in Venezuela, battles in Libya, attacks on oil facilities in Saudi Arabia – that it must be considered an inherent feature of the fossil economy.[215] Every time it happens, profits for the companies that can still sell their goods soar as well. Market price volatility is a function of the profile of the stock and a regularly (or irregularly) renewed source of super-profits. The bottlenecks of the late Covid-19 period belonged to the same category, and new instantiations are sure to come as the world moves deeper into overshoot territory.

As record-breaking as the bonanza of the early 2020s was, it was but a concatenation of these exceedingly generic sources. From the depths of Covid-19, oil prices climbed to a peak in the summer of 2022; but they never reached the heights from the Libyan civil war or the Iranian Revolution. They caused unheard-of windfalls only in combination with low-cost wells. ExxonMobil and all the rest had been adept at fine-tuning the latest recovery technologies and keeping the more hard-worked fields in abeyance.[216] In the wake of the bonanza, Equinor, Total, BP, Shell all re-emphasised this dual strategy: trimming operational costs to the utmost *and* making the most of the choicest fields, from Barents Sea to the Gulf of Mexico, so as to maximise the margin at any given price level.[217] This was how the winners had won the most in the years of fading pandemic and flaring war. It was and is and will be the formula for maximum profits in the primitive accumulation of fossil capital, in the 2020s as much as in the 1970s or 2050s. It remains the ever-present promise throwing the deficiencies of the flow into the sharpest relief.

For the flow appears without labour. It would be as impossible and redundant to mobilise labour for making the sun shine or the wind blow as for making humans exercise their lungs to breathe: these things come naturally, by themselves. Like mushrooms in the forest, here the fuel is ripe for picking prior to and in proud disregard of any process of production. 'Value is labour,' Marx spells out; 'value itself is defined as social labour,' Adorno reaffirms.[218] It follows that the flow *cannot have value*. In the parable of cheap water power – the paramount manifestation of the flow in nineteenth-century Britain – in the third volume of *Capital*, Marx makes as much clear: 'the waterfall, like the earth in general and every natural force, has no value, since it represents no objectified labour and hence no price, this being in the normal case nothing but value expressed in money. Where there is no value, there is *eo ipso* nothing to be expressed in money.'[219] And where there is nothing to be expressed in money, *there can be no profit*. The baseline for profitability caves in: where companies have made inordinate profits on selling coal and oil and gas for centuries, there is simply nothing to lay hands on. No one has yet made a penny from producing sunlight or wind, because they cannot be conjured up as commodities, and so cannot be sold to cover the subsistence of workers, and so likewise cannot be turned into a surplus product. The owners of a wind farm may sell electricity to consumers. So may the owners of a power plant fuelled by gas. But the difference, whose implications can hardly be exaggerated, is that *gas can be sold to the power plant as a commodity*, whereas no company has ever been seen delivering wind *qua* fuel on a barge or truck. 'Coal has value and water-power does not': the most general contrast between stock and flow.[220]

In the early twenty-first century, that contrast was indirectly demonstrated by the oil and gas giants all coming around to spurning solar and wind but accepting *biofuels* as the one passably promising sort of renewable energy; for, unlike the former, the latter can come about only through labour.[221] Biofuels are constituted like any other commodity from half a millennium of agrarian capitalism. Like sheep or sugarcane, they must be tended by humans doing work, or else they will not crop up. Moreover, biofuels can be processed in refineries and burnt in coal plants or combustion engines with only minor modifications to the infrastructure. After its second dumping of solar and wind, BP maintained at least a nominal commitment to biofuels; even ExxonMobil for many years flaunted its research into algae as the next-generation feedstock for gasoline and jet fuel.[222] The green goo grown in laboratories would eventually become grain for tanks. Most hard-nosed of the primitives, ExxonMobil deigned to traffic in the promise of *this* one renewable fuel that worked *almost* like oil, plastering social and other media with commercials for its plans to become an 'energy farmer', a notion that would have been oxymoronic for solar or wind. (But this hype, too, came to an early end. In the midst of the bonanza, ExxonMobil pulled the plugs on the research, because it remained 'extremely challenging to produce large quantities of algae bio-fuels at a profit' – incidentally, just as the research was yielding real results, some strains of algae growing in ponds in concentrations that approached fuel quality.[223] The ponds were then closed down as hopelessly unprofitable in comparison.)

But consider again the flow as a source of electricity. Imagine a market dominated by gas-fired power plants. Now some intrepid entrepreneurs enter it, armed with PV and wind farms that generate electricity at lower cost. They sell the same commodity – an identical current of electricity – to the prevailing price, make a good profit, seize a fair share of the market and expand. Nothing yet deviates from the standard capitalist situation. A problem, however, will soon arise: during sunny and windy hours, the pioneers will have so much electricity on their hands as to flood the grid with supplies. This might sound like bad news for the operators of the gas plants, who cannot benefit from any similar moments of free bounty and so cannot pocket the difference; but if the pioneers keep expanding and attract more firms to their green line of business, the gains will rather backfire on them. The market will begin to drown in electricity of no value. Insofar as fossil fuels are crowded out of some corners of the market, the logic of energy free of labour takes over.[224] Revenues to the pioneers dry up, as the commodity they sell – electricity – is sucked back into the hole of the non-commodity on which it's based: there is so much of it around; it is so abundant and needless of labour that prices collapse. This phenomenon is known in the technical literature as the 'revenue decline' or – a synonym – 'value decline' for wind and solar. It correlates perfectly with penetration levels; empirical observations from some of the most highly developed markets suggest that the decline sets in early, particularly for solar, and then accelerates once the two sources, together or alone, generate one quarter to one third of the electricity.[225] That is when the overabundance really starts depressing prices and, consequently, revenues. After that point, both tumble 'until they reach a floor (either close to zero or mildly negative)' – meaning, contrary to the assumptions of the IAMs, that *there is no floor*, only a labourless void.[226]

One temporary solution to this problem would be to smooth out fluctuations. If the excess supplies from sunny and windy hours can be bottled up in batteries or reservoirs or offloaded on distant regions, an artificial scarcity may be created – up to a point. If a grid relies to 100 per cent on the flow, or even just approaches that goal, there will be no way to protect against the cost-free nature of it.[227] Solutions to the time problem would, in any case, facilitate the penetration of solar and wind and so eventually restart the revenue decline. In the early 2020s, just as *and because* these two sources of energy became so superiorly cheap, specialists predicted that 'we will soon enter a regime of accelerating value decline', with a perfectly predictable corollary.[228] 'There will come a point when it will no longer make economic sense to invest in a renewable source' – or, in the words of the *MIT Technology Review*, 'it could become difficult to convince developers and investors to continue building ever more solar plants if they stand to make less money or even lose it': the slide along the downwards blade of the scissors.[229]

Under capitalist property relations, the flow hits a glass ceiling of sorts. It can grow to some extent, as an addition in the margins, but a *transition propelled by agents seeking to maximise profit* would be a self-defeating enterprise. The more developed the productive forces of the flow, the more proficient their capture of a kind of energy in which no labour can be objectified, the closer the price and the value and the profit all come to zero. Then investors will have to lose interest. This would resemble the Marxian law of the tendency of the rate of profit to fall, in that both extrapolate a disappearance of labour; but the differences are perhaps greater than the similarities. In the latter, it is the immanent laws of motion of the capitalist mode of production that play out across the full spectrum of commodities. In the former, the falling rate of profit is rather a function of the flow being *an alien presence in this mode of production*, because it begins and ends as a non-commodity; any profits that might be made in between those points – after 0 but long before 100 per cent – would be ephemeral. The classical Marxian law hints at a terminal crisis of the capitalist mode. The law of 'accelerating value decline' suggests that this mode will defeat prospects for a transition away from fossil fuels: a sign not of its weakness, but of its overpowering strength.

It follows, furthermore, that on this downward trajectory, the flow cannot offer anything like the three sources of super-profits in hydrocarbons. Consider first technological development. This cannot happen in the extraction of the fuel, because there is none. Capturing solar and wind is like cupping hands in a river and raising them to the mouth: an essentially passive act of borrowing something that rushes by, which means that no one in this business can go below any average socially necessary labour time – except in manufacturing and installation. The branches where technologies develop and profits might accrue are those of producing the *instruments* – the panels, the turbines – and constructing them on site. In the oil business, the equivalent situation would be one where the services of Halliburton and Schlumberger were in demand, whereas the places of Saudi Aramco and its peers would be taken up by a yawning nothing. The former players would then likely be rather insignificant, as their fortunes are pegged to those of the latter. (Somewhat analogously, the production of rain barrels and gutters has always allowed for the making of profit, but not in any sensational amounts, because rainfall comes without labour; no profitable industry can grow around its extraction.) Now a photovoltaic module is a commodity like any other, obviously. It has, as we have seen, been subject to extreme advances in labour productivity. While profits remain possible in this sideline business, its progress hastens the maturation of the central field. It cheapens the utilisation of solar and wind, lubricate their diffusion and more fully realise their condition as free gifts of nature.[230] The exchange value of the flow is an empty husk: technological development peels it away.

Seen from another angle, the only costs of solar and wind are for upfront acquisition and installation (plus negligible maintenance).[231] They cover the fixed capital required to get the process going – again: the panels, the turbines – which might well have a lifetime of half a century but will never become the kind of milch cows found on the other side of the fence. No first-order commodity will ever leave this farm, as from an oil field. The puniest cells and the largest concentrated solar power parks are alike in this regard: fixed capital but no surplus value. As for the second source of super-profits in fossil fuels, there is no equivalent either. An unusually bounteous field does not allow the owner to carve out an exception to any average labour time. It just means even more of the free stuff. Last but not least, because the flow cannot be commodified, it cannot be traded, and so it cannot have any world market, and so it will not be vulnerable to geopolitical disruptions the way oil and gas in particular are.[232] There cannot even be any OPEC – the most quotidian arrangement for manipulating hydrocarbon prices – for solar or wind. The possibility of a cartel for limiting and stockpiling fuel falls away.[233]

But what about the materials critical for building the instruments? Their supply chains might well be global, as we have seen, and here shocks would remain possible: Chile, say, could shut down its lithium exports. (In fact, the leftist government of that country nationalised its lithium industry in early 2023; as of that date, it held the largest reservoirs in the world, although more were constantly discovered and developed. The measure did not stop lithium prices from continuing their fall from a brief peak in 2022.[234]) Similar things could happen to silver, copper, cobalt, gallium, aluminium, potentially boosting the profits of their producers. Cartels may be formed; inter-imperialist rivalries – notably between China and the US – could cause ructions. But materials for building the instruments have no capacity to generate trade and profits on anything like the scale of fossil fuels. We have seen what a chasm a transition away from them would open up in the global circulation of commodities; not even under the tightest schedule could materials make up for it, either in value or tonnage.[235] Would not the cost of the flow, though, swell if their supplies were to be disrupted, for some reason or other? Producers of panels (and turbines) would then be squeezed, but it would have zero effect on the cost of those already in operation.[236] Yet a lasting and steady rise in the value of materials would threaten to bend the blade upwards: and this is the rare factor with a potential to close, if only ever so slightly, the scissors of prices.[237] For obvious reasons, it would not make the production of solar and wind *per se* any more profitable.

From whichever angle one looks at it, the flow appears incapable of ever spawning profit opportunities close to those of the stock – the incapacity so profusely illustrated during the early decades of energy addition. The void was then flanked and fed by dwarves. In 2022, the world's largest manufacturer of solar panels was a Chinese firm called Longi: had anyone heard the name? It managed to defend a position on the Fortune 500 list *for China*, ranking 177.[238] Put differently, the world's largest manufacturer of solar panels was nowhere close to entering the renowned global league of corporations with the highest revenues – which remained, as ever, filled up by primitive fossil capital.[239] Neither did the largest manufacturer of wind turbines (Vestas) or that of offshore wind farms (Ørsted) make it onto Fortune Global 500. Solar was unique not only for seeing the most spectacular price collapse ever recorded for an energy technology, but also for producing no mania, no quickly rising stars; for this tech, there was no Microsoft or Apple or Facebook. More broadly, there was no Boulton & Watt of the flow, no Edison Machine Works, no Ford factories, *no ascendant clusters of capital accumulation* riding this wave. The best candidate, as of the early 2020s, would be Tesla. But Tesla produced *automobiles*, the premier commodity of golden-age capitalism, of a type that happened to run on electricity (from whatever sources). Elon Musk was not the face of an emerging flow economy. He was, rather, the personification of the all-of-the-above approach; when anointed richest man in the world, his main sources of fortune, besides Tesla, were one company for shooting rockets into space and one for building tunnels to ease car traffic.[240] In the early third decade, after a long period of gestation inside the capitalist mode of production, the technologies of the flow evinced no talent for powering the accumulation of capital.

Other hypotheses have been put forth to explain this. One suggests that renewables cannot give a good profit because they are weighed down by too much and too perfect competition: too many small firms cutting prices so effectively as to eradicate any gains. The theoretical premise here is that profit requires monopoly; the empirical contrast is the oil and gas industry, where such an arrangement allegedly holds. Brett Christophers has made this a centrepiece of his explanation. 'For capital to be able to realize the benefits of cost reductions, a degree of monopoly power is necessary.'[241] The diffusion of a cost-cutting technology stagnates and comes to an end unless there is monopoly power. A strange inversion of the Baran-and-Sweezy school of monopoly capitalism – which, rather, postulates that *monopoly* is the cause of stagnation – it shares with it the idea that competition fizzles out when firms are big and few. But this is a fallacy. Competition can be just as intense between a handful of giants, as was indeed the case in the oil and gas industry in the early twenty-first century, the shareholder pressure we have inspected only one of its mechanisms.[242] Moreover, if the presence of hundreds of small firms on a market by definition blocks the making of profit and the ascent of novel technologies, plenty of episodes in the history of capitalist development – the British cotton industry the most canonical case – would be inexplicable. Another version of the hypothesis says that the renewables industry of the early twenty-first century was burdened by throat-cutting overcapacity and oversupply; but again, there was nothing exceptional about this predicament.[243] *Competition as such cannot explain the unprofitability of the flow.* The former does not cause the latter, but merely lets it 'be *seen*'.[244] From the flow, competition strips away value, because there is no social labour at its core; in commodity production, competition is compatible with value because there is social labour; in each sphere, it does not establish but *executes* the underlying law of value.[245] The flow is not unprofitable because it conforms to some idealised image of perfect competition. It is so because it cannot fit into the procrustean bed of the commodity form.

But there is an obvious and naïve objection to all of this: if solar and wind are so fantastically cheap, why do not consumers just call them forth from the big companies? Surely suppliers will respond to their demand and rise to the occasion of a transition? The idea that supply reflects demand would appear to be at the core of bourgeois economics, but even that branch of thinking makes room for anomalies. A certain deviant category of goods will not show up merely on the cue of consumer preference. Paul Samuelson famously defined them as 'collective consumption goods', the individual consumption of which 'leads to no subtraction from any other individual's consumption of that good' – in the formulaic language of this type of economics, they are 'non-rivalrous'.[246] A clearer definition says that 'goods are non-rivalrous when they can be consumed by any number of people *without being depleted*.'[247] If one person Z avails herself of a thing X, without thereby making person Y any less able to do so, the good in question makes for a poor business prospect. Everyone can just swim in it. Abnormal, freakish even, goods of this sort are by nature 'public'; however much consumers may want them, they are 'insufficiently profitable to be provided by the private sector. Therefore, in the absence of government provision, these goods or services would be produced in relatively small quantities or, perhaps, not at all.'[248]

We can easily see that the stock offers highly rivalrous goods: the consumption of one barrel of oil or one wagon-load of coal means that no one can ever consume it again. Every piece of fossil fuel burns once and once only. But supplies of sunlight and wind are in no way affected by any one consumer's use.[249] It is not so much that they are *renewed* upon consumption – biofuels would be a better match for that phrasing – as them being unmoved and unruffled by it. A producer of oil has something to sell that will be extinguished in the moment of use; a seller of solar-powered electricity offers a good that will last for another 4 or 5 billion years. Even in terms of bourgeois economics, then, solar and wind power would appear to end up in the category of public goods, which will be *underprovided* by profit-seeking enterprises.[250] This, of course, is but another way to say that the flow has no labour and no value.

But why then, we might ask, was there any investment in renewables at all? For plainly there was. At least five factors account for it. First, much of the investment in renewables was induced by states, through various politically motivated pro-grammes for supporting this sort of energy technology, by means of subsidies, feed-in tariffs, mandatory quotas or other measures. When states removed these props, investment often fell sharply, as Christophers has extensively demonstrated in *The Price Is Wrong*. Second, much of the investment came from end consum-ers – think Germans or Gazans buying PVs and putting them on rooftops – who appreciated the flow precisely for its use value. Unlike centralised companies, these people did not need to make a profit. Third, rates of profit on renewable energy production will, as we have argued, be high early on and then fall over time, which implies an inducement to invest by private profit-maximising firms, up to a certain point. Fourth, even low rates of profit might justify investments from such firms, again up to a certain point. Fifth, even fossil fuel companies might, as we have seen, have reason to invest in the construction of renewable energy capacity for the production of cheap inputs to their commodities. How these factors – not the least the first and the second – develop in the coming decades will determine investment trends; so far, they have been sufficient to animate an energy addition on the margins. But in the absence of states committing to transition and taking control, or, to speak with Looney, a 'replumbing and rewiring' of the energy system so that it comes to be based on decentralised consumption – two ways to privilege use value over exchange value – renewables will run ever more violently into the problem of diminishing returns.

What is in stock is then determined by the role that profit is allowed to play. All production takes time, and capital will commit to the expenditures involved only if there is reason to expect profit down the line: 'production is always initiated on the basis of prospective profit.'[251] Supply and demand may strut around on the stage, but behind them, 'profit is pulling the strings' and regulating the appearances of both.[252] The higher the profit from the production of one set of commodities, the more capital will stream to it and the more of it will take place.[253] The amount of fossil fuels in circulation is a function of their profitability; conversely, unprofitability puts a lid on the development of solar and wind, *insofar as the actors who decide about investment in the means of energy production are guided by profit* rather than some other principle or pursuit.[254] And it is the latter possibility that defies the imagination.

Nature of the Demon

By the early 2020s, then, the climate crisis was, to a large extent, a scissors crisis. There were no signs that the gap between the blades was about to close rather than widen further.[255] As 1.5°C was in the wind – and, shortly behind it, 1.7°C ... – the demon that held the world in its grip was value in general and self-expanding value in particular. This was the force that piled up assets that must not be stranded and produced ever more value that must not be destroyed, like the pharaoh made pyramids arise in Egypt. It ran the world out of control. This is the truth intuited in the pages of fossil fuel fiction, from Munif to Kelly, when they describe the demonic and monstrous character of the machinery. And rarely have the demonological passages in Marx himself been more apposite: there is 'a mechanical monster whose body fills whole factories, and whose demonic power, at first hidden by the slow and measured motions of its gigantic members, finally bursts forth'; fixed capital is an '*animated monster*', capitalist wealth a 'monstrous objective power', capital 'a Moloch demanding the whole world as a sacrifice belonging to it of right'.[256] As David McNally has stressed, these passages should not be read as mere rhetorical flourishes.[257] They should be taken deadly seriously.

Value rules like a demon because it is an invisible, immaterial quality.[258] But it must, again like a demon, inhabit material commodities as its body. The means for such inhabitation is dead labour. Over the living, the dead labour comes to exercise a mute and stifling power: the pipelines, the platforms, the mines, the gas stations, the airports now certifiably having an iron grip on people (and other forms of life).[259] There really is something demonic about the forces at work here – which does not necessarily mean that they are of a metaphysical, supernatural character, any more than the inner demons that rule a person suffering from neurosis or psychosis or any other mental disorder. Something in the life of an individual can bring forth powers in her psyche to which she submits in compulsive, repetitive obedience. She is then not in control of herself. Freud tells us that

> there are people in whose lives the same reactions are perpetually being repeated uncorrected, to their own detriment, or others who seem to be pursued by a relentless fate, though closer investigation teaches us that they are unwittingly bringing this fate on themselves. In such cases we attribute a 'daemonic' character to the compulsion.

Analogously, society is no longer in control of itself, when the emergent properties of capitalist property relations have it under their spell.[260] Foremost among them is the compulsion to produce value, a quality prone to infinite expansion in a world marked by physical limits; value drives the world towards destruction under 'the demonic impulse that everything solid should melt into air', in McNally's own phrase, suggestive of the core business of fossil capital.[261] Or, again in the passive voice of Looney of BP, 'we're going to be driven by value.'

Often couched in terms of 'the Anthropocene' versus 'the Capitalocene', there was in the early twenty-first century an academic debate about whether the human enterprise in general or the capitalist mode of production in its extraordinary specificity drove the climate crisis. When 1.5°C appeared on the horizon, the question was settled on the ground. It was, very specifically, the profit motive that revved the engine; in this moment of truth for climate politics, no one could reasonably blame the acceleration on the technical deficiencies of renewables or some universal human propensity (or, for that matter, on the Soviet Union).[262] A bit of *apokalyptein*, then, in the original sense of the term, and to the same extent that the climate crisis was a scissors crisis, it was also a crisis of rationality. In, say, the year 2000, it might still have been considered somewhat rational to stick with fossil fuels, because they were at least cheaper; but when they became definitively dearer, this last veil of reason was ripped from the mode of production. 'There is something crazy at work' in the law of value.[263] In Adorno's words, that law 'is the summation of all the social acts taking place through exchange. It is through this process that society maintains itself and, according to Marx, continues to reproduce itself and expand despite all the catastrophes that may eventuate.'[264] It was also, as we have seen, that same law that most deeply structured the IAMs that incubated overshoot ideology. This is the law that 'determines how the fatality of mankind unfolds', and it is, at its very root, irrational: 'exchange value, merely a mental configuration when compared with use value, dominates human needs and replaces them; *illusion dominates reality.*' Yet at the same time, '*this illusion is what is most real*, it is the formula used to bewitch the world.'[265] If this was how the overshoot conjuncture began, it is not a wild guess that illusion will be a dominating force deeper into it too.

Who Fires the Overshoot Gun?

In late July 2021, *Nature Communications* published a study on how CO_2 emissions cause people to die. The scientist behind it, R. Daniel Bressler of Columbia University, had come up with a way to calculate just how many lives would be ended during the rest of this century by 1 million metric tons – or just tonnes, for short – of CO_2 emitted in 2020: namely, 226 lives. Through a lifetime of consumption, three and a half Americans would then cause emissions sufficient to end one life, while it would require 146 citizens of Nigeria to do the same.[266]

These figures could be questioned on a number of grounds. First, they were, as Bressler was quick to point out, serious underestimates, since he only factored in mortality from a single climate hazard: exposure to heat as such. He did not count deaths from floods, storms, hurricanes, diseases, droughts or any other consequence of global heating. Second, insofar as the climate crisis intensifies over time, the figures would have to be revised upwards. Third, are individual consumption choices really to blame? In an interview with the *Guardian*, Bressler advised readers not to 'take their per-person mortality emissions too personally. Our emissions are very much a function of the technology and culture of the place that we live [in].'[267] And, evidently, not everyone – not even in the US – has a say in what technologies get developed in whichever culture we live in.

Consider the case of EACOP. Take Bressler's 'mortality cost of carbon' – 226 lives – and multiply it with the tonnes of CO_2 released from the oil to be pumped through that pipeline. We then get the following figure: when operational, EACOP will each year cause the death of 7,661 human beings (deaths spread out during the rest of this century).[268] The assumptions being the same as in Bressler's study, the estimate is exceedingly low; 7,661 people killed per year would be an unrealistic minimum. Who, in the early 2020s, was responsible for planning this mass killing? Was it the average commuter in the US or France? Or was it rather the owners of Total – the single largest private company headquartered in France – and its partners and backers, including President Macron, who gave wholehearted support to this opportunity to 'increase the French economic presence' in south-eastern Africa?[269] The act in question really should be considered one of violence: fossil fuels taken out of the ground are projectiles fired indiscriminately into humanity, primarily the part of it living in the global South – carbon bombs, literally.

By putting the matter in such terms, one is immediately exposed to the objection from weapons manufacturers: guns don't kill, people do. This line has long been peddled by the National Rifle Association and the companies that litter the US with guns to absolve themselves of any responsibility for the massacres committed with these instruments. It has also been pushed by oil and gas companies, everyone from ExxonMobil to the Norwegian state washing its hands of any emissions.[270] Fossil fuels, on this view, do not self-combust; consumers put them on fire. Both lines are patently spurious. But the National Rifle Association has the advantage of greater realism. Guns can, in fact, lie unused for decades, or be used merely for display. A fraction of the total weapons output in a country like the US ever gets involved in any actual killing. Crude oil works differently. It is not stored for decades, nor can it be used through parading or patrolling; no one buys oil from Total to manifest power by painting the crude on faces or smearing it on opponents. There is one sole purpose of the product: combustion upon delivery. Total would never drill a drop of oil from the fields around Lake Albert without the expectation that it could be transported straight to fireplaces around the world – just what the pipeline would be there to ensure. Any particular quantum of fossil fuels from those fields would make it into combustion only because Total has explored and drilled in them. That is how every projectile of this kind is detonated: the fuse lit in the moment of extraction.[271]

A temporal arch determines the magnitude of this violence. If the science of anthropogenic climate change is accepted, the recognition is logically unavoidable: fossil fuels kill people, and they do so *in greater numbers the longer business as usual continues.* One thousand tonnes of coal dug up in 1850 or barrels of oil pumped in 1950 did not cause any measurable deaths, because the atmosphere was not yet oversaturated with CO_2. When it is, the lethality of any additional quantum of fossil fuels tends to rise; or, the higher the temperatures, the larger the 'mortality cost' of any given such quantum extracted. In 2021, at the height of the preparations for EACOP, Total boasted of producing 3 million barrels of oil per day.[272] On Bressler's assumptions, this would mean that Total during this single year sentenced 106,412 people to death, by its own activities in the oil sector only; but again, that figure merely gives a rough sense of the proportions, and it will be higher in 2031 *insofar as the atmosphere contains more CO_2 by then,* even if the 3 million barrels stay constant.[273]

But what about the purpose of the act? Surely someone who extracts coal or oil, be it in 1850 or in 2050, is not doing it with any intent of ending lives? The goal is another: to make money. Once it becomes fully established and commonly known that this form of money-making kills multitudes, however, the absence of intention begins to fill up. Henceforth mass casualties are an ideologically and mentally processed, de facto accepted result of accumulation.[274] 'If you're doing something that hurts somebody, and you know it, you're doing it on purpose,' prosecutor Steve Schleicher said in his closing argument against Derek Chauvin, later convicted for the murder of George Floyd; mutatis mutandis, the same applies here. Indeed, the violence of fossil fuel production becomes more lethal *and more purposeful* for every passing year.[275] (A process going on for quite some time: one of the more sensational findings in the field of climate history in 2021 was that Total knew about the 'catastrophic consequences' of a rising atmospheric concentration of CO_2 as early as 1971.[276] For half a century, the corporation then continued to do something that hurt somebody and knew it.) These were, and are ever more so, acts of commission, not omission. Their compulsive character does not alter this moral status. 'We are very simple guys,' said the CEO of Total in March 2023: 'when we see giant resources, we *cannot avoid* trying to get in.'[277] A serial killer might say something similar in court – 'I cannot stop myself from doing what I do' – but it would not get him acquitted (possibly forensic psychiatric care). Less concerned with law and ethics, Marxian political economy considers it a truism that capitalists are both captives and beneficiaries of their mode of production. If anybody gains from it, it is the people who own the means of production and decide in what to invest. It is to them the profits accrue.

But after all this is said, even if 'production is the real point of departure', it still remains the case that 'a product becomes a real product only by being consumed. For example, a garment becomes a real garment only in the act of being worn; a house where no one lives is in fact not a real house'; a piece of fossil fuel becomes CO_2 in the atmosphere 'only through consumption. Only by decomposing the product does consumption give the product the finishing touch,' and so the process may as well be studied in that moment – if nothing else, to test the claims from Chevron et al. that they produce on behalf of the poor.[278] If that were the case, the whole matter would stand in a different, less unambiguous light. The causation of overshoot would be rather more complicated. But a quick scan of the landscape of consumption is enough to shatter such claims.

In the late 2010s and early 2020s, the poorer half of humanity accounted for one tenth of all emissions, seen from this end-use perspective – the CO_2 released from the exhaust pipes of cars and aeroplanes, the chimneys of houses, the goods used in households. The richest one tenth was responsible for *half*. Regardless of methodology and data set, studies kept finding this distribution: one half of humanity, one tenth of emissions; one tenth, one half.[279] The poorest one seventh, scraping by under the extreme poverty level, caused less than 2 per cent of emissions – no surprise, since such deprivation by definition means meagre consumption.[280] But was it perhaps the emancipation from such conditions that drove consumption to higher levels in this moment in history? Might the *growth* in emissions have been largest on the threshold from wood-fuelled huts to apartment blocks, so that poverty alleviation formed the reverse side of any overshoot?

This is, as we know well by now, a question of cumulative emissions. It took some 140 years for humanity to emit 750 gigatonnes; and then in only twenty-five years, between 1990 and 2015, it emitted as much again; and it was this later spurt that brought it so close to depleting the carbon budget for 1.5°C. Now during that quarter of a century of gorging on fossil fuels, the richest 1 per cent, on one count, managed to stack up cumulative emissions *more than twice as large as those of the poorest half* – this tiny bracket, in other words, exercising upwards of double the causative power behind the looming overshoot.[281] On another count, it was rather responsible for 21 per cent of per capita emissions growth between 1990 and 2019, as against 16 from the poorest half.[282] On yet another, the total emissions of the former grew more than three times as much as those of the latter.[283] Whichever figure comes closest to the exact events in those decades, there can be no doubt that the budget for 1.5°C was '*squandered in the service of increasing the consumption of the already affluent*, rather than lifting people out of poverty'.[284] Indeed, from the vantage point of 2020, the richest one tenth would have fully depleted the budget over the next decade *on their own*, had they continued on their ways while the emissions of everyone else in that year dropped to zero.[285]

One can study overshoot from a still more individualised perspective. If equally shared, a 1.5°C budget would entitle every human being to releasing around two tonnes of CO_2 per year between 2020 and 2050 – equivalent to one ticket for one round-trip flight between London and New York.[286] Much of humanity did not use up this allotment when the third decade dawned. This undershooting segment of the species was concentrated to sub-Saharan Africa and Asia outside of China.[287] The rich, on the other hand, were the advance guard of overshoot, stridently led by the richest of the rich: in 2018, the Russian oligarch Roman Abramovich (who built his fortune on oil) emitted an estimated 31,000 tonnes of CO_2 from his mansions, jets and yachts, other appliances uncounted. Bernard Arnault of Louis Vuitton released 10,000 tonnes in that year from his dwellings and vehicles, Bill Gates 7,000 tonnes, Elon Musk a comparatively austere 2,000 – only one thousand times more than the ideal individual allotment for 1.5°C (he owned no yacht).[288] Zooming out to the richest 0.1 per cent of the US population, it generated emissions from consumption 57 times higher than the bottom decile in that country, 597 times higher than average household in a low-income country. Some American individuals produced more carbon dioxide in a week than plenty of people in the global South would do in a lifetime. But these figures were, as usual, based on conservative assumptions.[289]

Poverty alleviation was not a concomitant or cause of overshoot and could be accomplished without it.[290] It was not the masses of the global South that, suicidally, tipped the world into 1.5°C. In fact, not even the working classes of the North were party to the process: between 1990 and 2019, per capita emissions of the poorest half of the populations of the US and Europe *dropped by nearly one third*, due to 'compressed wages and consumption'.[291] The overshoot conjuncture was the creation of the rich, with which they capped their victory in the class struggle. Needless to say, circles of consumption extended far beyond those of investment decisions: this is in the nature of the relation. But in two respects, the moments evinced a tight overlap. The consumption that did most to bring about overshoot was concentrated to the summit of class society, and so were the most egregious and excessive practices of consumption, similar to wanton acts of killing. With the original focus and methodology of Bressler, as well as the common sense of climate science in general, there is no escaping the conclusion that the worst mass killers in this rapidly warming world are the billionaires, merely by dint of their lifestyles. In these respects, then, one could speak of an *'immediate identity'* between production and consumption in the critical decades: the same classes that controlled the former also concentrated around them the latter, like someone who posts a letter to himself.[292] Or to stick with the ballistic language, the carbon bombs were both detonated and enjoyed by the same officer corps, not by some plebeian infantry, the violence of production and consumption of a piece.

The pendant to the profits from fossil fuels was the record sale and use of private jets in 2021, surpassed in 2022.[293] In those years, the habit of the ultra-rich to fly on their own planes for thirteen or seventeen minutes between locations comfortably reached by other means began to baffle a broader public.[294] Why on Earth did they do it? Just as mystifying was the appearance of 'ghost flights': planes with no passengers in them, sailing through the skies for no obvious reason. 'Why ghost flights operate remains unclear. Only airlines know the reasons but they do not publish data that explains the practice,' as if they wished to stage a performance of the demonic impulse and crisis of rationality.[295] On the ground, the sales figures for SUVs never flagged, pointing only upwards, the car fleet of the world so increasingly gas-guzzling as to wipe out the gains from electrification.[296] New models appeared in the luxury segments. One American company unveiled 'Vengeance', an SUV designed like a futuristic armoured vehicle, fully equipped with bulletproof glass, helmets, gas masks and wing mirrors that could shoot pepper spray and pipes that could fire thick smoke onto adversaries; an interior dressed like a Cadillac, an engine of 700 horsepower, nearly as much as a tank – all for the price of 250,000 US dollars.[297] Things were out of control.

PART III

Into the Long Heat

The overshoot conjuncture confronts us with a series of new and old conditions in climate politics. These can be tentatively summarised in ten theses. The one question that then remains to be answered is, of course, the most classic and difficult one: what on Earth is to be done?

7

Ten Theses on the
Overshoot Conjuncture

One. Overshoot in all its forms is capital protection. Its entrance onto the climate stage has been scandalously unnecessary, sub-optimal, irrational and poorly conceived, except in this regard: it permits the accumulation of capital to continue unimpeded and crash through any limits to warming. The closer we come to those limits, the stronger the overshoot logic becomes.

Two. The assets built on fossil fuels form an armour around the drivers of the warming. Its apparent thickness is the main reason for the enormity of the passivity on the mitigation front and hence for the immensity of the disasters unfolding. The armour is held in place as much by forces in the base as by ideologies operating in the superstructure.

Three. More extreme warming is now virtually guaranteed to move alongside more trench wars to uphold the value of fossil capital, without any apparent feedback between them. The feed-back that does exist increasingly takes one and one form only, and that is overshoot.

Four. The overshoot conjuncture arises out of the failure to fulfil the historical task of asset stranding, or the political destruction of fossil capital. But sooner or later, one way or another, some anthropogenic forces will have to face up to this task and execute it. The alternative is unlimited combustion, or the maximum realisation of the potentials for catastrophe.

Five. Logic admits of three further alternatives to the political destruction of fossil capital: adaptation, carbon dioxide removal and geoengineering, alone or in combination. It follows that fossil capital will do everything in its might to make these stand in for its own downfall – in other words, to turn overshoot management into a substitute for mitigation. Unlimited combustion would then proceed under thin covers.

Six. Geopolitical conflicts – hot wars in particular – fan the flames of business as usual. Global warming is a conflagration inside which wars burn like local fires, exacerbating the former. The war in Ukraine was the first instantiation of this dialectic of the conjuncture, but hardly the last. A fractal pattern of destruction runs from the smallest scale of a hospital or home to the largest of the biosphere.

Seven. The hotter the planet becomes, the easier it will be – in the strictly technical sense – to stop more fossil fuels from being poured on the fire. The productive forces for powering all the world economy with the flow will mature progressively. The contradiction between these forces and the property relations holding them back will then grow ever more intense.

Eight. In this version of the contradiction between forces and relations, an amalgam of profits and stock is pitted against one of needs and flow. The property relations that compel owners of the means of production to always and ever strive for maximum profit have come to be one with fossil fuels. To make room for an economy running to 100 per cent on the flow, these fetters would need to be burst asunder and energy production transferred to public control, a rule of the associated producers, that gives priority to the satisfaction of needs – the need for survival, first and foremost. It is because this task is so herculean that overshoot continues to make its way through the world.

Nine. Overshoot springs from the refusal of the dominant classes to adjust to this reality – a refusal identical to self-preservation – a fundamental madness likely to shape the rest of the conjuncture too. We should count on psychopathology rather than reason to determine the governing responses to it. Climate politics will run into diversifying derivatives and self-amplifying spin-offs from the initial break with a revolutionary reality.

Ten. If it is now too late to prevent warming of 1.5°C, it is then too late, above all, and for that very reason, to give up on the struggle against fossil capital.

8

Induce the Panic

The analysis developed here has concrete implications for late climate politics. The subject and dynamic of any transition will be marked by the constellation of forces coalescing in fossil capital, its inertia and integrative power setting the parameters for intervention. We shall offer some brief pointers next.

Against Popular Front Politics

How will the subject shape up? In an essay in *New Left Review*, summing up the long debate over ecological strategy in that journal, Thomas Meaney has invoked popular front politics as the most promising, or even the sole available, model: progressive forces must ally with their natural friends in capital. That is where the impetus to transition will come from. 'In the near future, the most plausible agency of an adequate decarbonization – sufficient to keep the planet cool enough for other purposes later on – is the coordination of three global forces: Wall Street, the European Union, and Communist China,' a troika beginning with Wall Street, metonym for a part of the capitalist classes undergoing genuine greening. 'The rest of the world may then find itself in the tight spot of an Indian revolutionary in the 1940s, caught between the British Empire and Japanese fascism: choose your imperialism. Green capital or brown?' The green will have interests strong enough to motivate a life-and-death struggle with the brown forces, just as Anglo-America took on fascism. 'Put this way, the choice is bleak but clear. Mere survival is not victory; yet the former is surely the condition of the latter.'[1] A popular front with Wall Street – clearly bowing to its lead – is, on this reading of the world situation, the path to survival.

From the above, a different conclusion follows. Fossil fuels may have similarities to fascism, but it is highly doubtful that any significant fractions of the capitalist classes look upon them as some of their interwar forebears regarded Hitler. Does a distinct 'green capital' even exist? In the early 2020s, Ørsted was not on a warpath with ExxonMobil; it willingly agreed to power its oil fields in Texas.[2] Goldman Sachs did run its own Renewable Power LLC: it provided the solar panels to the Lost Hills of Chevron.[3] As opposed to energy addition, capital had yet to demonstrate any appetite for transition. What was green was rather happily also brown. And there would seem to be scant reason to expect one part of capital to ever develop an existential interest in defeating the fossil part of itself, as it once geared up for battle against the Axis powers, *as long as there is money to be made in it*. Indeed, the analysis outlined above suggests the opposite: the longer the time goes on, the *less* inclined Wall Street will be to break off from its fossil axis, in the absence of a hostile intervention, and the more it will pivot in the direction of adaptation, removal, geoengineering. As we have seen, fixed capital formation integrates oil and gas companies ever more deeply into the common capital of the class – the loans, the bonds, the equities; the banks, the asset managers, the stock exchanges without which the infrastructure could not be drummed up. To destroy ExxonMobil would be to destroy JPMorgan. To knock out Glencore would be to hit BlackRock as collateral damage. *The same factors that make fossil fuels more expensive also make them more difficult to pull out without risking the whole capitalist edifice crashing down.* If these fuels were easy to get, there would be less need for entanglement in financial circuits; then their deletion would not also endanger the major representatives of money capital. But in the world of the scissors crisis, fossil fuels become more integral to capital in general to the same degree as they are outperformed by solar and wind.

The tight spot of a revolutionary in the 2020s and coming decades is another: we are alone in this. We have no reliable friends in the capitalist classes. To say this is not to suggest that a united front of the forces of the climate revolution would sail to victory; to the contrary, it is to take the measure of the powers driving the world into unliveable temperatures, and to face the reality – 'bleak but clear' – that any path to survival runs through their defeat.

A Frontal Assault with Light Sabres

But do we have the time for such conquests? In his exquisite ethnography of wind power in Spain, *Who Owns the Wind? Climate Crisis and the Hope of Renewable Energy*, David McDermott Hughes restates the by now classic case against revolutionary climate politics: this 'crisis will not wait for a long march to strip elites of assets they hold and defend'.[4] Hence anything like socialism must be considered off the table. The adjective that settles the case is 'long'. And, indeed, this crisis will not wait for a *long* march to strip elites of assets, as in a march over decades or half a century – in Gramscian lingo, we have no time for a war of position. When carbon budgets run out and temperature limits come into sight, a war of manoeuvre is the option that remains. There has to be a lightning strike, a *coup de main* that knocks down enemy defences *precisely because there is so little time left* (if any).

The temporal compressions of overshoot upend the original argument from Gramsci. In the fragments of the *Prison Notebooks* where he reflects on the distinction, Rosa Luxemburg and Trotsky are the adventurists of the war of manoeuvre, Lenin the sage of the war of position. Only 'Ilyich' understood that when revolutionary politics moved from East to West, it had to hunker down and prepare for slow, gruelling campaigns. Structures in the East were 'embryonic and unsettled,' and so they could be seized in a sudden assault; but in the West, they had developed, expanded, accreted for so long that 'a succession of sturdy fortresses and emplacements' stood in the way of the revolutionaries, who had to dig in for the long haul.[5] In the climate crisis, au contraire, it is the overdevelopment of fossil capital that rules out a war of position. That could have worked when the structures were embryonic. As things now stand, the crisis will not wait for *anything less* than a blitz to strip elites of the assets they hold and defend.

To pick Trotsky – we shall return to Luxemburg in a moment – the temporality of the crisis vindicates and radicalises the vision in the original Transitional Programme. Even the most minimal of minimum demands is now a transitional demand. In the lexicon of the Second and Third Internationals, a minimum demand corresponded to the most immediate needs of the day – say, a 2 per cent wage hike – while a maximum demand was a call for bringing down the capitalist order in toto and replacing it with something completely different, such as a workers' government.[6] In normal, tranquil times, the former had scant connection to the latter. A minimal demand could be conceded by the capitalist class without structural injury. But in critical conditions of crisis, that class could not afford to accommodate such a demand without also putting its own power in peril and so would resist it with all its force. In this sharpened atmosphere, even the humblest requests for improving the lives of workers would, in the words of Clara Zetkin, point '*in the direction of the limitation of capitalist private property*' as a precondition for their fulfilment.[7] The distinction between minimum and maximum demand would then collapse: the former would become a *transitional* demand, according to a sliding dynamic first developed by the Comintern in the years of the united front and later conceptualised by Trotsky. 'Insofar as the old, partial, "minimal" demands of the masses clash with the destructive and degrading tendencies of decadent capitalism,' he wrote in the Programme, there develops 'a system of *transitional demands*, the essence of which is contained in the fact that ever more openly and decisively they will be directed against the very bases of the bourgeois regime.'[8] This is objectively the political situation in overshoot.

At 1.1°C or thereabouts, the slogan '1.5°C to stay alive' sets in motion a system or series of transitional demands: consider

the moratorium sought in the knife report from the International
Energy Agency. An end to all new oil and gas and coal installations would put fossil capital on its death bed. Locking away half
of the fields and seams already in production would constitute a
more instantaneous death sentence. For reasons by now familiar, this could become a mass mortality event in the accounts of
capital – money dying, to an extent potentially threatening the
very life force of the bourgeois regime. Not at 0.5°C, but at 1.6°C
or so, '2°C to make do' would set off the same dynamic. Note that
the basic slogans here are minimalist in the extreme: they concern
survival, not improvement of life; staying alive and making do
with what is left of the planet, not higher wages (let alone a
workers' government).[9] But they are 'in sharp contradiction to
the capitalist class interests'.[10] In the constrained conditions of
overshoot, every meaningful mitigation points in the direction
of the limitation of private property, and in this specific respect, a
key line from the Transitional Programme has a validity no lesser
than in 1938, the year of its composition: 'without a socialist
revolution, in the next historical period at that, a catastrophe
threatens the whole culture of mankind.'[11] But is this anything
more than a restatement of the case for voluntarism, attached
to whatever temperature target? And what is that if not wildly
magical thinking on the Left, a deep red version of the belief in
unicorns and light sabres?

Paying the Polluters or Having Them Pay

The problem of asset stranding can, of course, be approached in
a non-revolutionary manner. Among the visionaries, Griffith has
the advantage over Jacobson in acknowledging stranded assets
as the main obstacle to a 100 per cent flow economy – the latter
at most alludes to it – and sketching something like an action
plan. Griffith's ancestors introduced coke to Australia. His own
first job was in the steel industry of that country. From a lineage
of winners, appreciating 'the marvels coal has given the world',
Griffith is, now that the time has come to stop using fossil fuels,

in a position to spread some love.[12] 'The best strategy may be to treat the owners of these assets, the fossil-fuel industry, as friends rather than enemies; after all, they did provide us with reliable vehicles and warm homes for a century.' Regard them not as opponents, but as 'the best allies' in the building of a fossil-free future. Stroke their egos, love-bomb them, 'celebrate them as having done an incredible job bringing us the energy we so obviously have enjoyed using'; and after that 'celebratory toast to them for a job well done, let's invite them to be a driving force' in decarbonisation.[13] Yeah, right: if all they needed to come over to this side were a little bit of sympathy and affection, the Obamas of the world, the preachers of 'stubborn optimism', the economists in the Nordhaus school, the organisers of the COPs would have accomplished this long ago.

But could there be another reason for them to swallow the bait? Griffith also advocates a proposal for overcoming the inertia: buy the owners out. Compensate for all stranding, with trillions of dollars to the holders of the liquidated property – first in line, the fossil fuel companies. They would thereby receive 'a huge amount of clean capital' to invest in renewables. '*Their margins would increase* as they built infrastructure spanning supply- and demand-side technologies, and they could leverage the initial capital investment to build businesses with *valuations far exceeding* that of their stranded assets.'[14] An offer they cannot refuse: full compensation for losses, to be ploughed into renewables that yield greater profits and more value than the ways of old. Why Griffith would know better about the profitability differentials than the companies themselves, he does not explain. There is no reason to expect them to jubilantly receive the toast and the trillions and run off to make more money from the flow. Had profit rates been higher on that side, they would, again, have defected long ago – their business acumen is not to be mistrusted.

If both flattery and profitability sound like weak attractors in a peaceful transition, what, though, of the compensation? Could the capital be bailed out rather than destroyed? There is a precedent for this code of conduct: enslavers in the Americas were handsomely compensated for the loss of their assets. The

British government, for example, resolved to make up for the financial wounds of abolition by paying them the full value of their slaves. Such 'generous compensation helped to buy off opposition from the West Indian proprietors' and maintained the sanctity of private property.[15] In the US, Lincoln sought to solve the equation in the same honourable manner, paying off the enslavers in the District of Columbia and drawing up a plan for payments to the entire planter class of the South, at $400 per slave; but the scheme fell through, the Civil War left that class without indemnities and the US became the main exception to the pattern of 'compensated emancipation'.[16] What would the precedent be here? One might think that if anyone should have been paid anything to make amends for suffering endured, it ought to have been those enslaved. The case for reparations to their descendants stands to this day. Analogously, one might think that, if monetary compensation should be part of the transition away from fossil fuels, the money ought to go to *anyone but* their owners.[17] Their career in planetary destruction should not be topped off with one final act of 'unjust enrichment'.[18] It would violate the polluter pays principle: the polluters would be *paid* for what they have done and wish to continue doing.[19] Crimes generally should not be rewarded – not celebrated with a toast, nor ended with 'a huge amount of clean capital' as consolation.

If these are some ethical objections to compensation for stranded assets, practical ones arise as well: how high would the price tag be? In the case of slavery in the Caribbean, the British and French states could 'finance compensation schemes without great strain, essentially because slaveholding represented only a tiny fraction of total imperial wealth'.[20] Do fossil fuels make up a tiny fraction of total capitalist wealth today? We have seen that the losses could amount to anything between 4 and 185 trillion dollars. Griffith proposes offering 'our friends' in the fossil fuel industry 9 trillion; other estimates of compensation sums have been in a similar range.[21] The figure could be compared to the 1.6 trillion of the entire US military budget anno 2022.[22] Even if states would be able cough up such heaps of money, it would create two serious problems. First, the transition would become far more expensive for taxpayers than if the losses were born by the propertied class itself.[23] (Better use of the money would be to pay for the rollout of flow technologies.[24]) Second, the moment a compensation scheme is announced – or even just put on the table – it would offer investors a financial shield and give them the perverse incentive to *keep investing* in fossil fuels: a promise of no losses, only gains.[25] The assets to be stranded would then grow larger still. Even the editors of *Financial Times* have understood that this idea is terrible: governments and the populations funding them through taxes are 'being asked to bear all the risk associated with assets rendered less valuable or worthless by necessary climate action. The prospect of "bailing out" fossil fuel projects risks disincentivising the steps needed now', sending the signal that 'fossil fuels cannot lose.' In short, 'if profits are to be private, so too should losses.'[26] The editors of *Financial Times* would here stand closer to Lenin than to Lincoln: when entering this battle, we should not even think of compensating the capitalists.

The ethical and practical approach to the problem of stranded assets, in other words, is mercilessly confrontational – you have created this mess, you carry the losses – and aim at unredeemed destruction of capital. It shades into expropriation, or the practise of depriving proprietors of their property without redress.[27] As Jacob Blumenfeld has recently argued, this is 'the *content* of revolution' as conceived in the Marxist tradition; more specifically, a move to 'expropriate the expropriators' proceeds in two steps.[28] It follows, first, the tendencies of ongoing capitalist development: the stock and credit markets entangle all enterprises in one web of connections. Then it makes that development trip over itself, by some kind of intervention that initiates a takeover.[29] Everything is ready to be swept up under public control. Might there be a similar dialectic between objective and subjective factors in asset stranding?

Hail the Meltdown

The more inflated a balloon, the easier to pierce it. The double logic of inertia and exposure pushes fossil capital towards just this point: close to a temperature limit, *any* intervention may induce the panic. Any measure significant enough to suggest that the fears harboured for so long are about to come true could pop the bubble. In this 'climate Minsky moment', the stampede would be frenzied and unstoppable, due to the extent of the financial connections: the strength would turn out a weakness, the columns of tanks a paper tiger. The contradiction of the last moment would flip over into an epic crash, when the valorisation of value has been disrupted. It is, as we have seen, the potential for such a crash that has informed the discourse of stranded assets from the very beginning; divestment campaigners sought to spook investors by talking about it. That did not work. What remains is the option of actually inducing the panic: the first strike in a war of manoeuvre, along the lines of Luxemburg, as per Gramsci's reading. She considered the tendencies of capitalist development – including the ever-greater sway of money capital – to be productive of convulsive economic crises. For her camp, these would be opportune moments. Crises 'open a breach in the enemy's defences, after throwing him into disarray and making him lose faith in himself, his forces, and his future', allowing for 'one's own troops to break through' (a shock doctrine from the left, as it were). Gramsci, of course, polemicised against this mix of economist and spontaneist 'prejudice' in Luxemburg; he did not think crises had the potential to sap the confidence of the capitalist class any longer – developments had advanced too far for that.[30] But the strategic horizon opened up by the problem of asset stranding is hers.

What intervention could trigger the crash? There is no way of knowing; the repertoire of possible 'supply-side policies' is wide. Subsidies to fossil fuel companies could be removed, taxes slapped on extraction, exports restricted, state-owned lands and waters closed for explorationists, production subjected to quotas to be traded and steadily withdrawn, permits revoked, credits and other kinds of funds to new projects choked off. Moratoria and bans could be imposed on anything from oil drilling to coal plants.[31] Such measures might be implemented by a province or nation or club of states, or extrapolated all the way to a 'Coal Elimination Treaty' for prohibiting coal use in the world by 2030, or a 'Fossil Fuel Non-Proliferation Treaty' – that deliciously apt label – covering all fossil fuels and all their infrastructure everywhere, first ending their expansion (non-proliferation) and then phasing them out (disarmament).[32] Any of these could push the pin into the balloon. Any could send the appropriate shockwaves through the system, if it is taut enough, stretched between an excess of assets and one of heat: a type of stress inherent in the conjuncture. Any could be a transitional demand.

In global diplomacy, precedents for comprehensive crackdowns on supply run from Tolba's Montreal Protocol to the Ottawa Treaty on landmines.[33] From Kyoto to Paris and beyond, climate politics has been woefully fixated on the demand side, when focusing on supply makes so much more sense: the points of production are fewer and more easily shut down than those of consumption; popular enthusiasm would be of a different order; the problem would be addressed at the source.[34] But perhaps most importantly, supply side politics *emanates from local resistance*, from which it works its way up to the higher scales. Moreover, insofar as they succeed, blockades and other types of place-based resistance *count as policies in and of themselves*, on a par with anything a government might legislate.[35] If a popular campaign strikes a blow hard enough, more than its immediate object could topple over. Luxemburgian masses would here be present at the moment of crisis creation, along the lines we have inspected above; once the crash is underway, they would have to push through the breach and enforce a conclusive expropriation. In countries where primitive fossil capital is organised through private property rather than state-owned enterprises – in Anglo-America, western Europe, the peripheral frontiers where supermajors prowl – this would take the form of nationalisation: companies taken into public control and forced to terminate production of fossil fuels.[36] Such a coup de grace would be preceded by a collapse in their valuation. But it might have to extend to bankrupt banks and asset managers and other kinds of enterprises too. Where it would end, no one could know.

A fever dream, a catastrophic endgame, obviously: but its possibility inheres in the problem. Policy measures that block supplies are not inconceivable (we shall shortly examine one remarkable case). Nor is a financial crisis ensuing from a pierced carbon bubble. But would it not be exceedingly reckless to *seek* a mega-meltdown of the world economy, likely to be worse than in 2008 or 1929? People do not thrive in times of suchlike crises. They do not necessarily rally behind causes of a progressive kind.[37] Crossing the bridge of transition at ultra-fast speed clearly comes with the risk of many things falling apart. Yet things have now gone so very far that other scenarios for sufficiently rapid and system-wide change have receded from view. No risk-free pathways remain, and the question of realism has become one of priorities: total breakdown of capital or climate; assets or lives; profits or planet. If the scenario we have just sketched comes across as chimerical, it is because the destruction of the biosphere has been accepted as the epitome of realism.

To be clear, this strategic sequence of intervention – systemic panic – expropriation should not be taken as writ. The two of us have no recipes for the kitchens of this hot future. Nor has anyone else, on this decisive count: neither the Green New Deal nor degrowth or any other programme in circulation has a plan for how to strand the assets that must be stranded. They are as rich in proposals for scaled-up renewables and reduced throughput as they are poor in ideas for the inescapable destructive aspect of the transition, the tearing down of the ancient temples, as against the wishes of their guardians and holy men: but this, we submit, is the nub of the matter. It is the point where strategic thinking and practise should be urgently concentrated in the years ahead. It is here that any alternative to the triple substitutes – adaptation, removal, geoengineering – must be forged. The overshoot conjuncture leaves little room for other futures, and so it behoves us to elaborate on a future other than that captured by the IAMs:

one where we bite the bullet of asset stranding, if not before
1.5°C, then before the limit coming up next.

Yasuní Victorious

Sometimes things happen, even on the climate front, that seem
too good to be true. We mentioned Ecuador in passing above,
as one of the two tributaries of the stranded assets discourse in
the early 2010s. For decades, this impoverished country had
by then been dependent on oil export and subjected to foreign
companies – notably Texaco, later Chevron – leaving behind
utter wastelands; on one count, a stretch of the Ecuadorian
Amazon was among the most contaminated industrial sites in
the world.[38] But the country was also home to a vibrant ecosystem of movements social, environmental and indigenous.
The fightback against primitive fossil capital came, in the early
twenty-first century, to focus on the rainforest known as Yasuní:
a biodiversity hotspot ne plus ultra, with more tree species in
one single hectare than in all of the landmass of Canada and
the US combined, the highest richness of amphibian, reptile, bat
and insect species in the world, plus a multitude of rare vertebrates ranging from the giant otter to the harpy eagle.[39] Yasuní
stood on top of the second largest oil reserves in Ecuador. It was
also the ancestral lands of two indigenous peoples in voluntary isolation, hunter-gatherers in the habit of killing unwanted
intruders with spears, including some two dozen employees of
oil companies.[40]

When Rafael Correa became president of Ecuador in 2007,
as part of the pink tide, he picked up a proposal from Acción
Ecológica, CONAIE and other movements: leave the oil under the
ground of Yasuní, undrilled, unextracted in perpetuity. But this
could not come for free. To compensate for revenues foregone,
Correa turned to the international community and asked for 3.6
billion dollars, roughly half of the estimated value of the oil. If his
state did not receive the sum, he would unleash the companies.
The Yasuní-ITT initiative became a cause célèbre, championed by

figureheads like Desmond Tutu and Prince Charles, who sought to persuade 'the international community' to pay up, so the oil could remain underground and the local poverty alleviated in more sustainable ways: but the money was not forthcoming. Prospective donors such as the German government were wary of the precedent for compensation. If Ecuador received 3.6 billion dollars, what would Saudi Arabia expect on judgement day? In 2013, after less than 0.4 per cent of the money had been raised, Correa called off the initiative: 'the logic that prevails is not that of justice, but of power', he lamented.[41] It was a dispiriting defeat for movements trying to 'keep it in the ground'. The local struggle, however, did not come to an end. As soon as Correa gave a green light to drilling, the campaigns to save Yasuní reorganised and tried out another tactic: calling for a national referendum.

Knocking on doors during half a year in 2014, an army of volunteers from the young Yasunidos collected nearly 800,000 signatures, far more than needed to trigger a referendum in Ecuador. The Correa government declared them null and void, leading to a protracted battle in the Supreme Court. At long last, in May 2023, the judges gave their verdict: the referendum must be held. The Yasuní rainforest had at that point been stabbed with 12 oil platforms and 230 wells, producing 57,000 barrels per day, some 12 per cent of total production in the country; Petroecuador, the state-owned enterprise, had invested more than 2 billion dollars in the project.[42] Its subsidiary Petroamazonas was the only registered operator inside the block. BlackRock held upwards of 1 billion in bonds from it.[43] On the eve of the referendum, Fitch, one of the top credit rating agencies that adjudicate on whether countries deserve investments, downgraded Ecuador to junk status, in anticipation of the 'approval of the vote to halt oil extraction'.[44] After all, more than one third of this country's export then consisted of oil. Fitch admonished it to come to its senses or face serious difficulties in obtaining credit and other forms of capital; but movements kept canvassing. On 21 August 2023, the results were announced: 58 per cent, or 5.2 million people, had voted in favour of a ban on oil extraction in the Yasuní, as against 41 per cent and 3.6 million. 'Today is a historic day,' exulted Nemonte Nenquimo, a leader of the Waorani tribe – 'finally, we are going to kick oil companies out of our territory! This is a major victory for all Indigenous peoples, for the animals, the plants, the spirits of the forest and our climate!'[45]

The outcome of the referendum obliged Petroecuador to remove machinery from within the Yasuní within a year. Platforms, wells, pipelines and other infrastructure would have to be dismantled, all operations discontinued, the grounds reforested, the oil reserves left untouched for eternity. The company complained of 16 billion dollars in losses over the next two decades: no compensation would be paid.[46] Some 1 billion barrels of oil would stay outside of the active carbon cycle (slightly more than one third of what's in Johan Sverdrup). Using the same mortality cost and mathematics as above, subtracting the daily production figures from 2023 – below the expected peak – the ban would save 2,011 lives per year; but we know this to be an underestimate.

One cannot take for granted that the referendum would be honoured. As we write these words, in late October 2023, the boyish heir of a banana fortune and bandsman of the bourgeoise has been elected interim president; his intentions for Yasuní remain unclear.[47] But regardless, history was made in Ecuador in August of that year: for the first time ever, a people democratically elected to strand fossil fuel assets.[48] It was a radicalised version of the original Yasuní-ITT initiative, as the ban would be implemented *after* extraction had begun – with all the forms of fixed capital in place – but without any mechanism of compensation. It was the culmination of decades of mass mobilisation against oil in Ecuador, spanning the tactical spectrum from litigation to sabotage – perhaps a Gramscian war of position more than a war of manoeuvre, but one that would need to be copied and condensed in the latter form.[49]

The referendum did not spark a 'climate Minsky moment'. Nor did Costa Rica when it banned oil exploration and extraction in 2002, or Belize offshore activities in 2017, or Ireland drilling in 2019.[50] None of these supply-side measures involved *a major producer of fossil fuels*, be it a nation or corporation.[51] The strategic task would be to win something like the Yasuní victory in a country like the US, Canada, Australia, the UK, Norway; or run a mass sabotage campaign against Total or nationalise ENI (expropriation could be starting point as much as endpoint) – anything damaging enough for sufficiently vested interests to touch off the panic. Ecuador offers the finest model so far, in its nearly perfect sequence from grassroots activism via referendum to (hopefully) execution by the state. Worldwide emulation might be highly improbable. Unchecked overshoot is likelier. But Ecuador, at the very least, showed what *could* be done, even in a year as thoroughly deranged as 2023.

9

Chronicle of One More Year of Madness

As we write this in October 2023, the world has just experienced its warmest September on record. This followed equivalent back-to-back records for June, July, August, with particularly out-of-the-way sea temperatures in the North Atlantic and off the coast of Antarctica setting off alarm bells in research centres around the world.[1] But it was the size of the September anomaly that caused most jaws to drop. Climate commentators were – not for the first time, not for the last – scrambling for the appropriate superlatives. The month was a full 1.75°C warmer than the pre-industrial average, 'smashing', in the uncharacteristic words of the World Meteorological Organization, the previous record by no less than 0.5°C.[2] Temperature records are more typically – and typical they have become – exceeded by one tenth of a degree, not one half. No one had seen a jump of this size before. In fact, September 2023 was warmer than *any* month in the historical record, including those months that normally top the charts – July and August – and so in this one month, normal only in its abnormality, the planet reached half the distance between the two limits of 1.5°C and 2°C. One scientist briefly ascended to viral fame when he called it 'gobsmackingly bananas'.[3]

September also saw the deadliest climate disaster so far in the third decade. A low-pressure system formed in the Mediterranean on the fourth day of the month; on the next, it was upgraded into a cyclone and named Storm Daniel. It first struck the Balkan

Peninsula, hitting the central Greek plain of Thessaly with one year's worth of rain in twenty-four hours, covering this former breadbasket in a thick coat of silt, which, locals expected, would keep it unfertile for at least three years.[4] The deluge came on the heels of a series of wildfires in Greece, including the largest ever measured in the EU.[5] The mayor of one town called it a 'biblical catastrophe'.[6] But Storm Daniel became catastrophic on another scale when it reached eastern Libya on the tenth.[7] During its twenty-four hours' visitation, it dropped a load of water around seventy times larger than the average amount for the month of September.[8] The city of Derna was located at the mouth of a river, running through a wadi towards the sea, normally within narrow banks, if indeed it ran at all. This was desert country. But now suddenly the river rose, burst through two dams and crashed into Derna, the water, sediment, debris forming a bulldozer that ripped and roared through the city in the middle of the night to 11 September – a force of such speed and violence as to drive structures and streets into the Mediterranean and turn the former centre into a brownish, muddy bog.[9] Reports described how random furniture and body parts poked up through pulverised buildings.[10] 'Corpses still litter the street, and drinkable water is in short supply. The storm has killed whole families' – 'a catastrophe unlike anything we have ever seen', according to one native of the town. 'The residents of Derna are searching for the bodies of their loved ones by digging with their hands and simple agricultural tools.'[11]

These scenes formed a striking prefiguration of those unfolding further to the east on the Mediterranean littoral one month later, in Gaza; in Derna, however, the anonymous shower of death from the sky was not an air force, but the cumulative carbon bombing of years past. (There was also a more intimate connection between Derna and Gaza: at least a dozen Palestinians who had fled from the latter enclave to the former town were killed in the floods. 'Two catastrophes took place, the catastrophe of the displacement (in 1948) and the storm in Libya,' commented one relative, who survived the storm because he had stayed in Gaza.)[12] Using the refined methodologies of weather attribution, researchers soon concluded that the floods on 11 September had been made fifty times more likely by the average warming of 1.2°C – a mathematical formula for identifying the cause of the disaster.[13] During the preceding summer months, the waters off North Africa had been no less than 5.5°C warmer than the average from the previous two decades.[14] Warm water holds heat energy that can get packed into a storm like fuel into a missile.[15] Some 11,300 people were killed by Storm Daniel in Libya. This made it the most intense event of mortality from carbon so far in the decade, possibly the century.[16]

But it was not, of course, the first of its kind in that year. In February and March, Cyclone Freddy circled through Mozambique and Malawi – fourth and eighth poorest countries in the world – making a first round, retreating into the Indian Ocean and then returning for a second, its thirty-four days of gyrations making it the longest-lasting tropical cyclone so far. Some 1,000 died, mostly in Malawi. People dug through mud for survivors with shovels and bare hands.[17] In April, temperature records snapped like dry twigs in Laos, Thailand, Myanmar, Bangladesh, forcing schools to close and pushing people to seek shelter inside air-conditioned shopping malls, where such existed.[18] In May, torrential rains killed some 600 in Rwanda, Uganda, the Democratic Republic of Congo.[19] In July, torrential rains killed several dozens in Japan and South Korea.[20] But in that month, it was the heatwaves that dominated: 'we live as if we are inside an oven, struggling to breathe due to the heat inside the tent. If it weren't for the water we sprinkle on the tent, we would have died from the intense heat,' said one Syrian woman in a refugee camp in the Idlib province.[21] Over in Tunis, migrants from sub-Saharan Africa did not even have tents to stay in, but slept on hard ground that 'rarely cools, leaving their bodies with little chance to recover from the intense heat of the day, increasing the risks of heat exhaustion and stroke'; with temperatures reaching 50°C – seasonal average 33°C – life in the city had 'slowed to a crawl'.[22] Deadly blazes broke out from Algeria to Croatia, and in August, the turn came to the US.[23] Dried out, heated up, the Hawaiian island of Maui ignited in the nation's worst wildfire in over a century, by the number of lives lost (97): cars torched, homes turned to ashes, the ground black, grey and white, even in American paradise.[24]

Nor could the most affluent parts of Europe claim ignorance. Over the course of 2022 and 2023 – two short years in history – Switzerland lost one tenth of the volume of its glaciers.[25] In late August, a Swiss weather balloon had to climb a record 5,300 meters into the air before temperatures fell to 0°C.[26] Cacti spread on the Alpine hills.[27] The red fire ant, a pest that can form 'super-colonies' crowding out local species and devouring crops, made a first landing in Sicily.[28] Paris was fumigated for tiger mosquitos: a vector for viruses like dengue and zika, the species was making its way north and had now reached the trees and ponds of the French capital.[29] But needless to say, it was the global South that bore the brunt of climate suffering. One year after the killer monsoon, four in ten children in the affected parts of Pakistan had stunted growth, their parents struggling to access basic necessities; desperate farmers were caught in the nets of debt bondage.[30] A drought held eastern Africa – Ethiopia, Somalia, Kenya, Djibouti – in a silent, uncommented, crippling grasp.[31] A drought left the Uruguayan capital Montevideo bereft of drinking water.[32] A drought reduced the Amazon River to its lowest level in over a century.[33] A cyclone battered Rio Grande do Sul and killed thirty-nine, the deadliest climate disaster so far in the Brazilian province: record after record after record; and as we write these words, news has just come in of a hyper-intense hurricane in south-western Mexico, swamping streets, flooding hospitals.[34] This is what warming looked like in 2023, at the dawn of the overshoot conjuncture, with 1.2°C behind us and no limits in sight. What traces might a year like this leave on trends in politics and investment?

We Need to Start Working Together in Harmony

Also during 2023, preparations were underway for COP number twenty-eight. It would feature a novel element. The man appointed to direct this instalment of the show, to be held in the United Arab Emirates, was Sultan Al Jaber. He was the CEO of the Abu Dhabi National Oil Company, ADNOC.[35] That enterprise was then in the midst of a thrust of expansion, planning to pour more than 1 billion dollars into oil and gas projects *per month* until 2030 (by comparison, we might recall that Petroecuador had invested a total of 2 billion in Yasuní when the referendum took place).[36] In July 2023, ADNOC hit an output of 4.5 million barrels of oil per day, a milestone on the way to 5 million, the goal for 2027, the Emirates now the third largest producer in OPEC.[37] The jewel in this particular crown was Upper Zakum, second biggest offshore field in the world, shared between ADNOC and ExxonMobil. With partners across the circuit of primitive fossil capital, the company expanded right and left and rigorously modernised its drilling and refining machinery: and the man overseeing it all was Sultan Al Jaber, who would also be the president of negotiations at COP28.[38]

The US climate envoy John Kerry thought it was a 'terrific choice'.[39] (So did Tony Blair.[40]) Others were somewhat more circumspect about the idea of giving the top job in international climate politics to an oil boss, a scepticism that lead to a trail of minor scandals on the way to Expo City in Dubai: members of Al Jaber's team edited Wikipedia pages so that readers learned that he was 'precisely the kind of ally the climate movement needs'.[41] A battery of fake accounts spread similar messages on social media.[42] ADNOC employees handled the email correspondence of the COP28 office.[43] PR executives from the company also made PR for the summit.[44] Al Jaber himself toured the world in amazement that anyone would question his qualifications, made the case for the indispensability of fossil fuels and defended the strategy of ADNOC: 'investing in our reserves here, it's investing in our economy.' A *Guardian* journalist found 'only one point when he's nonplussed. Does he intend to put Adnoc out of business? He stops still and turns to stare at me in astonishment. "Why would I want to do that?"'[45] Sultan Al Jaber had another preference for how to deal with the crisis. 'If we're serious about mitigating climate change and reducing in a practical manner emissions we must scale up carbon capture technologies.'[46] Or, as he put it to a journalist from Reuters: 'are we after decarboniza-tion, or are we after some ideological idea against oil and gas? We are after emissions, so let's stay focused on that, that is our enemy; let's fight that, let's not fight an industry that has helped shape the world we are all in today.'[47] Thus, in 2023, active capital protection had been insinuated into the highest echelons of climate governance, the irreal turn coming full circle, the theatre now a tragedy and farce wrapped into one, overshoot ideology the official decor. And true to script, Al Jaber was optimistic. 'We need to stop the finger-pointing. We need to stop this polarisation. We need to flip the page and start focusing on being optimistic, positive and working together in harmony.'[48]

When we write this, there is one month to go before the doors are opened to Expo City, and so we cannot know the outcome of COP28; but betting on the absence of a breakthrough at this twenty-eighth edition of the summit does not seem overly bold. No trends in 2023 pointed beyond the incantatory rituals. No climate realities were admitted entry into the central halls and corridors. A coincidence or part of a broader trend, the EU appointed a climate chief in October 2023, one Wopke Hoekstra from the Netherlands: he had worked for Shell.[49] The new thing in European climate politics in this year was 'greenlash', or the pushback against the mild reforms initiated so far.[50] Emmanuel Macron called for a 'pause' to environmental regulations in the Union.[51] The UK government postponed the phasing out of fossil gas and internal combustion engines, clamped down on the installation of solar panels and effectively terminated the construction of wind turbines, onshore and offshore.[52] (The invaded country of Ukraine built more onshore turbines in 2022 than England did.[53]) Sweden, long regarded as a green light to the nations, undertook a Trump-like deconstruction of existing climate policy and planned for a massive increase in domestic emissions throughout the decade.[54] This was the way the wind blew in Europe in 2023, despite plenty of lip service to the need for a change of direction. Meanwhile, in May the WMO revised its assessment of the probability of one of the next five years breaching 1.5°C, from a one-in-two chance to two-in-three.[55] Then the summer months did just that.[56]

Year of the Demon

A tinge of sadness and disappointment in the bureaucratic prose, the International Energy Agency noted that – contrary to the recommendation in its knife report, and despite there being no need for it – 'what has happened in practice is that oil and gas investment has risen'.[57] It now stood at double the levels compatible with the Agency's scenario for 1.5°C.[58] 'The world is racing through the available budget,' the combustion of each of the three fossil fuels hitting a new all-time high in 2023 and expected to do the same again next year.[59] The bonanza slowed down one notch, as prices fell in the first half of 2023; but profits stayed very healthy indeed.[60] Expenditure on oil and gas was on the up, flows of money capital – equity, bond – into the circuit speeding up, stocks of the companies going up.[61] 'Betting on oil is becoming a favourite trade on Wall Street,' observed one analyst.[62] The price was also turning back up in the summer, and then on 7 October, the Palestinian resistance in Gaza launched the Toufan al-Aqsa operation against military bases and colonies in the south of the country. The Israeli occupation forces responded with overwhelmingly devastating bombardment of the refugee camps and towns of Gaza. Because this happened in the Middle East, and because the US rushed to protect its investments in the state of Israel with guns pointing towards Iran, prices shot up further; in October 2023, they were back at the heights of the same month one year prior.[63] We know what this augured.

For the first time, oil was the number one commodity leaving the United States in 2023.[64] Mitigation now meant closing down the single largest export product from the single most powerful nation in history – or, in the words of Biden, 'the essential nation', the leader that 'holds the world together'.[65] This president handed over the northern slopes of Alaska to ConocoPhillips. Slated to run for at least three decades, peak output 180,000 barrels per day, the Willow project required an unusual type of fixed capital: giant chillers to refreeze the thawing permafrost in summertime, lest pipelines and other infrastructure sink into the soggy soil.[66] One dozen major banks – JPMorgan, Citigroup, Barclays, Wells Fargo – had devoted funds to the project.[67] The Biden administration also showered the industry with leases in the Gulf of Mexico, fast-tracked the Mountain Valley Pipeline slicing through Appalachian forests, removed legal hurdles to fresh oil and gas projects and oversaw a non-stop surge in volumes produced.[68] In the UK, the gift of the year was Rosebank. To the west of the Shetland Islands, this oil field was given to Equinor and Ithaca Energy, the Israeli upstart; the Norwegians promised to electrify extraction.[69] 'Today's announcement is a welcome shot in the arm for the UK energy sector which will give investors, operators and the wider supply chain confidence as they strive to provide the power we need,' the CEO of the chamber of commerce in the Scottish capital of oil rejoiced.[70] In the South, hot new frontiers for primitive fossil capital from the North included the Okavango delta in Namibia, the Nile delta in Egypt, South Africa, Suriname.[71] Amid these trends, at least one luminary of climate diplomacy offered a mea culpa of sorts. This was Christiana Figueres, the priestess of 'stubborn optimism', who wrote: 'I thought fossil fuel firms could change. I was wrong.'[72] But how they should be crushed, now that it was clear that they could not be changed, she did not say.

A more substantial exception came from Latin America, the sole continent where forces of the left retained a capacity to seize state power. Colombia declared an end to exploration. No more licenses would be given; fracking operations were off.[73] The first left-wing president in the country's history, former guerrilla fighter Gustavo Petro, had the phase-out of fossil fuels as a plank of his electoral campaign, and then he made good on the promise.[74] 'We are convinced that strong investment in tourism, given the beauty of the country, and the capacity and potential that the country has to generate clean energy, could, in the short term, perfectly fill the void left by fossil fuels,' he swore in the church of the World Economic Forum in Davos.[75] He 'says oil is poisoning the earth and warns that climate change could lead to the extinction of the human race', the *Wall Street Journal* reported, a smug disdain in its voice.[76] Colombia was at this point the third largest producer of oil in Latin America.[77] The 749,000 barrels per day were expected to hit zero within ten years without new discoveries. If this prognosis was correct, and if the ban on exploration withstood the immense pressures from bourgeoisies local and global, Colombia would be on track to have zero output by 2034 – the task *rich* countries ought to have shouldered. 'Yet no other significant oil producer is following the Petro government's lead,' the *Journal* noticed with relief.[78]

But the energy addition carried on: 2023 was a shining year for solar power. In a record on the bright side, some 500 gigawatts were added. Growth in capacity stayed exponential and cell technologies broke new boundaries.[79] The International Energy Agency predicted that the price scissors would extend into the foreseeable future.[80] Rising interest rates, however, threatened to throttle the flow of private capital into the solar sectors, as they made low profits less bearable, and the stocks of solar companies were going down, all gains on US exchanges since 2020 wiped out by late October 2023.[81] Trying to strike an upbeat note, the Agency counted on fossil fuel combustion to keep growing until around 2030. Oil and gas would then hit a peak and *stay at a plateau* until mid-century, or as long as the forecast stretched, meaning that no reductions were on the horizon; but coal would decline, largely due to the progress of alternatives in the steel industry.[82] For the time being, solar and coal boomed in parallel in the People's Republic, bastion of both. The installation of solar *and* the approval of coal plants accelerated in 2023.[83]

Was there anything sustainable in these trajectories? 'We need to stop burning fossil fuels,' said one climatologist at Lord Stern's research institute. 'Now. Not some time when we've allowed companies to make all the money they possibly can.'[84] But the CEO of ENI was of a different mind. 'We cannot shut down everything and rely just on renewables and that is the future, no. It's not like that. We have infrastructure, we have investment that we have to recover and we have the demand that is still there.'[85] Meanwhile, in the sphere of consumption, the sale of private jets was on course for another record, and one British company came up with a brand new niche product: customers could pay to bring their dogs on board luxury flights between London and Dubai, drinking champagne with one hand and nuzzling their golden retrievers with the other.[86] The owner of the company said that it 'believes pet family members deserve to travel in comfort and style alongside their owners. We couldn't be more excited to kick off this new route, just in time for the holidays, so guests can celebrate with their loved ones (including pets) in style.'[87] Just in time, indeed.

Malmö, 28 October 2023

Acknowledgements

This book builds on work that each of us have carried out in various forms over the past years, and the causal chain of conversations, inputs and criticisms that led to this point therefore extends far beyond what we can do justice to here. A few names, however, need to be named. Parts of the argument were developed in conversation with the group of interns from the Lund University master's programme in Human Ecology during the Fall Semester of 2022. More of their research input will make its way into the sequel to this book: acknowledgements to be continued. Special thanks to Troy Vettese and Ståle Holgersen for incisive comments on an earlier version of the full manuscript. Jana Tsoneva sent cutting-edge reading recommendations at an astounding rate. Very special thanks, as always, to Sebastian Budgen, whose work on all fronts is inestimable. Friends, family and colleagues showed patience while we prioritised finalising this book over other things. To Anna from Wim: writing this would not have been possible without your endless support and understanding, and now we have committed to another instalment of this. I am so grateful for the adventures large and small that you and I have embarked on during the writing of this book – let there be many sequels to that as well! The project that this book is based on was generously funded by the Swedish Research Council for Sustainable Development, Formas, grant number 2018-01686. All remaining errors are ours alone.

Notes

Preface

1 See e.g. Mathis Wackernagel, Niels B. Shulz, Diana Deumling et al., 'Tracking the Ecological Overshoot of the Human Economy', *PNAS* (2002) 99: 9266–71.
2 It should be stressed again here that we understand mitigation in the traditional sense, as emissions cuts. In recent nomenclature, used, for example, by the IPCC, carbon dioxide removal is subsumed under the heading of mitigation, but we keep the two separate in this book.
3 International Energy Agency [hereafter IEA], 'Global Methane Tracker 2023', February 2023; Marielle Saunois, Ann R. Stavert, Ben Poulter et al., 'The Global Methane Budget 2000–2017', *Earth System Science Data* (2020) 12: 1594.
4 The discrepancy between the gigatonnes in this graph and in chapter 1 below is due to the former counting *carbon* and the latter *carbon dioxide*. One gigatonne of carbon equals nearly four of carbon dioxide.

1. Chronicle of Three Years Out of Control

1 Estimates of the size of the fall vary. 5.1 per cent: IEA, *Global Energy Review: CO_2 Emissions in 2021*, March 2022, 3; 5.4 per cent: Global Carbon Project, 'Global Carbon Budget', n.d.; 6.3 per cent: Zhu Liu, Zhu Deng, Biqing Zhu et al., 'Global Patterns of Daily CO_2 Emissions Reductions in the First Year of COVID-19', *Nature Geoscience* (2022) 15: 615–20; 6.4 per cent: Jeff Tollefson, 'COVID Curbed Emissions – But Not by Much', *Nature* (2021) 589: 343. The highest estimate, deviating from all others, is 7 per

cent: Corinne Le Quéré, Glen P. Peters, Pierre Friedlingstein et al., 'Fossil CO_2 Emissions in the Post-Covid-19 Era', *Nature Climate Change* (2021) 11: 197–9.

2   Göran Therborn, 'The World and the Left', *New Left Review* (2022) 137: 41.

3   IEA, *Global Energy*, 3. Already in December: Liu et al., 'Global', 616.

4   Chris Mooney, 'To Truly Grasp What We're Doing to the Planet, You Need to Understand this Gigantic Measurement', *Washington Post*, 1 July 2015. Only about 400,000 such elephants existed by the early 2020s.

5   IEA, *Global Energy*, 3. This figure for the addition in 2021 is slightly higher than those reported in Steven J. Davis, Zhu Liu, Zhu Deng et al., 'Emissions Rebound from the COVID-19 Pandemic', *Nature Climate Change* (2022) 12: 412–14; Friedlingstein et al., 'Global Carbon', 4811–4900. Following the latter two would mean placing the exceedance of the 2019 levels some months later, the surge spanning the years 2021 and 2022. The variations in these estimates are minor; for a detailed discussion of methodologies and data discrepancies, see Friedlingstein et al., 'Global Carbon'. Note that the IEA here gives 36 gigatonnes as the total for CO_2 emissions from fossil fuel combustion and industry in 2021 (a figure with which Friedlingstein et al. are in accord); with emissions from deforestation and other forms of land-use change included, the total reaches 40. Ibid., 4814.

6   Zeke Hausfather and Pierre Friedlingstein, 'Global CO_2 Emissions from Fossil Fuels Hit Record High in 2022', *Carbon Brief*, 11 November 2022; Friedlingstein et al. 'Global Carbon', 4814, 4827.

7   Sandro W. Lubis, Samson Hagos, Eddy Hermawan et al., 'Record-Breaking Precipitation in Indonesia's Capital of Jakarta in Early January 2020 Linked to Northerly Surge, Equatorial Waves, and MJO', *Geophysical Research Letters* (2022) 49: 1–11; Paige Van de Vuurst and Luis E. Escobar, 'Climate Change and the Relocation of Indonesia's Capital to Borneo', *Frontiers in Earth Science* (2020) 8: 1–6. As the latter article details, the relocation of the capital to Borneo threatens to cause a disaster for the biodiversity remaining on this island.

8   Danielle Celermajer, Rosemary Lyster, Glenda M. Wardle et al., 'The Australian Bushfire Disaster: How to Avoid Repeating this Catastrophe for Biodiversity', *WIREs Climate Change* (2021) 12: 1–9; Michelle Ward, Ayesha I. T. Tulloch, James Q. Radford et al., 'Impact of 2019–2020 Mega-Fires on Australian Fauna Habitat', *Nature Ecology and Evolution* (2020) 4: 1321–6; Sergey

Khaykin, Bernard Legras, Silvia Bucci et al., 'The 2019/20 Australian Wildfires Generated a Persistent Smoke-Charged Vortex Rising Up to 35 km in Altitude', *Communications Earth and Environment* (2020) 1: 1–12; Lisa Cox, 'Smoke Cloud from Australian Summer's Bushfires Three-Times Larger than Anything Previously Recorded', *Guardian*, 2 November 2020. The 3 billion animals were mammals, birds and reptiles, invertebrates excluded.

9 Shin-Chan Han, Khosro Ghobadi-Far, In-Young Yeo et al., 'GRACE Follow-On Revealed Bangladesh Was Flooded Early in the 2020 Monsoon Season Due to Premature Soil Saturation', *Proceedings of the National Academies of Sciences* [hereafter *PNAS*] (2021) 118: 1–9.

10 Walter Leal Filho, Ulisses M. Azeiteiro, Amanda Lange Salvia et al., 'Fire in Paradise: Why the Pantanal is Burning', *Environmental Science and Policy* (2021) 123: 31–4; Jose A. Marengo, Ana P. Cunha, Luz Adriana Cuartas et al., 'Extreme Drought in the Brazilian Pantanal in 2019–20: Characterization, Causes, and Impacts', *Frontiers in Water* (2021) 3: 1–20; Renata Libonati, João L. Geirinhas, Patrícia S. Silva et al., 'Assessing the Role of Compound Drought and Heatwave Events on Unprecedented 2020 Wildfires in the Pantanal', *Environmental Research Letters* (2022) 17: 1–12; Maria Luca Ferreira Barbosa, Isadora Haddad, Ana Lucia da Silva Nascimento et al., 'Compound Impact of Land Use and Extreme Climate on the 2020 Fire Record of the Brazilian Pantanal', *Global Ecology and Biogeography* (2022) 31: 1960–75.

11 James P. Schultz, Ryan C. Berg, James P. Kossin et al., 'Convergence of Climate-Driven Hurricanes and COVID-19: The Impact of 2020 Hurricanes Eta and Iota on Nicaragua', *Journal of Climate Change and Health* (2021) 3: 1–5; John L. Beven II, 'The 2020 Atlantic Hurricane Season: The Most Active Season on Record', *Weatherwise* (2021) 74: 33–43; Philip J. Klotzbach, Kimberly M. Wood, Michael M. Bell et al., 'A Hyperactive End to the Atlantic Hurricane Season, October–November 2020', *Bulletin of the American Meteorological Society* (2022) 103: 110–28; John J. Kennedy, Kate M. Willett, Robert J. H. Dunn et al., 'Global and Regional Climate in 2020', *Weather* (2021) 76: 364.

12 See references in the five preceding notes. Needless to say, this is but a handful of the extreme weather events of 2020; for a fuller chronicle, see Kennedy et al., 'Global and Regional'.

13 Damian Carrington, 'Rain Falls on Peak of Greenland Ice Cap for the First Time on Record', *Guardian*, 20 August 2021; Min Xu, Qinghua Yang, Xiaoming Hu et al., 'Record-Breaking Rain Falls at Greenland Summit Controlled by Warm Moist-Air Intrusion',

Environmental Research Letters (2022) 17: 1–7; Jason E. Box, Adrien Wehrlé, Dirk van As et al., 'Greenland Ice Sheet Rainfall, Heat and Albedo Feedback Impacts From the Mid-August 2021 Atmospheric River', *Geophysical Research Letters* (2022) 49: 1–11.

14 Alan H. Taylor, Lucas B. Harris and Carl N. Skinner, 'Severity Patterns of the 2021 Dixie Fire Exemplify the Need to Increase Low-Severity Fire Treatments in California's Forests', *Environmental Research Letters* (2022) 17: 1–5.

15 *BBC*, 'Israel Battles Huge Wildfire Near Jerusalem', 17 August 2021; Al-Haq, 'COP26: Jerusalem Wildfires in Summer 2021, when Climate Change-Related Wildfires Expose the JNF's Colonial Afforestation Project', 6 November 2021.

16 Vincent Ni and Helen Davidson, 'Death Toll Rises and Thousands Flee Homes as Floods Hit China', *Guardian*, 21 July 2021 ('unseen': Chinese media quoted in this report); Munyaradzi Makoni, 'Southern Madagascar Faces "Shocking" Lack of Food', *Lancet* (2021) 397: 2239. For a chronicle of the year, see UNFCCC, 'State of Climate in 2021: Extreme Events and Major Impacts', 31 October 2021.

17 Rahmat Tunio, '"It Seems This Heat Will Take Our Lives": Pakistan City Fearful After Hitting 51C', *Guardian*, 25 May 2022; Zoha Tunio, 'In Jacobabad, One of the Hottest Cities on the Planet, a Heat Wave Is Pushing the Limits of Human Livability', *Inside Climate News*, 20 June 2022.

18 Ayesha Rascoe, 'How Melting Glaciers Contributed to Floods in Pakistan', *NPR*, 4 September 2022; Zulfiqar A. Bhutta, Shereen Zulfiqar Bhutta, Shabina Raza and Ali Tauqeer Sheikh, 'Addressing the Human Costs and Consequences of the Pakistan Flood Disaster', *Lancet* (2022) 400: 1287–8; Muhammad Ahmed Waqas, 'Pakistan's Floods Flow from Climate Injustice', *Science* (2022) 378: 482; Nina Lakhani and Shah Meer Baloch, 'Rich Nations Owe Reparations to Countries Facing Climate Disaster, Says Pakistan Minister', *Guardian*, 4 September 2022; The National Disaster Management Authority of Pakistan, 'Floods (2022)', 18 November 2022. Sindh was reported to have received nearly eight times more rain than the average: Reuters, 'Pakistan Floods Have Affected Over 30 Million People: Climate Change Minister', 27 August 2022.

19 Ruth Maclean, 'Nigeria Floods Kill Hundreds and Displace Over a Million', *New York Times*, 17 October 2022.

20 NASA Earth Observatory, 'Worst Drought on Record Parches Horn of Africa', 14 December 2022.

21 Susan Schulman, 'Between Two Rivers: The Iraq Drought', *RUSI*

Journal (2022) 167: 130–47; Simona Foltyn, '"The Green Land is a Barren Desert": Water Scarcity Hits Iraq's Fertile Crescent', *Guardian*, 7 September 2022; Ghaith Abdul-Ahad, 'Death in the Marshes: Environmental Calamity Hits Iraq's Unique Wetlands', *Guardian*, 29 January 2023.

22 Jon Henley, '"The New Normal": How Europe Is Being Hit by a Climate-Driven Drought Crisis', *Guardian*, 8 August 2022; Jon Henley, 'Europe's Rivers Run Dry as Scientists Warn Drought Could be Worst in 500 Years', *Guardian*, 13 August 2022; Copernicus Programme, '2022 Was a Year of Climate Extremes, with Record High Temperatures and Rising Concentrations of Greenhouse Gases', 9 January 2023.

23 Observations by Malm; see further e.g. Tobias Jones, 'Quiet Flows the Po: The Life and Slow Death of Italy's Longest River', *Guardian*, 10 July 2022.

24 E.g. 1.1°C: World Meteorological Organization [hereafter WMO], '2021 One of the Seven Warmest Years on Record, WMO Consolidated Data Shows', 19 January 2022; 1.2°C: Copernicus Programme, '2022'.

25 WMO, '50:50 Chance of Global Temperature Temporarily Reaching 1.5°C Threshold Next 5 Years', 9 May 2022. Similar estimations could be found in e.g. Jochem Marotzke, Sebastian Milinski and Christopher D. Jones, 'How Close Are We to 1.5 DegC or 2.5 DegC of Global Warming?', *Weather* (2022) 77: 147–8; Leon Hermanson, Doug Smith, Melissa Seabrook et al., 'WMO Global Annual to Decadal Climate Update', *Bulletin of the American Meteorological Society* (2022) 103: 1117–29. The WMO announcement was widely reported, e.g. Matt McGrath, 'Climate Change: "Fifty-Fifty Chance" of Breaching 1.5C Warming Limit', *BBC*, 10 May 2022; Sam Meredith, '"It's a 50–50 Call": The Earth is Close to Crossing a Key Temperature Threshold within 5 Years', *CNBC*, 10 May 2022.

26 More precisely, the baseline is 1850–1900, the first period of reliable global measurements, a moment not before the birth of fossil capital – it had matured in Britain by 1850 – but rather before such capital had grown large enough to impact the energy balance on Earth.

27 Valérie Masson-Delmotte, Panmao Zhai, Hans-Otto Pörtner et al. (eds), *Global Warming of 1.5 °C: An IPCC Special Report* (Cambridge: Cambridge University Press, 2018), 53, 56–9, 81, 184, 338; Marotzke et al., 'How Close', 147; Celermajer et al., 'The Australian Bushfire', 3; Hermanson et al., 'WMO Global', 119.

28 WMO, '50:50 Chance'; WMO, 'United in Science: We Are Heading in the Wrong Direction', 13 September 2022.

29 Friedlingstein et al., 'Global Carbon', 4814; cf. Zhu Liu, Zhu Deng, Steven J. Davis et al., 'Monitoring Global Carbon Emissions in 2021', *Nature Reviews Earth and Environment* (2022) 3: 218. This forecast derived from an annual total of 40 gigatonnes in 2022: see footnote 5 ch. 1. Others, however, concluded that only six and a half more years of emissions á la 2022 would finish up the budget. Piers Forster, Debbie Rosen, Robin Lamboll and Joeri Rogelj, 'What the Tiny Remaining 1.5C Carbon Budget Means for Climate Policy', *Carbon Brief*, 11 November 2022.

30 E.g. IPCC, 'Summary for Policymakers', in Masson-Delmotte et al., *Global Warming*, 12; and further e.g. Jeff Tollefson, 'Clock Ticking on Climate Action', *Nature* (2018) 562: 172–3; Jonathan Watts, 'We Have 12 Years to Limit Climate Change Catastrophe, Warns UN', *Guardian*, 8 October 2018; Coral Davenport, 'Major Climate Report Describes a Strong Risk of Crisis as Early as 2040', *New York Times*, 7 October 2018.

31 Friedlingstein et al., 'Global Carbon', 4814; Forster et al., 'What the Tiny'; Le Quéré et al., 'Fossil CO_2', 197; Liu et al., 'Global Patterns', 615–16.

32 IEA, *Net Zero by 2050: A Roadmap for the Global Energy Sector*, May 2021, 20; see further e.g. 21, 23, 26, 51, 99, 100–3, 152, 162, 175.

33 Ibid., 160. Still in the preface to this report, the director of the agency, Fatih Birol, defined one of its 'core missions' to be promoting 'secure and affordable energy supplies to foster economic growth' – an entirely honourable bourgeois mission. Ibid., 3.

34 Thom Allen and Mike Coffin, 'Paris Maligned: Why Investors Should Assess the Climate Alignment of Oil and Gas Companies', Carbon Tracker, December 2022, 12.

35 David Hodari, 'Stop New Oil Investments to Hit Net-Zero Emissions, IEA Says', *Wall Street Journal*, 18 May 2021; Dave Jones quoted in David Vetter, 'End New Fossil Fuel Development, IEA Demands in Groundbreaking Net Zero Plan', *Forbes*, 18 May 2021. Cf. e.g. Leslie Hook and Anjli Raval, 'Energy Groups Must Stop New Oil and Gas Projects to Reach Net Zero by 2050, IEA Says', *Financial Times*, 18 May 2021.

36 See e.g. Simon Evans, 'New Fossil Fuels "Incompatible" with 1.5C Goal, Comprehensive Analysis Finds', *Carbon Brief*, 23 October 2022; Dan Tong, Qiang Zhang, Yixuan Zheng et al., 'Committed Emissions from Existing Energy Infrastructure Jeopardize 1.5°C Climate Target', *Nature* (2019) 572: 373, 376.

37 Dan Welsby, James Price, Steve Pye and Paul Ekins, 'Unextractable Fossil Fuels in a 1.5°C World', *Nature* (2021) 597: 230–4.

38 Dan Calverley and Kevin Anderson, 'Phaseout Pathways for Fossil

Fuel Production within Paris-Compliant Carbon Budgets', Tyndall Centre, University of Manchester, March 2022.

39 Greg Muttit, James Price, Steve Pye and Dan Welsby, 'Socio-Political Feasibility of Coal Power Phase-Out and its Role in Mitigation Pathways', *Nature Climate Change* (2023) 13: 140–7.

40 Kelly Trout, Greg Muttit, Dimitri Lafleur et al., 'Existing Fossil Fuel Extraction Would Warm the World Beyond 1.5°C', *Environmental Research Letters* (2022) 17: 1–12. Initiation here includes 'a financial and regulatory commitment to extraction': ibid., 2. This paper thus confirmed Tong et al., 'Committed Emissions', and superseded Christopher J. Smith, Piers M. Forster, Myles Allen et al., 'Current Fossil Fuel Infrastructure Does not yet Commit Us to 1.5°C Warming', *Nature Communications* (2019) 10: 1–10.

41 Trout et al. 'Existing Fossil', 7, 9; cf. e.g. Saphira A. C. Rekker, Katherine R. O'Brien, Jacquelyn E. Humphrey and Andrew C. Pascale, 'Comparing Extraction Rates of Fossil Fuel Producers against Global Climate Goals', *Nature Climate Change* (2018) 8: 489–92.

42 Stanley Reed and Clifford Krauss, 'Too Much Oil: How a Barrel Came to Be Worth Less than Nothing', *New York Times*, 20 April 2020.

43 Kenneth Rapoza, 'Despite Historic Production, Oil Is Dying', *Forbes*, 15 April 2020. As Gabe Eckhouse points out, the same hasty obituary was written by more well-meaning observers too, such as Chris Saltmarsh, who announced in *Jacobin* that 'the oil industry is dying right now' – 'this crisis isn't a momentary blip' – 'significantly contracting oil output is now inevitable.' Chris Saltmarsh, 'The Oil Industry Is Dying Right Now. Don't Resuscitate It', *Jacobin*, 21 April 2020. Cf. Gabe Eckhouse, 'Covid-19, Global Oil Market Volatility, and the Renewable Energy Transition', *Environment and Planning A* (2022) 54: 1648–9.

44 Javier Blas, 'In a World Fighting Climate Change, Fossil Fuels Take Revenge', *Bloomberg*, 10 October 2021. See further Eckhouse, 'Covid-19', 1649.

45 'Higher Prices Triggered a Record Increase in Revenues of Oil and Gas Producers in 2021': IEA, *World Energy Investment 2022*, June 2022, 109.

46 Darren Woods, CEO of ExxonMobil, in 'ExxonMobil (XOM) Q4 2021 Earnings Call Transcript', *Motley Fool*, 1 February 2022.

47 Mike Wirth in 'Chevron (CVX) Q4 2021 Earnings Call Transcript', *Motley Fool*, 28 January 2022.

48 IEA, *World Energy Investment*, 62–4; Caroline Kuzemko, Mathieu Blondeel, Claire Dupont and Marie Claire Brisbois, 'Russia's War on Ukraine, European Energy Policy Responses and Implications

for Sustainable Transformations', *Energy Research and Social Science* (2022) 93: 1; Isabella M. Weber, 'Big Oil's Profits and Inflation: Winners and Losers', *Challenge* (2022) 65: 152; Gabe Eckhouse and Anna Zalik, 'Whither Hydrocarbons? The Rescaling of Oil and Gas Markets Amidst Covid-19 and a Contested Tradition. Introduction', *Environment and Planning A* (2022) 54: 1642.

49 Global Witness, 'Crisis Year 2022 Brought $134 Billion in Excess Profit to the West's Five Largest Oil and Gas Companies', February 2023; Oliver Milman, '"Monster Profits" for Energy Giants Reveal a Self-Destructive Fossil Fuel Resurgence', *Guardian*, 8 February 2023; Sam Meredith, 'Big Oil Rakes in Record Profit Haul of Nearly $200 Billion, Fueling Call for Higher Taxes', *CNBC*, 8 February 2023.

50 ExxonMobil, '4Q Earnings Release', 31 January 2023; Exxon-Mobil, 'ExxonMobil (XOM) Q4 2022 Earnings Call Transcript', *Motley Fool*, 31 January 2023. That is to say, ExxonMobil was the largest of the private supermajors; Saudi Aramco was larger but mostly state-owned, as were the Chinese giants.

51 Murray Auchincloss in BP, 'BP (BP) Q4 2021 Earnings Call Transcript', *Motley Fool*, 8 February 2022.

52 Patrick Lavery, 'Saudi Aramco Makes Largest Ever Quarterly Profit for Listed Company', *International Flame Research Foundation*, 29 August 2022. See further Kalyeena Makortoff, 'Saudi Aramco Profits Soar by 90% as Energy Prices Rise', *Guardian*, 14 August 2022. One report that made a splash in July 2022 claimed that the profits accruing to oil and gas companies that year were likely to measure up to twice the annual average for the past half-century, the total for those fifty years estimated as $52 trillion. Damian Carrington, 'Revealed: Oil Sector's "Staggering" $3 Bn-a-Day Profits for Last 50 Years', *Guardian*, 21 July 2022. It was later published as Aviel Verbruggen, 'The Geopolitics of Trillion US$ Oil and Gas Rents', *International Journal of Sustainable Energy Planning and Management* (2022) 36: 3–10, but, unfortunately, the composition, language and general quality of this paper are distinctly subpar.

53 This seems more than a coincidence, given the outsized climatic power of Saudi Aramco, the oil company responsible for most CO_2 emissions (at the point of combustion) between 1988 and 2015. Paul Griffin, 'The Carbon Majors Database: CDP Carbon Majors Report 2017', CDP, July 2017, 14. Saudi Arabia has exported more than just oil and Wahhabism to Pakistan (on the former, see further below).

54 Mark Sweney, 'Saudi Aramco's $161bn Profit Is Largest Recorded by an Oil and Gas Firm', *Guardian*, 12 March 2023.

55 Joanna Partridge, 'ExxonMobil's Record-Breaking $20bn Profit Nearly Matches Apple's', *Guardian*, 28 October 2022; Ian Palmer, 'Oil and Gas Profits Very High Once Again – What this Feels Like to Energy Consumers', *Forbes*, 4 November 2022.

56 Lex, 'Saudi Aramco: Crude Colossus Takes the Crown', *Financial Times*, 17 May 2022.

57 Krisztian Bocsi, 'Big Coal Miners' Profits Triple as Demand Surges', *Financial Times*, 28 December 2022.

58 Javier Blas and Jack Farchy, *The World for Sale: Money, Power, and the Traders Who Barter the Earth's Resources* (Oxford: Oxford University Press, 2021), 21, 183, 258, 273.

59 Bocsi, 'Big Coal'; *Fortune*, 'Glencore', Fortune Global 500, 2023; *Mining Technology*, 'Glencore Logs Record Profit in 2022 on Coal Boom', 16 February 2023.

60 Chris Moore, 'EU Coal: The Good, the Bad and the Ugly', Carbon Tracker, September 2022, 41, 50. On German lignite and its dirtiness, see e.g. Kristina Juhrich, 'CO$_2$ Emission Factors for Fossil Fuels', UmweltBundesamt (German Environment Agency), June 2016, e.g. 18–25, 45–7.

61 Cecilia Nikpay, 'Svenske oljemiljardären ökar förmögenheten mest – efter rekordåret', *Affärsvärlden*, 6 April 2023; Sofia Clason, 'Svenske miljardärens rekordår – på rysk olja', *Expressen*, 8 April 2023; *Forbes*, 'Profile: Torbjorn Tornqvist', 12 October 2023; and on Törnqvist and Gunvor, see further Blas and Farchy, *The World*, 20, 207–16.

62 April Merlaux, Caleb Schwartz, Ruth Breech et al., 'Banking on Climate Chaos: Fossil Fuel Finance Report 2023', Rainforest Action Network, Banktrack, Indigenous Environmental Network, Oil Change International, Reclaim Finance, Sierra Club and Urgewald, 2023, 4, 8.

63 Nick Ferris, 'Weekly Data: One Million Kilometres of New Fossil Pipelines Poses Stranded Asset Risk', *Energy Monitor*, 14 February 2022.

64 Robert Rozansky and Baird Langenbrunner, *Gas Bubble 2022*, Global Energy Monitor [hereafter GEM], October 2022, 3.

65 Ryan Driskell Tate, Christine Shearer and Andiswa Matikinca, 'Deep Trouble: Tracking Global Coal Mine Proposals', GEM, June 2021, 3; GEM, CREA, E3G et al., 'Boom and Bust Coal: The Global Coal Plant Pipeline', April 2022, 3. Note that the first figure refers to early 2021 and would have been higher in the following year.

66 It bears pointing out that the profit boom of 2021 immediately spurred an uptick in investments, returning them to pre-pandemic levels: IEA, *World Energy Investment*, 71.

67 ExxonMobil, '4Q Earnings', 1.

68 Lomarsh Roopnarine, 'Managing Guyana's Newfound Oil and Gas Wealth: Controversy and Challenge', *Caribbean Studies* (2022) 50: 3–33; Reuters, 'Exxon Strikes Oil Again in Guyana with Two New Discoveries', 26 October 2022; Melissa Cavcic, 'ExxonMobil's New Oil Discovery off Guyana Could Underpin Future Development', *Offshore Energy*, 26 January 2023.

69 ExxonMobil, 'ExxonMobil (XOM) Q4 2022 Earnings Call Transcript'.

70 Allen and Coffin, 'Paris Maligned', 21.

71 Partridge, 'ExxonMobil's Record-Breaking'. IEA put the figure at one third in *World Energy Investment*, 66.

72 Matthew DiLallo, 'Chevron Plans to Significantly Boost Spending in 2023. Should Investors Worry?', *Motley Fool*, 10 December 2022.

73 IEA, *World Energy Investment*, 18–19, 66, 70. IEA here put the *average* increase in investment from the US majors at 30 per cent. On Aramco's expansion plans, largest in absolute terms, cf. David Tong, Romain Ioulalen and Kelly Trout, 'Investing in Disaster: Recent and Anticipated Final Investment Decisions for New Oil and Gas Production Beyond the 1.5°C Limit', Oil Change International, November 2022, 9–10.

74 Allen and Coffin, 'Paris Maligned', 22.

75 Ibid., 27–33. Details on ExxonMobil's Uaru field in Guyana: Kemol King, 'At Over US$12.5 Billion; Exxon's 5th Guyana Project Will Be the Largest Ever to Date', *Oil Now*, 9 January 2013.

76 Allen and Coffin, 'Paris Maligned', 33; Tong et al., 'Investing in Disaster', 10; David Tong and Kelly Trout, 'Big Oil Reality Check 2023: TotalEnergies', Oil Change International, May 2023.

77 East Africa Crude Oil Pipeline, 'Final Investment Decision (FID) Announced', 1 February 2022.

78 Les Amis de la Terre France and Survie, 'A Nightmare Named Total: An Alarming Rise in Human Rights Violations in Uganda and Tanzania', December 2020; Les Amis de la Terre France and Survie, 'EACOP: A Disaster in the Making', October 2022; Richard Heede, 'East Africa Crude Oil Pipeline: EACOP Lifetime Emissions from Pipeline Construction and Operations, and Crude Oil Shipping, Refining, and End Use', Climate Accountability Institute, 21 November 2022. Emissions here included construction of the pipeline, operation of it, transportation of the oil, refining and, above all, end use, compared to the annual emissions of Uganda and Tanzania in 2022. On the logic of including end use, see further chapter 6 below. Total had 62 per cent of the stakes in the pipeline consortium; partners included the China

National Offshore Oil Corporation (8 per cent) and Uganda National Oil Company (15). Les Amis de la Terre France and Survie, 'EACOP', 6.

79 DRC hydrocarbons minister Didier Budimbu quoted in Les Amis de la Terre France and Survie, 'EACOP', 8.

80 Ruth Maclean and Dionne Searcey, 'Congo to Auction Land to Oil Companies: "Our Priority Is Not to Save the Planet"', *New York Times*, 24 July 2022.

81 TotalEnergies, 'Strategy, Sustainability and Climate: Presentation', 21 March 2023, 11; and further TotalEnergies, 'Strategy, Sustainability and Climate: Transcript', 21 March 2023, 7.

82 Eni, 'Africa', 2023; David Tong and Kelly Trout, 'Big Oil Reality Check: ENI', Oil Change International, May 2023.

83 Tong et al., 'Investing in Disaster', 6–7; Mei Li, Gregory Trencher and Jusen Asuka, 'The Clean Energy Claims of BP, Chevron, ExxonMobil and Shell: A Mismatch between Discourse, Actions and Investments', *PLoS One* (2022) 17: 15–16, 19.

84 GEM, 'The Scramble for Africa's Gas', December 2022.

85 An anonymous former executive at state-owned oil company Sonatrach, quoted in Ali Bokhlef, 'Algeria: The Trans-Saharan Pipeline, A Nigerian Alternative to Russian Gas?', *Middle East Eye*, 10 September 2022; further Lamine Chikhi, 'Algeria, Niger and Nigeria Sign MoU for Saharan Gas Pipeline', Reuters, 28 July 2022; Jerome Onoja Okojukwu-Idu, 'Nigerian Welders Actively Engaged on Trans-Sahara Gas Pipeline Project – Welding Institute', *Majorwaves Energy Report*, 26 November 2022; Martina Schwikowski, 'African Countries Seek to Revive Sahara Gas Pipeline', *Deutsche Welle*, 8 December 2022; and for some background on the project, see Benjamin Augé, 'The Trans-Saharan Gas Pipeline: An Illusion or a Real Prospect?', Institut Français des Relations Internationales, June 2010.

86 Daniel Avis, 'Israel Plans Gas Exports to Europe as Output Surges by 22%', *Bloomberg*, 24 August 2022; Lahav Harkov, 'Israel, Egypt, EU Sign Initial Gas Export Agreement', *Jerusalem Post*, 15 June 2022; Steven Scheer, 'Energean Starts Gas Production at Israel's Karish Site', Reuters, 26 October 2022; *Times of Israel*, 'Israel Exports Crude Oil for First Time, with Shipment Heading to Europe', 16 February 2023.

87 GEM, 'When Is Enough, Enough? The State of Play with Europe's New LNG Terminal Projects in Response to the Energy Crisis', December 2022; Kuzemko et al., 'Russia's War', 4; Rozansky and Langenbrunner, 'Gas Bubble', 7; Habeck in Shotaro Tani and Guy Chazan, 'Qatar to Supply Germany with LNG as EU Seeks Secure Energy Options', *Financial Times*, 29 November 2022.

88 Jonah Fisher, 'UK Defies Climate Warnings with New Oil and Gas Licences', *BBC*, 7 October 2022. See further Scott Zimmerman and Hanna Fralikhina, 'Hooked on Hydrocarbons: The UK's Risky Addiction to North Sea Oil and Gas', GEM, October 2022.

89 Josh Lamb, 'Shell and BP Involved in UK's First Oil and Gas Licensing Round Since 2019', *Proactive Investors*, 17 January 2023.

90 Reuters, 'Ithaca Energy Set for London's Biggest IPO in 2022', 2 November 2022.

91 Zimmerman and Fralikhina, 'Hooked', 9–12. Of Tornado, Shell owned half.

92 *Offshore Engineer*, 'Talos Brings Online Tornado Attic Well in the Gulf of Mexico: Output Beats Expectations', 27 July 2021.

93 Jim Pickard, George Parker, Attracta Mooney and Rachel Millard, 'Grant Shapps Vows to "Max Out" UK's North Sea Oil and Gas Reserves', *Financial Times*, 23 July 2023.

94 Silje Ask Lundberg, 'The Aggressive Explorer: How Norway's Rapid Ramp-Up of Oil and Gas Licensing Is Incompatible with Climate Leadership', Oil Change International, February 2022.

95 Terje Solsvik, 'Norway Plans to Expand Arctic Oil and Gas Drilling in New Licensing Round', Reuters, 17 March 2022.

96 Terje Aasland in ibid. and Regjeringen, 'Great Interest in Further Exploration Activity on the Norwegian Shelf', 10 January 2023.

97 Rozansky and Langenbrunner, 'Gas Bubble', 3–5.

98 Oliver Milman, 'How the Gas Industry Capitalized on the Ukraine War to Change Biden Policy', *Guardian*, 22 September 2022. Cf. e.g. Hiroko Tabuchi, 'U.S. Oil Industry Uses Ukraine Invasion to Push for More Drilling at Home', *New York Times*, 26 February 2022.

99 Steven Mufson, 'Administration Awards Gulf of Mexico Drilling Leases to Oil Giants', *Washington Post*, 14 September 2022.

100 David Blackmon, 'Permian Basin Drives the U.S. Oil Industry Despite Limits on Growth', *Forbes*, 19 November 2022; Stephanie Kelly, 'U.S. Permian Oil Output Hit Record in December, but Gains Are Slow', Reuters, 14 November 2022.

101 Oliver Milman, 'Burning World's Fossil Fuel Reserves Could Emit 3.5tn Tons of Greenhouse Gases', *Guardian*, 19 September 2022.

102 Romain Ioulainen and Kelly Trout, 'Planet Wreckers: How 20 Countries' Oil and Gas Extraction Plans Risk Locking in Climate Chaos', Oil Change International, September 2023, 4, 9, 15.

103 Ioulainen and Trout, 'Planet Wreckers', 4–5, 14–17. This share was far larger than that of Russia, Iran, China, Brazil, the

United Arab Emirates, Argentina, Iraq, Turkmenistan, Saudi Arabia, Guyana, Qatar, Mexico, Nigeria, India and Kazakhstan combined.

104 Ioulainen and Trout, 'Planet Wreckers', 14.

105 CREA and GEM, 'China Permits Two New Coal Power Plants per Week in 2022', February 2023; Tate et al., *Deep Trouble*, 4, 8. Note that the expansion sped up already in 2020 and 2021, before reaching the extreme velocity of 2022: GEM, 'Why China's Coal Mine Boom Jeopardizes Short-Term Climate Targets', May 2022.

106 Stockholm Environment Institute [hereafter SEI], International Institute for Sustainable Development, Overseas Development Institute et al., 'The Production Gap: Governments' Planned Fossil Fuel Production Remains Dangerously Out of Sync with Paris Agreement Limits', October 2021, 40; Brototi Roy and Anke Schaffartzik, 'Talk Renewables, Walk Coal: The Paradox of India's Energy Transition', *Ecological Economics* (2021) 180: 1–12; Patrik Oskarsson, Kenneth Bo Nielsen, Kuntala Lahiri-Dutt and Brototi Roy, 'India's New Coal Geography: Coastal Transformations, Imported Fuel and State-Business Collaboration in the Transition to More Fossil Fuel Energy', *Energy Research and Social Science* (2021) 73: 1–10.

107 Ankur Paliwal, '"It Was a Set-Up, We Were Fooled": The Coal Mine that Ate an Indian Village', *Guardian*, 20 December 2022.

108 Katrina Beavan, 'Adani's First Carmichael Mine Coal Export Shipment Imminent after Years of Campaigns Against It', *ABC News*, 29 December 2021.

109 Tate et al., 'Deep Trouble', 4, 8, 19.

110 Kate Connolly, 'Germany Puts Coal Power Plant Back on the Network after Gas Supply Cut', *Guardian*, 1 August 2022; Vanessa Dezem, 'Germany Bolsters Coal-Fired Power to Meet Winter Demand', *Bloomberg*, 21 October 2022; Robert Bryce, 'The Iron Law of Electricity Strikes Again: Germany Re-Opens Five Lignite-Fired Power Plants', *Forbes*, 28 October 2022; Moore, 'EU Coal', 41–6.

111 Kuzemko et al., 'Russia's War', 5; Fiona Harvey, 'UK's First New Coalmine for 30 Years Gets Go-ahead in Cumbria', *Guardian*, 7 December 2022. Oil and gas from the North Sea, however, was in no way minor: the UK accounted for one fifth of all hydrocarbon projects due to commence operations in Europe between 2023 and 2027. Global Data Energy, 'The UK Leads Upcoming Oil and Gas Projects Starts in Europe by 2027', *Offshore Technology*, 6 February 2023.

112 Ottomar Edenhofer, Jan Christoph Steckel, Michael Jakob and Christopher Bertram, 'Reports of Coal's Terminal Decline May

be Exaggerated', *Environmental Research Letters* (2018) 13: 7. Cf. e.g. Christian Hauenstein, 'Stranded Assets and Early Closures in Global Coal Mining under 1.5°C', *Environmental Research Letters* (2023) 18: 1.

113 IEA, *World Energy Investment*, 16–18, 43–5, 99–102; Tate et al., 'Deep Trouble', 4–5, 9, 13.

114 Rystad Energy, 'Energy Services Sector Set to Grow to $1 Trillion in 2025', 30 January 2023. Cf. e.g. Ron Bousso and Sabrina Valle, 'Big Oil's Good Times Set to Roll on after Record 2022 Profits', Reuters, 17 January 2023.

115 Woods and Jeff Miller in Clifford Krauss, 'What Exxon and Chevron Are Doing with those Big Profits', *New York Times*, 1 February 2023.

116 Cf. e.g. Carole Nakhle, 'Oil and Gas: The Investment Gap Dilemma', *GIS Reports*, 3 February 2023.

117 Jeffrey Rothfeder and Christopher Maag, 'How Wall Street's Fossil-Fuel Money Pipeline Undermines the Fight to Save the Planet', *Fortune*, 23 February 2023.

118 Andy Serwer and Dylan Croll, 'Jamie Dimon Sounds Off On . . . Almost Everything: Morning Brief', *Yahoo! Finance*, 13 August 2022; Jamie Dimon, 'Chairman and CEO Letter to Shareholders', JPMorgan Chase and Co, 2022.

119 Merlaux et al., 'Banking on', 4–5, 10, 14, 48; Tong et al., 'Investing in Disaster', 9. For the European banks, cf. Jillian Ambrose, 'Europe's Banks Helped Fossil Fuel Firms Raise More than €1tn from Global Bond Markets', *Guardian*, 26 September 2023.

120 InfluenceMap, 'Finance and Climate Change: A Comprehensive Climate Assessment of the World's Largest Financial Institutions', March 2022, 12–14. See also GEM, 'The Hidden Financial Pipelines Supporting New Coal', October 2022.

121 Merlaux et al., 'Banking on', 19.

122 Theodor Adorno, *Philosophical Elements of a Theory of Society* (Cambridge: Polity Press, 2019 [1964]), 127.

123 Or, 'there continues to be a disconnect between the production outlook of different countries and corporate entities and the necessary pathway to limit average temperature increases.' Welsby et al., 'Unextractable Fossil', 230.

124 UN, 'Secretary-General's Video Message on the Launch of the Third IPCC Report', 4 April 2022. This became a theme of Guterres' throughout the year: see e.g. Matthew Taylor, '"Grotesque Greed": Immoral Fossil Fuel Profits Must be Taxed, Says UN', *Guardian*, 3 August 2022.

125 Kjell Kühne, Nilsk Bartsch, Ryan Driskell Tate et al., '"Carbon Bombs": Mapping Key Fossil Fuel Projects', *Energy Policy*

(2022) 16: 1–10. See also Damian Carrington and Matthew Taylor, 'Revealed: The "Carbon Bombs" Set to Trigger Catastrophic Climate Breakdown', *Guardian*, 11 May 2022.

126 SEI et al., 'The Production Gap', e.g. 3–4, 14–15.

127 Tate et al., 'Deep Trouble', 12.

128 Data from the Global Registry of Fossil Fuels, as reported in Milman, 'Burning World's'.

129 Philip Oltermann, 'Stop Dismantling German Windfarm to Expand Coalmine, Say Authorities', *Guardian*, 26 October 2022; Lawrence Richard, 'Coal Mine Demolishes Neighboring Wind Farm to Boost Country's Energy Supply, Drawing Ire of Climate Activists', *Fox Business*, 27 October 2022; Tod Gillespie and Petra Sorge, 'German Power Plants Warn of Coal Shortage Amid Low River Levels', *Bloomberg*, 2 September 2022; Henley, 'Europe's Rivers'; Henley, '"The New Normal"'.

130 Mubarak Zeb Khan, 'Agreement with Kabul Reached to Facilitate Coal Trade for Power Generation', *Dawn*, 24 July 2022; Tribal News Network, 'Energy Crisis Has Pushed Pakistan to Seek Coal Imports from Afghanistan', 20 July 2022; UK Energy Research Centre, 'The EU's Global Scramble for Gas', 12 May 2022; Kuzemko et al., 'Russia's War', 5.

131 Mariyam Suleman Anees, 'Why Do Floods Devastate Balochistan So Intensely?', *Diplomat*, 17 August 2022; Khalid al-Falih quoted in *TankTerminals.com*, 'Saudi Arabia to Install $12 Billion Aramco Oil Refinery Unit in Gwadar', 31 October 2022; Khurshid Ahmed, 'Masterplan for Pakistani Oil City, Including Aramco Refinery, to Be Ready by Year-End', *Arab News*, 2 March 2021; Zafar Bhutta, 'Riyadh Renews $10b Refinery Project', *Express Tribune*, 25 October 2022.

132 CREA and GEM, 'China Permits'; Michael Le Page, 'Heatwave in China Is the Most Severe Ever Recorded in the World', *New Scientist*, 23 August 2022 (historian: Maximiliano Herrera); Yuan Ye, 'As Drought Worsens, Guangdong to Face "Severe" Water Shortages', *Sixth Tone*, 15 February 2022. Cf. also Benjamin Parker and Chloe Cornish, 'India Boosts Coal Production to Tackle Power Crunch Amid Searing Heatwave', *Financial Times*, 4 May 2022. Not that there were no alternatives to coal for producing electricity, however: see further chapter 6 below.

133 Adam Tooze, 'Welcome to the World of the Polycrisis', *Financial Times*, 28 October 2022; Rebeca Grynspan, 'Weathering a "Perfect Storm" of Cascading Crisis', UNCTAD, 25 May 2022; Andreas Malm, *Corona, Climate, Chronic Emergency: War Communism in the Twenty-First Century* (London: Verso, 2020). For the most clarifying dissection of 'crisis' as concept and reality

to appear in several decades at the least, see Ståle Holgersen, *Against the Crisis: Economy and Ecology in a Burning World* (London: Verso, 2024).

134 Sigmund Freud, 'The Loss of Reality in Neurosis and Psychosis', in *The Standard Edition of the Complete Psychological Works of Sigmund Freud, Volume XIX* (London: Vintage, 2001 [1924]), 185.

2. When Is It Too Late?

1 Observations by Carton; UN, 'Secretary-General's Remarks to High-Level Opening of COP27', 7 November 2022.

2 At the last minute they also put the word 'unabated' before 'coal', a reference to carbon capture and storage as supposed solution to the problem of emissions from coal: see further below.

3 *Economist*, 'Goodbye 1.5°C', 3 November 2022.

4 Jeff Tollefson, 'Top Climate Scientists Are Sceptical that Nations Will Reign in Global Warming', *Nature* (2021) 599: 22–4.

5 Scientist Rebellion, 'Calling Academics of All Disciplines!', n.d.

6 Bill McGuire of Scientist Rebellion in Chelsea Harvey, 'The World Will Likely Miss 1.5 Degrees C – Why Isn't Anyone Saying So?', *Scientific American*, 11 November 2022. See further e.g. Bill McGuire, 'Why We Should Forget about the 1.5C Global Heating Target', *Guardian*, 15 September 2022; James Dyke, 'We Need to Stop Pretending We Can Limit Global Warming to 1.5°C', *Byline Times*, 6 July 2022.

7 H. Damon Matthews and Seth Wynes, 'Current Global Efforts Are Insufficient to Limit Warming to 1.5°C', *Science* (2022) 376: 1404–9 (quotations from 1406–7).

8 Bill McGuire, 'The 1.5C Climate Target Is Dead – To Prevent Total Catastrophe, COP27 Must Admit It, *Guardian*, 12 November 2022. Cf. the account of the debate in Laurie Laybourn, Henry Throp and Suzannah Sherman, '1.5°C – Dead or Alive? The Risks to Transformational Change from Reaching and Breaching the Paris Agreement Goal', Institute for Public Policy Research and Chatham House, February 2023.

9 As argued, if in an undertone, in Matthews and Wynes, 'Current Global'.

10 Birol in Fiona Harvey, 'Giving Up on 1.5 Climate Target Would be Gift to Carbon Boosters, Says IEA Head', *Guardian*, 30 November 2022. This logic was also acknowledged in e.g. James Dyke and Julia K. Steinberger, 'Climate Breakdown: Even if We Miss the 1.5°C Target We Must Still Fight to Prevent Every Single Increment of Warming', *Conversation*, 11 May 2022.

11 We Mean Business Coalition, 'COP27: Business and Civil Society Is All in for Delivery', 12 November 2022. On the Coalition, see Sarah Benabou, Nils Moussu and Birgit Müller, 'The Business Voice at COP21: The Quandaries of a Global Political Ambition', Stefan C. Aykut, Jean Foyer and Edouard Morena (eds), *Globalising the Climate: COP21 and the Climatisation of Global Debates* (Abingdon: Routledge, 2017), e.g. 60, 67.

12 Fiona Harvey, 'Boris Johnson to Attack "Corrosive Cynicism" on Net Zero at Cop27', *Guardian*, 6 November 2022; Salma El Wardany, Yousef Gamal El-Din and John Ainger, 'COP27 Latest: Sunak Says UK to Triple Funding for Adaptation', *Bloomberg*, 7 November 2022; cf. Sky News, 'COP26: "Let's Keep 1.5 Alive" Boris Johnson Tells Climate Change Summit – Video', 2 November 2021. On Johnson's career path from climate denialism to something else, see Andreas Malm and the Zetkin Collective, *White Skin, Black Fuel: On the Dangers of Fossil Fascism* (London: Verso, 2021), 69–81, 96–7, 528–9.

13 Reuters, 'G20 Agrees to Pursue Efforts to Limit Temperature Rise to 1.5C', 16 November 2022; AFP, 'Leaders Plant Mangroves on Final Day of G20', YouTube, 16 November 2022.

14 *Transnational Law and Contemporary Problems*, 'Living History Interview with Dr. Mostafa Kamal Tolba' (1992) 2: 259–71; Hamed Ead and Tarek Y. S. Kapiel, 'Mostafa Kamal Tolba, Egyptian Scientist, Environmental Expert, and the Founder of Environmental Diplomacy (1922–2016)', *Egyptian Journal of Botany* (2022) 62: 595–610.

15 UN, treatise no. 2639, 'Montreal Protocol on Substances that Deplete the Ozone Layer (with annex). Concluded at Montreal on 16 September 1987' (quotations from 30).

16 Shardul Agrawala, 'Context and Early Origins of the Intergovernmental Panel on Climate Change', *Climatic Change* (1998) 39: 605–20. Tolba used Montreal as the model for a climate protocol in his op-ed 'For a World Campaign to Limit Climate Change', *International Herald Tribune*, 15 March 1988.

17 Mostafa Tolba, 'Highways to Nowhere: Statement to Berlin Environment Forum, November 1982', in *Sustainable Development: Constraints and Opportunities* (London: Butterworth, 1987), 30; for his concerns about climate change, also from 1982, see e.g. 4, 47.

18 Mostafa Tolba, 'Climatic Effects of Carbon Dioxide and other Trace Gases: Statement to the Joint WMO/UNEP/ICSU Assessment of the Role of Carbon Dioxide and other Radiatively Active Constituents in Climate Variations and their Associated Impact', in *Sustainable Development*, 196. On this workshop, see Agrawala,

'Context and Early', 607–8; Mostafa K. Tolba, *Global Environmental Diplomacy: Negotiating Environmental Agreements for the World, 1973–1992* (Cambridge, MA: MIT Press, 1998), 90.

19 Tolba, 'Climatic Effects', 196.

20 Agrawala, 'Context and Early', 609.

21 F. R. Rijsberman and R. J. Swart (eds), *Targets and Indicators of Climatic Change* (Stockholm: The Stockholm Environment Institute, 1990), 9.

22 Ibid., viii–ix, 24. The later figures: NASA Earth Observatory, 'Tracking 30 Years of Sea Level Rise', 11 August 2022; National Oceanic and Atmospheric Administration, 'Carbon Dioxide Now More than 50% Higher than Pre-Industrial Levels', 3 June 2022; Masson-Delmotte et al., *Global Warming*, 51. For more on the pioneering target practice of the Advisory Group at Villach and Bellagio, see Shardul Agrawala, 'Early Science-Policy Interactions in Climate Change: Lessons from the Advisory Group on Greenhouse Gases', *Global Environmental Change* (1999) 9: 158, 162; Agrawala, 'Context and Early', 610; Rik Leemans and Pier Vellinga, 'The Scientific Motivation of the Internationally Agreed "Well Below 2°C" Climate Protection Target: A Historical Perspective', *Current Opinion in Environmental Sustainability* (2017) 26–7: 137–8; Piero Morseletto, Frank Biermann and Philipp Pattberg, 'Governing by Targets: Reductio ad Unum and Evolution of the Two-Degree Target', *International Environmental Agreements* (2017) 17: 658.

23 Rijsberman and Swart, *Targets and Indicators*, viii, x.

24 Agrawala, 'Context and Early', 612–14.

25 Rijsberman and Swart, *Targets and Indicators*, 145, 148.

26 See Agrawala, "Context and Early'; 'Early Science-Policy'; and cf. Samuel Randalls, 'History of the 2°C Climate Target', *WIREs Climate Change* (2010) 1: 601.

27 UN, 'United Nations Framework Convention on Climate Change' (1992), 9.

28 But the spirit of the Montreal Protocol is evident in the text, which often has the character of an appendix to or copy of that agreement: see e.g. 4, 10–13.

29 Morsoletto et al., 'Governing by Targets', 660; Randalls, 'History of the 2°C', 598–602; Leemans and Vellinga, 'The Scientific Motivation', 135; *Carbon Brief*, 'Two Degrees: The History of Climate Change's Speed Limit', 8 December 2014.

30 George Monbiot, *Heat: How to Stop the Planet Burning* (London: Penguin, 2006), 15, 17. 'Two degrees is important because it is the point at which some of the larger human impacts and critical positive feedbacks are expected to begin.' Ibid., 15. On the movement

supporting 2°C, see Randalls, 'History of the 2°C', 602, 605.

31 UN, 'Draft Decision -/CP.15. Proposal by the President: Copenhagen Accord' (2009), 1–2; Randalls, 'History of the 2°C', 598,
602; Morselotto et al., 'Governing by Targets', 662–4.

32 E.g. Steinar Andresen, 'International Climate Negotiations: Top-
Down, Bottom-Up or a Combination of Both?', *International
Spectator* (2015) 50: 20.

33 The baseline year for these cuts was 1990, i.e. countries needed
to reduce emissions to a level 5 per cent lower than in that year
and do so by 2012.

34 The formula appears three times: UN, 'United Nations', 2, 9, 10.

35 Cf. Joanna Depledge, 'The "Top-Down" Kyoto Protocol? Explor-
ing Caricature and Misrepresentation in the Literature on Global
Climate Change Governance', *International Environmental
Agreements* (2022) 22: 679.

36 Aubrey Meyer, *Contraction and Convergence: The Global Solution to Climate Change* (Totnes: Green Books, 2000); and see
further e.g. Michel den Elzen, Paul Lucas and Detlef van Vuuren,
'Abatement Costs of Post-Kyoto Climate Regimes', *Energy Policy*
(2005) 33: 2138–51; Renaud Gignac and H. Damon Matthews,
'Allocating a 2°C Cumulative Carbon Budget to Countries',
Environmental Research Letters (2015) 10: 1–9. A similar principle was laid down in the Montreal Protocol, article 5 of which
stipulated that developing countries should be allowed to delay
compliance by ten years. UN, 'Montreal Protocol', 34.

37 Tom Athanasiou and Paul Bear, *Dead Heat: Global Justice and
Global Warming* (New York: Seven Stories Press, 2002); Paul Baer,
Tom Athanasiou and Sivan Kartha, *The Right to Development
in a Climate Constrained World: The Greenhouse Development
Rights Framework* (Berlin: Heinrich Böll Foundation, 2007). On
the sub-national scale, a buzzword in those years was 'carbon
rationing', the idea that everyone would get the same per capita
ration for emissions and use it like a credit card swiped for
every transaction, with those consuming more than their per-
sonal allowance having to pay those who consumed less for
additional points; in 2006, David Miliband proposed such a
scheme – which, two decades later, seems fantastically radical –
in the UK. David Adam and David Batty, 'Miliband Unveils
Carbon Swipe-Card Plan', *Guardian*, 19 July 2006. Cf. e.g.
Monbiot, *Heat*, 43–9.

38 David Ciplet, J. Timmons Roberts and Mizan R. Khan, *Power in
a Warming World: The New Global Politics of Climate Change
and the Remaking of Environmental Inequality* (Cambridge, MA:
MIT Press, 2015), 65–74; David Ciplet and J. Timmons Roberts,

'Climate Change and the Transition to Neoliberal Environmental Governance', *Global Environmental Change* (2017) 46: 152.

39 Cf. Hélène Guillemot, 'The Necessary and Inaccessible 1.5°C Objective: A Turning Point in the Relations between Climate Science and Politics?', in Aykut et al., *Globalising the Climate*, 42.

40 Tolba, 'For a World'.

41 Atiq Rahman quoted in Athanasiou and Baer, *Dead Heat*, 23. Sea level rise also dominated much of the discussions about climate change and the Third World at the alternative summit in Berlin: observations by Malm.

42 UN, 'Draft Decision', 3; Guillemot, 'The Necessary and Inaccessible', 43; Lisette van Beek, Jeroen Oomen, Maarten Hajer et al., Jasmine E. Livingston and Markku Rummukainen, 'Taking Science by Surprise: The Knowledge Politics of the IPCC Special Report on 1.5 Degrees', *Environmental Science and Policy* (2020) 112: 11; Lisette van Beek, Jeroen Oomen, Martin Hajer et al., 'Navigating the Political: An Analysis of Political Calibration of Integrated Assessment Modelling in Light of the 1.5°C Goal', *Environmental Science and Policy* (2022) 133: 194. On the negotiating skills and strategies of the Alliance of the Small Island States, or AOSIS, see Timothée Ourbak and Alexandre K. Magnan, 'The Paris Agreement and Climate Change Negotiations: Small Islands, Big Players', *Regional Environmental Change* (2018) 18: 2201–7.

43 Cf. Guillemot, 'The Necessary and Inaccessible', 43.

44 Pier Vellinga and Rob Swart, 'The Greenhouse Marathon: A Proposal for a Global Strategy', *Climatic Change* (1991) 18: vii, ix. Crossed in 2015: *Climate Analytics*, 'Global Warming Reaches 1°C above Preindustrial, Warmest in More Than 11,000 Years', 2015.

45 Leemans and Vellinga, 'The Scientific Motivation', 139.

46 David Adam, 'World Will Not Meet 2C Warming Target, Climate Change Experts Agree', *Guardian*, 14 April 2009. For a critique of 2°C as too lenient, see James E. Hansen, 'A Slippery Slope: How Much Global Warming Constitutes "Dangerous Anthropogenic Interference"?', *Climatic Change* (2005) 68: 269–79.

47 David G. Victor and Charles F. Kennel, 'Ditch the 2°C Warming Goal', *Nature* (2014) 514: 31. 'The 2°C goal . . . has allowed some governments to pretend that they are taking serious action to mitigate global warming, when in reality they have achieved almost nothing.' Ibid., 30. For the discussion about the feasibility of 2°C around this time, see further the wonderfully titled Céline Guivarch and Stéphane Hallegatte, '2C or Not 2C?', *Global Environmental Change* (2013) 23: 179–92.

48 *Economist*, 'Facing the Consequences', 25 November 2010.

'Wishful dream' was a phrase taken from Bob Watson and evidently accepted by the anonymous pundits of the *Economist*.

49 Oliver Geden and Silke Beck, 'Renegotiating the Climate Stabilization Target', *Nature Climate Change* (2014) 4: 747–8.

50 At this point, Saudi Arabia and Kuwait objected most loudly to the idea of 1.5°C: Livingston and Rummukainen, 'Taking Science', 11.

51 Ciplet et al., *Power in a Warming*, 87–90.

52 Beek et al., 'Taking Science', 195; Ourbak and Magnan, 'The Paris Agreement', 2203. In fact, the small island states sought a target *below* 1.5°C: see e.g. Ian Fry, 'The Paris Agreement: An Insider's Perspective', *Environmental Law and Policy* (2016) 46: 105–6. There was, however, one small island state that resisted 1.5°C as too rigorous, namely Singapore. Ibid., 106.

53 Beatrice Cointe and Hélène Guillemot, 'A History of the 1.5°C Target', *WIRES Climate Change* (2023), online first; Guillemot, 'The Necessary and Inaccessible', 45–6; and cf. Leon Sealey-Huggins, '"1.5°C to Stay Alive": Climate Change, Imperialism and Justice for the Caribbean', *Third World Quarterly* (2017) 38: 2444–63.

54 Fry, 'The Paris Agreement', 106.

55 Radoslav S. Dimitrov, 'The Paris Agreement on Climate Change: Behind Closed Doors', *Global Environmental Politics* (2016) 16: 4.

56 Cf. ibid.; Oliver Geden, 'The Paris Agreement and the Inherent Inconsistency of Climate Policymaking', *WIREs Climate Change* (2016) 7: 794.

57 UN, 'Paris Agreement' (2015), 3.

58 Fry, 'The Paris Agreement', 107.

59 Ciplet and Roberts, 'Climate Change', 153.

60 Cf. Aslak Brun, 'Conference Diplomacy: The Making of the Paris Agreement', *Politics and Governance* (2016) 4: 115.

61 The formulations in the text are exceedingly fluffy. Clause 9 of article 4 stipulates that 'each party shall communicate a nationally determined contribution every five years' – the so-called stock-take – while clause 11 specifies (if that is the right word) that 'a Party may at any time adjust its existing nationally determined contribution with a view to enhancing its level of ambition' – 'may' here being the softest verb imaginable. UN, 'Paris Agreement', 5.

62 Stefan. C. Aykut, 'Governing through Verbs: The Practice of Negotiating and the Making of a New Mode of Governance', in Aykut et al., *Globalising the Climate*, 31–2; Dimitrov 'The Paris Agreement', 3; Fry, 'The Paris Agreement', 107.

63 Dimitrov 'The Paris Agreement', 2; cf. e.g. Ciplet and Roberts, 'Climate Change', 148, 154; Geden, 'The Paris Agreement', 792.

64 William Hare, Claire Stockwell, Christian Flachsland and Sebastian Oberthür, 'The Architecture of the Global Climate Regime: A Top-Down Perspective', *Climate Policy* (2010) 10: 604–6, 609. The Kyoto Protocol, for all its flaws – and they were many – had a compliance rate of 100 per cent. Igor Shishlov, Romain Morel and Valentin Bellassen, 'Compliance of the Parties to the Kyoto Protocol in the First Commitment Period', *Climate Policy* (2016) 6: 768–82. This study purports to demonstrate that the 'hot air' – the collapse in the emissions of the Eastern bloc relative to 1990, induced by the transition to capitalism and with no relation to climate policy – was redundant to this compliance rate, i.e. the 5 per cent emissions cuts would have been realised even without the hot air.

65 Ciplet and Roberts, 'Climate Change', 152. See further 153–4; and cf. Ciplet et al., *Power in a Warming*, 2, 94; Dimitrov, 'The Paris Agreement', 8. This was the principle that replaced the Kyoto division between Annex B countries with mitigation duties and everyone else.

66 Dimitrov, 'The Paris Agreement', 8; cf. Geden, 'The Paris Agreement', 792.

67 Simon Donner, 'Why We Need the Next-to-Impossible 1.5°C Temperature Target', *Guardian*, 30 December 2015. Cf. Guillemot, 'The Necessary and Inaccessible', 39; Livingston and Rummukainen, 'Taking Science', 12–13.

68 Dimitrov, 'The Paris Agreement', 4.

69 Livingston and Rummukainen, 'Taking Science', 14; Daniel Mitchell, Rachel James, Piers M. Forster et al., 'Realizing the Impacts of a 1.5°C Warmer World', *Nature Climate Change* (2016) 6: 736.

70 Shinichiro Asayama, Rob Bellamy, Oliver Geden et al., 'Why Setting a Climate Deadline is Dangerous', *Nature Climate Change* (2019) 9: 571. This prim defence of the purity of science as against partisan contamination was earlier stated by Mike Hulme, '1.5°C and Climate Research after the Paris Agreement', *Nature Climate Change* (2016) 6: 222–4; for a rejoinder, see Mitchell et al., 'Realizing the Impacts'. On scandalised scientists, see further Guillemot, 'The Necessary and Inaccessible', 51; Livingston and Rummukainen, 'Taking Science', 12–15.

71 Cf. Cointe and Guillemot, 'A History', 4–6.

72 Masson-Delmotte et al., *Global Warming*, 175–312.

73 Colin J. Carlson, Gregory F. Albery, Cory Morow et al., 'Climate Change Increases Cross-Species Viral Transmission Risk', *Nature* (2022) 607: 555–62.

74 Masson-Delmotte et al., *Global Warming*, 179, 218, 256.

75 Ibid., 230, 254.

76 Ibid., 251.

77 Mika Rantanen, Alexey Yu. Karpechko, Antti Lipponen et al., 'The Arctic Has Warmed Nearly Four Times Faster than the Globe Since 1979', *Communications Earth and Environment* (2022) 3: 1–9; and see further Uta Kloenne, Alexander Nauels, Pam Pearson et al., 'Only Halving Emissions by 2030 Can Minimize Risk of Crossing Cryosphere Thresholds', *Nature Climate Change* (2023) 13: 9–11.

78 Masson-Delmotte et al., *Global Warming*, 206–7, 260; Matthias Mengel, Alexander Nauels, Joeri Rogelj and Carl-Friedrich Schleussner, 'Committed Sea-Level Rise under the Paris Agreement and the Legacy of Delayed Mitigation Action', *Nature Communications* (2018) 9: 1–10; Hans Joachim Schellnhuber, Stefan Rahmstorf and Ricarda Winkelmann, 'Why the Right Target Was Agreed in Paris', *Nature Climate Change* (2016) 6: 651; Kloenne et al., 'Only Halving', 9. The underestimations will be further dealt with in the second instalment of this study.

79 Masson-Delmotte et al., *Global Warming*, 191. The study: Alessandro Dosio, Lorenzo Mentaschi, Erich M. Fischer and Klaus Wyser, 'Extreme Heat Waves under 1.5°C and 2°C Global Warming', *Environmental Research Letters* (2018) 13: 1–10. Cf. e.g. Shingirai Nangombe, Tianjun Zhou, Wenxia Zhang et al., 'Record-Breaking Climate Extremes in Africa under Stabilized 1.5°C and 2°C Global Warming Scenarios', *Nature Climate Change* (2018) 8: 375–80.

80 Masson-Delmotte et al., *Global Warming*, 242, 260, 263.

81 Cf. Claire Fyson, Fahad Saeed, Robert Brecha et al., 'Stayin' Alive: Heatwave Makes Searing Case for 1.5°C', *Climate Analytics*, 15 August 2018; Jeff Tollefson, 'Clock Ticking on Climate Action', *Nature* (2018) 562: 172–3. Cf. e.g. two studies drawing a similar line for Europe: Andrew D. King, Markus G. Donat, Sophie C. Lewis et al., 'Reduced Heat Exposure by Limiting Global Warming to 1.5°C', *Nature Climate Change* (2018) 8: 549–51; Daniel Mitchell, Clare Heaviside, Nathalie Schaller et al., 'Extreme Heat-Related Mortality Avoided under Paris Agreement Goals', *Nature Climate Change* (2018) 8: 551–3.

82 Steven C. Sherwood and Matthew Huber, 'An Adaptability Limit to Climate Change Due to Heat Stress', *PNAS* (2010) 25: 9552–5 (quotations from 9552).

83 Jeremy S. Pal and Elfatih A. B. Eltahi, 'Future Temperatures in Southwest Asia Projected to Exceed a Threshold for Human Adaptability', *Nature Climate Change* (2016) 6: 197–9. One study gave a similar forecast for the Indus and Ganges basins:

Eun-Soon Im, Jeremy S. Pal and Elfatih A. B. Eltahir, 'Deadly Heat Waves Projected in the Densely Populated Agricultural Regions of South Asia', *Science Advances* (2017) 3: 1–7. And for the northern China plain: Suchul Kung and Elfatih A. B. Eltahir, 'North China Plain Threatened by Deadly Heatwaves Due to Climate Change and Irrigation', *Nature Communications* (2018) 8: 1–9. In one business-as-usual scenario, more than 3 billion people will experience mean annual temperatures in 2070 that only existed in patches of the Sahara in 2020. Chi Xu, Timothy A. Kohler, Timothy M. Lenton et al., 'Future of the Human Climate Niche', *PNAS* (2020) 117: 11350–55.

84 Colin Raymond, Tom Matthews and Radley M. Horton, 'The Emergence of Heat and Humidity Too Severe for Human Tolerance', *Science Advances* (2020) 6: 1–8. Hot days, it might be noted, warm somewhat quicker than the average: see Michael P. Byrne, 'Amplified Warming of Extreme Temperatures over Tropical Land', *Nature Geoscience* (2021) 14: 837–41.

85 Yi Zhang, Isaac Held and Stephan Fueglistaler, 'Projections of Tropical Heat Stress Constrained by Atmospheric Dynamics', *Nature Geoscience* (2021) 14: 133–7 (quotation from 133).

86 Oliver Milman, 'Global Heating Pushes Tropical Regions towards Limits of Human Livability', *Guardian*, 8 March 2021.

87 Sault, 'The Beginning and the End', *Untitled (Rise)*, 2020.

88 Xiang-Sheng Wang, Lei He, Xiao-Hu Ma et al., 'The Emergence of Prolonged Deadly Humid Heatwaves', *International Journal of Climatology* (2022) 42: 8607–18. See further Colin Raymond, Tom Matthews, Radley M. Horton et al., 'On the Controlling Factors for Globally Extreme Humid Heat', *Geophysical Research Letters* (2021) 48: 1–11; Luke J. Harrington, 'Temperature Emergence at Decision-Relevant Scales', *Environmental Research Letters* (2021) 16: 1–9; Nicolas Freychet, Gabriele C. Hegerl, Natalie S. Lord et al., 'Robust Increase in Population Exposure to Heat Stress with Increasing Global Warming', *Environmental Research Letters* (2022) 17: 1–10.

89 Fahad Saeed, Carl-Friedrich Schleussner and Moetasim Ashfaq, 'Deadly Heat Stress to Become Commonplace Across South Asia Already at 1.5°C of Global Warming', *Geophysical Research Letters* (2021) 48: 1–11. Another study indicated that the lethal threshold is actually lower than a wet-bulb temperature of 35°C but found that massive dangers kick in after average warming of 2°C: Daniel J. Vecellio, Qinqin Kong, W. Larry Kenney and Matthew Huber, 'Greatly Enhanced Risk to Humans as a Consequence of Empirically Determined Lower Moist Heat Stress Tolerance', *PNAS* (2023) 1–9.

90 Emma E. Ramsay, Genie M. Fleming, Peter A. Faber et al., 'Chronic Heat Stress in Tropical Urban Informal Settlements', *iScience* (2021) 24: 1–12.

91 David Armstrong McKay, Arie Staal, Jesse F. Abrams et al., 'Exceeding 1.5°C Global Warming Could Trigger Multiple Climate Tipping Points', *Science* (2022) 377: 1–11.

92 Karl Marx, 'Critique of Hegel's Philosophy of Right', in *Early Writings* (London: Penguin, 1975), 256 (emphases in original).

93 Cf. e.g. Cointe and Guillemot, 'A History', 5; Van Beek et al., 'Navigating the Political', 198; Asayama et al, 'Why Setting', 570.

94 Thomas Meaney, 'Fortunes of the Green New Deal', *New Left Review* (2022) 138: 82.

95 Maxwell Boykoff and Olivia Pearman, 'Now or Never: How Media Coverage of the IPCC Special Report on 1.5°C Shaped Climate-Action Deadlines', *One Earth* (2019) 1: 285.

96 Masson-Delmotte et al., *Global Warming*, 31. Cf. Karen O'Brien, 'Is the 1.5°C Target Possible? Exploring the Three Spheres of Transformation', *Current Opinion in Environmental Sustainability* (2018) 31: 153–60.

97 Or, in the more technical terms of the IPCC: 'if emission reductions do not begin until temperatures are close to the proposed limit, pathways remaining below 1.5°C necessarily involve much faster rates of net CO_2 emissions reductions'. Masson-Delmotte et al., *Global Warming*, 61. For other texts from the year of the Special Report with useful perspectives on this problem, see e.g. Henry Shue, 'Mitigation Gambles: Uncertainty, Urgency and the Last Gamble Possible', *Philosophical Transactions of the Royal Society A* (2018) 376: 1–11; Katarzyna B. Tokarska, 'Countdown to 1.5°C Warming', *Nature Geoscience* (2018) 546–9; H. Damon Matthews, Kirsten Zickfeld, Reto Knutti and Myles R. Allen, 'Focus on Cumulative Emissions, Global Carbon Budgets and the Implications for Climate Mitigation Targets', *Environmental Research Letters* (2018) 13: 1–8.

98 Note that the figure used for emissions growth in 2021 in this graph is the lower estimate: see footnote 5 ch. 1 above.

99 The phrase being 'late capitalism born out of the historical delay of world revolution': Ernest Mandel, *Long Waves of Capitalist Development: A Marxist Interpretation* (London: Verso, 1995), 82.

100 Kevin Anderson, 'Wrong Tool for the Job', *Nature* (2019) 573: 348. This Kevin Anderson has no known relation to the Marxist-humanist Kevin B. Anderson.

101 UN, 'Inadequate Progress on Climate Action Makes Rapid Transformation of Societies Only Option – UNEP', 27 October 2022.

The official term was 'wide-ranging, large-scale, rapid and systematic transformation': UNEP, *Emissions Gap Report 2022: The Closing Window – Climate Crisis Calls for Rapid Transformation of Societies* (Nairobi: UNEP, 2022), xxii, 38. See further e.g. Laybourn et al., *1.5 °C – Dead*, 7–9.

102 UNEP, *Emissions Gap*, 35–6.

103 UNEP quoted in Damian Carrington, 'Climate Crisis: UN Finds "No Credible Pathway to 1.5°C in Place"', *Guardian*, 27 October 2022.

104 Harvey, 'The World Will'.

105 *Economist*, 'The World Is Going to Miss the Totemic 1.5°C Climate Target', 5 November 2022.

106 IPCC, 'Summary for', 15. Indeed, this became a truism as soon as research on 1.5°C took off. Cf. e.g. Joeri Rogelj, Alexander Popp, Katherine V. Calvin et al., 'Scenarios Towards Limiting Global Mean Temperature Increase below 1.5°C', *Nature Climate Change* (2018) 8: 327.

107 Masson-Delmotte et al., *Global Warming*, 317. As if to soften the thrust of these words, the Panel inserted the bracket '(including inclusive markets)' after the word 'institutions'.

108 Ibid., 148, 279, 153. The term 'a marked shift in investment patterns' also appears on 33, 95; and cf. 155, 158.

109 Priyadarshi R. Shukla, Jim Skea, Andy Reisinger et al., *Climate Change 2022: Mitigation of Climate Change. Working Group III Contribution to the Sixth Assessment Report of the Intergovernmental Panel on Climate Change* (IPCC, 2022), 169; cf. e.g. 176.

110 Tiffany H. Morrison, W. Neil Adger, Arun Agrawala et al., 'Radical Interventions for Climate-Impacted Systems', *Nature Climate Change* (2022) 12: 1100–6. Cf. e.g. Thomas Wiedmann, Manfred Lenzen, Lorenz T. Keyßer and Julia K. Steinberger, 'Scientists' Warning on Affluence', *Nature Communications* (2020) 11: 1–10.

111 Cf. SEI et al., 'The Production Gap', 9–11.

112 Kühne et al., '"Carbon Bombs"', 5–6; cf. Trout et al. 'Existing Fossil', 9.

113 Leah Temper, Sofia Avila, Daniela Del Bene et al., 'Movements Shaping Climate Futures: A Systematic Mapping of Protests against Fossil Fuels and Low-Carbon Energy Projects', *Environmental Research Letters* (2020) 15: 1–23.

114 Chloé Farand, 'Philippines Declares Moratorium on New Coal Power Plants', *Climate Home News*, 28 October 2018; Philippine Movement for Climate Justice, 'Movement Stalls 8 Fossil Gas Projects', 26 August 2022. For the context of the

coal moratorium, see further Xu Chen and Denise L. Mauzer-
all, 'The Expanding Coal Fleet in Southeast Asia: Implications
for Future CO_2 Emissions and Electricity Generation', *Earth's
Future* (2021) 9: 1–17.

115 Climáximo, 'Victory: Portugal Free of Fossil Fuel Extraction
Projects as Company Abandons the Last Contract', 4 September
2000.

116 Tom Sims, 'Deutsche Bank Not Financing Controversial African
Oil Pipeline, Source Says', Reuters, 16 May 2022; Inclusive
Development International, 'World Bank Client Backs Out of
EACOP after Risk Assessment', 21 November 2022; 350.org,
'Global Protests against Multinational Banks Funding East
African Crude Oil Pipeline', 22 February 2023; #StopEACOP,
'Don't Bank on EACOP: Who's Backing the Pipeline and Who's
Ruled it Out?', November 2022. The three banks: Standard
Bank, Sumitomo Mitsui Bank and ICBC.

117 Climate Home News, 'DR Congo Delays Rainforest Oil Auctions', 31 January 2023.

118 Augé, 'The Trans-Saharan', 7–14.

119 See e.g. *China-Lusophone Brief*, 'Insurgency Imperils Exxon/
CNPC Investment in Mozambique', 30 March 2021; Matthew
Hill and Borges Nhamirre, 'Gas Bonanza in Limbo as Mozam-
bique Insurgency Simmers', *Bloomberg*, 3 April 2022; Matthew
Hill, 'Exxon Weighs Resuming Mozambican LNG Project with
Bigger Output', *Bloomberg*, 17 March 2023; Henrieke Butijn,
'Who Dares to Finance Eni and Exxon's Dangerous Rovuma
Gas Plans in Mozambique?', *BankTrack*, 20 March 2023. On
the social and ecological depredations caused by the arrival of
the gas industry, see JA!/Friends of the Earth Mozambique, 'The
Impacts of the LNG Industry in Cabo Delgado, Mozambique',
March 2020. On the discontent fuelling the insurrection, see
e.g. Emilinah Namaganda, Kei Otsuki and Griet Steel, 'Learning
from Africana Critical Theory: A Historicized Contextualization
of the Impacts of Mozambique's Natural Gas Project', *Extractive
Industries and Society* (2022) 10: 1–8; Martin Ewi, Liesl Louw-
Vaudran, Willem Els et al., 'Violent Extremism in Mozambique:
Drivers and Links to Transnational Organised Crime', Institute
for Security Studies, August 2022, 19–21. The US and its allies
sought to frame the conflict as one with Daesh. A *CNN* headline said it all: David McKenzie and Ghazi Balkiz, 'ISIS-Linked
Militants Are Threatening Huge Natural Gas Reserves the World
Needs Badly Right Now', *CNN*, 10 October 2022. An expert
writing for the *BBC* nuanced the picture of the ideological origins
of the uprising: 'protests were growing that the profits were all

going to an elite in the ruling party, Frelimo, and that few local jobs were being created. The coastal zone is historically Muslim. Local fundamentalist preachers said Sharia, or Islamic law, would bring equality and a fair sharing of wealth – effectively, a socialist message.' Joseph Hanlon, 'Mozambique Insurgency: Why 24 Countries Have Sent Troops', *BBC*, 23 May 2022. For the complicated Islamist factor, see Eric Morier-Genoud, 'The Jihadi Insurgency in Mozambique: Origins, Nature and Beginning', *Journal of Eastern African Studies* (2020) 14: 396–412. With military support from the EU, South Africa, Rwanda and others, the central state sought to re-establish order for foreign companies in the gas fields. As of early 2023, the northern Cabo Delgado province was under military occupation, primarily serving Total, which had become the de facto executive power, directing economic, humanitarian, legal and security operations. 'In the face of the weakening of the State, TotalEnergies comes to be a post-colonial version of the Niassa Company', the Portuguese company colonising these lands in the late nineteenth century. João Feijó, 'Return of the Populations and Reconstruction of the Northeast of Cabo Delgado: From the Weakening of the State to the Emergence of a Totaland', Observatório do Meio Rural, March 2023, 14. Such were the forms of political constitution of fossil fuel property in Mozambique in the early 2020s.

120 Cf. Jessica Jewell and Aleh Cherp, 'On the Political Feasibility of Climate Change Mitigation Pathways: Is It Too Late to Keep Warming below 1.5°C?', *WIREs Climate Change* (2022) 11: 2–3.

121 Cf. ibid., 8.

122 Stuart Jenkins, Chris Smith, Myles Allen and Roy Grainger, 'Tonga Eruption Increases Chance of Temporary Surface Temperature Anomaly Above 1.5°C', *Nature Climate Change* (2023) 13: 127–9; Damian Carrington, 'Warning of Unprecedented Heatwaves as El Niño Set to Return in 2023', *Guardian*, 16 January 2023.

123 Leon Trotsky, 'Before the Decision', in *The Struggle against Fascism in Germany* (New York: Pathfinder, 1971), 342–3.

124 Ibid., 347 (emphasis in original).

125 Cf. eg. Marotzke et al., 'How Close', 148; Laybourn et al., '1.5°C – Dead', 10.

126 McKay et al., 'Exceeding 1.5°C', 8. This possibility was already recognised by Schellnhuber et al., 'Why the Right', 650.

127 Tolba, 'For a World'. Although 'delay its onset' is a poor choice of words.

128 Capitalism as the quake behind the ecological tsunami: Steven Stoll, 'A Metabolism of Society: Capitalism for Environmental Historians', in Andrew G. Isenberg (ed.), *The Oxford Handbook of Environmental History* (Oxford: Oxford University Press, 2014), 369.

129 Timeless while predicated on the shortness of time: another symptom of the strange temporality of the climate crisis.

3. The Rise of Overshoot Ideology

1 For such a critique – of a soft-denialist inclination, characteristic of this scholar of late Kyoto prominence (and see further below) – see Richard S. J. Tol, 'Europe's Long-Term Climate Target: A Critical Evaluation', *Energy Policy* (2007) 35: 424–32.

2 One being Michel den Elzen and Malte Meinshausen, 'Meeting the EU 2°C Climate Target: Global and Regional Emission Implications', *Climate Policy* (2006) 6: 545–64.

3 'The role of the computer in the acceleration of scientific activity, the exponential growth rate of this activity, and its increasing socialization and capitalist organization are no less obvious' than e.g. the role of classical physics in the first stages of steam-powered industrial capitalism: Ernest Mandel, *Late Capitalism* (London: Verso, 1978 [1972]), 250.

4 In alphabetic (what else) order: Atmospheric General Circulation Model; Common but Differentiated Responsibilities and Respective Capabilities; Carbon Capture Utilisation and Storage; Clean Development Mechanism; Carbon Dioxide Removal; Emissions Trading System; Green Climate Fund; Greenhouse Gases; Global Warming Potential; Land Use, Land-Use Change and Forestry; Nationally Determined Contributions; Negative Emissions Technologies; Reducing Emissions from Deforestation and forest Degradation, and the role of conservation, sustainable management of forests and enhancement of forest carbon stocks in developing countries; Stratospheric Aerosol Injection; Social Cost of Carbon; Sea Level Rise; Special Report on Emissions Scenarios. As anyone with some familiarity with the institutions of COPs and the IPCC will know, this shortlist is indeed extremely short. It cannot be ruled out that these acronyms might serve as the raw material for some forms of Dadaist poetry.

5 Thus there was no sign of justice in the sphere of climate, but there was a supremely ugly acronym for referring to it: CBDR-RC.

6 For a primer, see Robert McSweeney and Zeke Hausfather, 'How Do Climate Models Work?', *Carbon Brief*, 15 January 2018; and

cf. Simon Evans and Zeke Hausfather, 'How "Integrated Assessment Models" Are Used to Study Climate Change', *Carbon Brief*, 2 October 2018.

7 Robert S. Pindyck, 'Climate Change Policy: What Do the Models Tell Us?', *Journal of Economic Literature* (2013) 51: 861–2; Béatrice Cointe, Christophe Cassen and Alain Nadaï, 'Organising Policy-Relevant Knowledge for Climate Action: Integrated Assessment Modelling, the IPCC, and the Emergence of a Collective Expertise on Socioeconomic Emission Scenarios', *Science and Technology Studies* (2019) 32: 37, 45; Lisette van Beek, Maarten Hajer, Peter Pelzer et al., 'Anticipating Futures through Models: The Rise of Integrated Assessment Modelling in the Climate Science-Policy Interface since 1970', *Global Environmental Change* (2020) 65: 2, 6. Understanding the distinctive features and problems of IAMs thus presupposes an analytical distinction between the natural and the social: cf. Andreas Malm, *The Progress of This Storm: Nature and Society in a Warming World* (London: Verso, 2018).

8 As critically described in e.g. Franck Ackerman, Stephen J. DeCanio, Richard I. Keppo and Kristen Sheeran, 'Limitations of Integrated Assessment Models of Climate Change', *Climatic Change* (2009) 95: 301, 308; Saskia Ellenbeck and Johan Lilliestam, 'How Modelers Construct Energy Costs: Discursive Elements in Energy System and Integrated Assessment Models', *Energy Research and Social Science* (2019) 47: 73–4; I. Keppo, I. Butnar, N. Bauer et al., 'Exploring the Possibility Space: Taking Stock of the Diverse Capabilities and Gaps in Integrated Assessment Models', *Environmental Research Letters* (2021) 16: 2; Salvi Asefi-Najafabady, Laura Villegas-Ortiz and Jamie Morgan, 'The Failure of Integrated Assessment Models as a Response to "Climate Emergency" and Ecological Breakdown: The Emperor Has No Clothes', *Globalizations* (2021) 18: 1180–3; Natalia Rubiano Rivadineira and Wim Carton, '(In)justice in Modelled Climate Futures: A Review of Integrated Assessment Modelling Critiques through a Justice Lens', *Energy Research and Social Science* (2022) 92: 3.

9 Ottomar Edenhofer, Ramón Pichs-Madruga, Youba Sokona et al. (eds), *Climate Change 2014: Mitigation of Climate Change. Working Group III Contribution to the Fifth Assessment Report of the Intergovernmental Panel on Climate Change* (Cambridge: Cambridge University Press, 2014), 422.

10 As pointed out by Hector Pollitt in Evans and Hausfather, 'How "Integrated"'.

11 Erich W. Schienke, Seth D. Baum, Nancy Tuana et al., 'Intrinsic

Ethics Regarding Integrated Assessment Models for Climate Management', *Science and Engineering Ethics* (2011) 17: 511–13; Mark Workman, Kate Dooley, Guy Lomax et al., 'Decision Making in Contexts of Deep Uncertainty: An Alternative Approach for Long-Term Climate Policy', *Environmental Science and Policy* (2020) 103: 81–2; Johannes Emmerling and Massimo Tavoni, 'Representing Inequalities in Integrated Assessment Modeling of Climate Change', *One Earth* (2021) 4: 177–80; Rivadineira and Carton, 'In(justice) in Modelled', 4; Asefi-Najafabady et al., 'The Failure of Integrated', 1182; Keppo et al., 'Exploring the Possibility', 5–6. Some IAMs have made feeble attempts to model inequalities between income groups within nations: see Emmerling and Tavoni, 'Representing Inequalities', 179. For a superb critique of the idea of a representative agent and neoclassical assumptions about hyper-rationality in general, see Anwar Shaikh, *Capitalism: Competition, Conflict, Crises* (Oxford: Oxford University Press, 2016), 78–83, 110, 114–15.

12 William D. Nordhaus, 'To Slow or Not to Slow: The Economics of the Greenhouse Effect', *Economic Journal* (1991) 101: 932. On Nordhaus as father of IAMs, see e.g. Pindyck, 'Climate Change', 861; Van Beek et al., 'Anticipating Futures', 4.

13 See further Ackerman et al., 'Limitations of Integrated', 298, 304–5; Rivadineira and Carton, '(In)justice in Modelled', 4.

14 Nordhaus, 'To Slow', 923. For a brilliant critique of the damage function in IAMs, see Adrienne Buller, *The Value of a Whale: On the Illusions of Green Capitalism* (Manchester: Manchester University Press, 2022), 48–50; and on the follies of Nordhaus, 89, 33, 36–42.

15 Nordhaus, 'To Slow', 930.

16 William D. Nordhaus, 'A Review of the *Stern Review on the Economics of Climate Change*', *Journal of Economic Literature* (2007) 45: 698.

17 Edenhofer et al., *Climate Change*, 421–2, 433; Richard A. Rosen, 'Critical Review of: "Making or Breaking Climate Targets: The AMPERE Study on Staged Accession Scenarios for Climate Policy"', *Technological Forecasting and Social Change* (2015) 96: 324–6; Wim Carton, 'Carbon Unicorns and Fossil Futures: Whose Emission Reduction Pathways Is the IPCC Performing?', in J. P. Sapinski, Holly Jean Buck and Andreas Malm (eds), *Has It Come to This? The Promises and Perils of Geoengineering on the Brink* (Brunswick, NJ: Rutgers University Press, 2021), 37; Silke Beck and Jeroen Oomen, 'Imagining the Corridor of Climate Mitigation: What Is at Stake in IPCC's Politics of Anticipation?', *Environmental Science and Policy* (2021) 123: 175. On

the assumptions about growth in IAMs, see Lorenz T. Keyßer and Manfred Lenzen, '1.5°C Degrowth Scenarios Suggest the Need for New Mitigation Pathways', *Nature Communication* (2021) 12: 1–16; and cf. e.g. Asefi-Najafabady et al., 'The Failure of Integrated', 1178; Keppo et al., 'Exploring the Possibility', 10–11. Omitted in the IAMs, on the other hand, was the very real prospect of a total financial gain from a transition to renewable energy, now consistently cheaper than fossil fuels: see further below. The idea of a carbon price reflects the understanding of the climate crisis as one of market failure – a deeply flawed understanding, of course. See further e.g. Marisa Beck and Tobias Krueger, 'The Epistemic, Ethical, and Political Dimensions of Uncertainty in Integrated Assessment Modeling', *WIREs Climate Change* (2016) 7: 634, 638.

18 See e.g. Elizabeth A. Stanton, Frank Ackerman and Sivan Kartha, 'Inside the Integrated Assessment Models: Four Issues in Climate Economics', *Climate and Development* (2009) 1: 166–84; Johannes Emmerling, Laurent Drouet, Kaj-Ivar van der Wijst et al., 'The Role of the Discount Rate for Emission Pathways and Negative Emissions', *Environmental Research Letters* (2019) 14: 1–11; Ackerman et al., 'Limitations of Integrated', 299, 308; Carton, 'Carbon Unicorns', 42; Beck and Oomen, 'Imagining the Corridor', 175.

19 See further Carton, 'Carbon Unicorns', 41–3.

20 Detlef P. van Vuuren, Michel G. J. den Elzen, Paul L. Lucas et al., 'Stabilizing Greenhouse Gas Concentrations at Low Levels: As Assessment of Reduction Strategies and Costs', *Climatic Change* (2007) 81: 131 (emphasis added). On the historical role of IMAGE, see Cointe et al., 'Organising Policy-Relevant', 41; Van Beek et al., 'Anticipating Futures', 5.

21 Van Vuuren et al., 'Stabilizing Greenhouse', 131.

22 See e.g. Robert S. Pindyck, 'The Use and Misuse of Models for Climate Policy', *Review of Environmental Economics and Policy* (2017) 1: 109–10; David L. McCollum, Ajay Gambhir, Joeri Rogelj and Charlie Wilson, 'Energy Modellers Should Explore Extremes More Systematically in Scenarios', *Nature Energy* (2020) 5: 104–7; Pindyck, 'Climate Change', 869; Rosen, 'Critical Review', 325; Beck and Oomen, 'Imagining the Corridor', 175. Defenders of IAMs would retort that the *aggregate* changes modelled still amounted to shifts in energy use on an historically unprecedented scale: e.g. Jessica Jewell, 'Clarifying the Job of IAMs', *Nature* (2019) 573: 349.

23 Ackerman et al., 'Limitations of Integrated', 305; Rivadineira and Carton, '(In)justice in Modelled', 4.

24 Nordhaus, 'To Slow', 930. The full insanity of Nordhausian economics is called out in Steve Keen, 'The Appallingly Bad Neoclassical Economics of Climate Change', *Globalizations* (2021) 18: 1149–77; for this particular point, see 1152. See also Jason Hickel, 'The Nobel Prize for Climate Catastrophe', *Foreign Policy*, 6 December 2018.

25 Nordhaus, 'To Slow', 930. The figure is for the US national output in 1981, then fantastically extrapolated to the whole world in 2050.

26 Ibid., 933. Moreover, early IAMs such as this one often predicted that global warming would be economically *beneficial* for at least the global North, on the basis of e.g. the fertilisation effect – higher concentration of CO_2 increasing agricultural productivity – but this prediction was later abandoned. Cf. e.g. ibid., 929; Ackerman et al., 'Limitations of Integrated', 306.

27 Tol in 2019, quoted in Keene, 'The Appallingly Bad', 1165.

28 IAM modellers quoted in Sean Low and Stefan Schäfer, 'Is Bio-Energy Carbon Capture and Storage (BECCS) Feasible? The Contested Authority of Integrated Assessment Modelling', *ERSS* (2020) 60: 3.

29 Evans and Hausfather, 'How "Integrated"'.

30 On this difference between climate models and IAMs, cf. Ackerman et al., 'Limitations of Integrated', 310.

31 Cointe et al., 'Organising Policy-Relevant', 37, 44–6; Van Beek et al., 'Anticipating Futures', 6, 9–10; Van Beek et al., 'Navigating the Political', 195–6.

32 Béatrice Cointe, 'Scenarios', in Kari De Pryck and Mike Hulme (eds), *A Critical Assessment of the Intergovernmental Panel on Climate Change* (Cambridge: Cambridge University Press, 2023), 137–47.

33 Naomi Oreskes, 'How Earth Science Has Become a Social Science', *Historical Social Research / Historische Sozialforschung* (2015) 40: 246–70; Silke Beck and Martin Mahony, 'The IPCC and the Politics of Anticipation', *Nature Climate Change* (2017) 7: 311–12; Silke Beck and Martin Mahony, 'The Politics of Anticipation: The IPCC and the Negative Emissions Technologies Experience', *Global Sustainability* (2018) 1: 1–8.

34 Detlef P. van Vuuren, Jae Edmonds, Mikiko Kainuma et al., 'The Representative Concentration Pathways: An Overview', *Climatic Change* (2011) 109: 5–31.

35 IPCC, *Towards New Scenarios for Analysis of Emissions, Climate Change, Impacts, and Response Strategies: IPCC Expert Meeting Report*, September 2007; Massimo Tavoni and Richard S. J. Tol, 'Counting Only the Hits? The Risk of Underestimating the

Costs of Stringent Climate Policy', *Climatic Change* (2010) 100: 769–78; Massimo Tavoni and Robert Socolow, 'Modeling Meets Science and Technology: An Introduction to a Special Issue on Negative Emissions', *Climatic Change* (2013) 118: 1–14.

36 Eva Lövbrand, 'Co-Producing European Climate Science and Policy: A Cautionary Note on the Making of Useful Knowledge', *Science and Public Policy* (2011) 38: 225–36; IPCC, *Towards New*.

37 Karl Marx, *Grundrisse* (London: Penguin, 1993 [1857–8]), 90. Note that it is exactly this process Marx is here referring to. For an excellent in-depth account of it, see David Beerling, *The Emerald Planet: How Plants Changed the Earth's History* (Oxford: Oxford University Press, 2007).

38 The graph is continuously updated at Global Monitoring Laboratory, 'Trends in Atmospheric Carbon Dioxide: Monthly Average Manua Loa CO_2'. The Southern hemisphere, of course, has a more limited landmass, and so it's the seasonal calendar in the Northern hemisphere that determines the atmospheric fluctuations over the year.

39 Van Vuuren et al., 'Stabilizing Greenhouse', 130.

40 Christian Azar, Kristian Lindgren, Eric Larson and Kenneth Möllersten, 'Carbon Capture and Storage from Fossil Fuels and Biomass: Costs and Potential Role in Stabilizing the Atmosphere', *Climatic Change* (2006) 74: 47–79.

41 Or, 'overshoot may, in particular, play an important role in making more ambitious climate targets feasible'. Michel G. J. den Elzen and Detlef P. van Vuuren, 'Peaking Profiles for Achieving Long-Term Temperature Targets with More Likelihood at Lower Costs', *PNAS* (2007) 104: 17931.

42 Van Vuuren et al., 'Stabilizing Greenhouse', 131 (emphasis added).

43 Ibid., 147.

44 Tom M. L. Wigley, 'Modelling Climate Change under No-Policy and Policy Emissions Pathways', OECD, 2003, 3. 'As a working-paper, this document has received only limited peer-review.' Ibid.

45 Ibid., 10 (emphasis in original).

46 Ibid., 11.

47 Ibid., 28.

48 Van Vuuren et al., 'Stabilizing Greenhouse', 123; Den Elzen and Van Vuuren, 'Peaking Profiles', 17932.

49 Den Elzen and Van Vuuren, 'Peaking Profiles', 17931.

50 Beck and Mahony, 'The Politics of Anticipation', 2; cf. Ellenbeck and Lilliestam, 'How Modelers', 72.

51 The question here is that of the 'discount rate'. It can be thought of as 'the inverse of an interest rate on savings: just as the value

of £100 in the present will increase in the future under different interest rates, so the value of £100 of damages in the future will decrease in the present under different social discount rates'. Catriona McKinnon, *Climate Change and Political Theory* (Cambridge: Polity, 2022), 57. Most IAMs used a high discount rate of around 5 per cent per year; lowering it to 2 would knock out half the overshoot and corresponding removal as economically suboptimal, according to Emmerling et al., 'The Role of the Discount'. Cf. e.g. Beck and Oomen, 'Imagining the Corridor', 175. Just how high the standard discount rate of IAMs is can be gauged from the following computations. At a rate of 5 per cent, one death next year will be more important than 1 billion deaths in 500 years. Or, it would be worth spending no more than $2,200 today to prevent $87 *trillion* in damages 500 years from now (the latter sum equivalent to total GDP in 2019). With this rate, the future is cheapened to the point of absurd nullity. Asefi-Najfabady et al., 'The Failure of Integrated', 1183.

52 IPCC, *Towards New*, 34.

53 Ibid., 42.

54 Cointe and Guillemot, 'A History', 2.

55 Kevin Anderson, 'Duality in Climate Science', *Nature Geoscience* (2015) 8: 899.

56 Van Beek et al., 'Taking Science', 8, 9.

57 Edenhofer et al., *Climate Change*, 433.

58 Beck and Mahony, 'The IPCC and the Politics', 312; Beck and Mahony, 'The Politics of Anticipation', 3–5; Van Beek et al., 'Anticipating Futures', 9.

59 Duncan McLaren and Nils Markusson, 'The Co-Evolution of Technological Promises, Modelling, Policies and Climate Change Targets', *Nature Climate Change* (2020) 10: 395.

60 Joeri Rogelj, Alexander Popp, Katherine V. Calvin et al. 'Scenarios towards Limiting Global Mean Temperature Increase below 1.5 °C', *Nature Climate Change* (2018) 8: 325–32.

61 Participant interviewed and quoted in Van Beek et al., 'Navigating the Political', 197. Cf. 195; Cointe and Guillemot, 'A History', 7.

62 Detlef P. van Vuuren, Elke Stehfest, David E. H. J. Gernaat et al., 'Alternative Pathways to the 1.5°C Target Reduce the Need for Negative Emission Technologies', *Nature Climate Change* (2018) 8: 392, 396.

63 Arnulf Grubler, Charlie Wilson, Nuno Bento et al., 'A Low Energy Demand Scenario for Meeting the 1.5°C Target and Sustainable Development Goals without Negative Emissions Technologies',

Nature Energy (2018) 3: 515–27. This scenario, it should be noted, relied on very optimistic assumptions about the spread of digital technologies and the effects of increased energy efficiency.

64 Cointe and Guillemot, 'A History', 4–5.

65 Cf. e.g. Oliver Geden and Andreas Löschel, 'Define Limits for Temperature Overshoot Targets', *Nature Geoscience* (2017) 10: 881–2.

66 Masson-Delmotte et al., *Global Warming*, 60. Cf. 63, 98.

67 Ibid., 61; see further e.g. 115, 121, 152, 395.

68 Ibid., 177.

69 Ibid., 18. It should be noted that the scenarios demanding emissions be halved before 2030 were in the fuzzy category of 'no or limited overshoot', the latter defined as overshoot up to 1.6°C. Ibid., 12, 24. But even the few absolute no-overshoot scenarios relied on some degree of carbon dioxide removal, if only to prevent the accumulation of CO_2 to take the world beyond the limit (rather than reversing it after the fact). See e.g. 277.

70 See further Holly Jean Buck, Wim Carton, Jens Friis Lund and Nils Markusson, 'Why Residual Emissions Matter Right Now', *Nature Climate Change* (2023) 13: 351–8; and for an informed analysis of 'net zero' and residual emissions generally, Holly Jean Buck, *Ending Fossil Fuels: Why Net Zero Is Not Enough* (London: Verso, 2021).

71 The website Net Zero Tracker (zerotracker.net) keeps a running record of these targets, as well as of corporate and regional pledges of the same sort. Net Zero Tracker, 2023.

72 Robin D. Lamboll, Zebedee R. J. Nicholls, Christopher J. Smith et al., 'Assessing the Size and Uncertainty of Remaining Carbon Budgets', *Nature Climate Change* (2023) online first, 1–10.

73 Matthews and Wynes, 'Current Global'.

74 E.g. Damian Carrington, '"Blah, Blah, Blah": Greta Thunberg Lambasts Leaders over Climate Crisis', *Guardian*, 28 September 2021.

75 Climate Action Tracker, 2023.

76 Van Vuuren et al., 'The Role of Negative', 16.

77 Jamie Allinson, *The Age of Counter-Revolution: States and Revolutions in the Middle East* (Cambridge: Cambridge University, 2022), 40. See further 34–46. This is undoubtedly the most brilliant study of these categories to appear in many decades, nourished by a revolutionary experience far more recent, of course, than that of 1917, namely the Arab Spring of 2011.

78 Ibid., 53.

79 Namely, for the situation in Putin's Russia. Ilya Budraitskis,

Dissidents among Dissidents: Ideology, Politics and the Left in Post-Soviet Russia (London: Verso, 2022), 36.

80 Simon Haikola, Anders Hansson and Mathias Fridahl, 'Map-makers and Navigators of Politicised Terrain: Expert Understandings of Epistemological Uncertainty in Integrated Assessment Modelling of Bioenergy with Carbon Capture and Storage', *Futures* (2019) 114: 8; Ellenbeck and Lilliestam, 'How Modelers', 72; Geden, 'The Paris Agreement', 793; Workman et al., 'Decision Making', 81; McLaren and Markusson, 'The Co-Evolution of Technological', 394.

81 Shukla et al., *Climate Change*, 1874 (emphasis added).

82 Anderson, 'Duality', 900.

83 This, of course, is also a bit of technical jargon. It's 'rather like calling the opposite of inhalation not "exhalation", but "negative inhalation", or referring to fasting as "negative eating"'. Henry Shue, *The Pivotal Generation: Why We Have a Moral Responsibility to Slow Climate Change Right Now* (Princeton: Princeton University Press, 2021), 93.

84 The ethical principle at stake here is that people have a 'fundamental right not to have their bodies damaged by the actions of others, when the damage is preventable.' Henry Shue, *Climate Justice: Vulnerability and Protection* (Oxford: Oxford University Press, 2014), 167. Shue invents a scenario of someone with the hobby of planting landmines that will go off far into the future, while speculating that these future victims 'will probably have terrific prosthetic limbs by then' – a most unacceptable pastime. Ibid., 162. We have modified the scenario so as better to reflect one central aspect of the push to delay climate mitigation: the defence of real, existing power relations in the economic sphere (rather than a flippant hobby).

85 Chris Huntingford and Jason Lowe, '"Overshoot" Scenarios and Climate Change', *Science* (2007) 316: 829.

86 Van Vuuren et al., 'The Role of Negative', 18.

87 William D. Nordhaus, 'Discounting in Economics and Climate Change', *Climatic Change* (1997) 37: 324.

88 Ibid., 317.

89 Stephen M. Gardiner, *A Perfect Moral Storm: The Ethical Tragedy of Climate Change* (Oxford: Oxford University Press, 2011), 273, 286; Tyler Cowen and Derek Parfit, 'Against the Social Discount Rate', in Peter Laslett and James S. Fishkin (eds), *Justice Between Age Groups and Generations* (New Haven: Yale University Press, 1992), 145; Simon Caney, 'Human Rights, Climate Change, and Discounting', *Environmental Politics* (2008) 17: 550–1; McKinnon, *Climate Change*, 59.

90 Simon Caney, 'Climate Change and the Future: Discounting for Time, Wealth, and Risk', *Journal of Social Philosophy* (2009) 40: 170; Simon Caney, 'Climate Change, Intergenerational Equity and the Discount Rate', *Politics, Philosophy and Economics* (2014) 13: 328.

91 Gardiner, *A Perfect*, 273, 285; Matthew Rendall, 'Discounting, Climate Change, and the Ecological Fallacy', *Ethics* (2019) 129: 460; Caney, 'Climate Change, Intergenerational', 330; Caney, 'Climate Change and the Future', 172–3. The only way to rebut this argument is to engage in climate denial, as indeed Bjorn Lomborg, one of the main proponents of bequeathing the mess to those who come after us, has consistently done.

92 Rendall, 'Discounting, Climate', 444–5, 458–9; Caney, 'Climate Change and the Future', 171–4; Caney, 'Human Rights', 551.

93 Cf. Emmerling and Tavoni, 'Representing Inequalities', 178–9; Rivadineira and Carton, '(In)justice in Modelled', 4.

94 E.g. Shue, *The Pivotal*, 15, 18–19. And this will also inevitably take the form of snowballing monetary costs: Caney, 'Climate Change, Intergenerational', 330–1.

95 Gardiner, *A Perfect*, 33–4. See further e.g. 150–4.

96 Shue, *The Pivotal*, 107–9. A case for this constituting domination rather than exploitation is made in Patrick Taylor Smith, 'The Intergenerational Storm: Dilemma or Domination', *Philosophy and Public Issues* (2013) 3: 207–44.

97 On this agreement, see McKinnon, *Climate Change*, e.g. 55.

98 Cf. Henry Shue, 'Climate Dreaming: Negative Emissions, Risk Transfer, and Irreversibility', *Journal of Human Rights and the Environment* (2017) 8: 214–15.

99 With the qualification that these children and grandchildren would have no other place to live. 'Move out' here simply means passing away.

100 E.g. McKinnon, *Climate Change*, 23; Shue, *The Pivotal*, 15.

101 Shue, *The Pivotal*, 10–11, 145.

102 Ibid., 145; Daniel Bensaïd, *Marx for Our Times: Adventures and Misadventures of a Critique* (London: Verso, 2009 [1995]) 77.

103 Leon Trotsky, 'Germany, the Key to the International Situation', in *The Struggle*, 124–5 (emphases in original).

104 The party leadership clung to this line – the belief that the downfall of the Nazi regime was imminent – as late as in the summer of 1934. Allan Merson, *Communist Resistance in Nazi Germany* (London: Lawrence & Wishart, 1985), 63.

105 Steve Rayner, 'What Might Evans-Pitchard Have Made of Two Degrees?', *Anthropology Today* (2016) 32: 1. Cf. Low and Schäfer, 'Is Bio-Energy', 1.

106 Cf. e.g. Shue, 'Climate Dreaming', 205, 212; Beck and Mahoney, 'The Politics of Anticipation', 2, 4; Beck and Oomen, 'Imagining the Corridor', 175; Workman et al., 'Decision Making', 78.

107 In the second instalment of this inquiry, we shall study BECCS and the very major obstacles to its large-scale realisation in detail.

108 Informants from the IAM community cited in Low and Schäfer, 'Is Bio-Energy', 4.

109 Roger Pielke Jr., 'Opening Up the Climate Policy Envelope', *Issues in Science and Technology* (2018) 34: 33.

110 Geden, 'The Paris Agreement', 794.

111 Quoted in Haikola et al., 'Map-makers and Navigators', 8.

112 Hillard G. Huntington, John P. Weyant and James L. Sweeney, 'Modeling for Insights, Not Numbers: The Experiences of the Energy Modeling Forum', *OMEGA: The International Journal of the Management Sciences* (1982) 10: 449–62; and the critique in Bill Keepin, 'A Technical Appraisal of the IIASA Energy Scenario', *Policy Sciences* (1984): 199–275 (quotation from 201); Brian Wynne, 'The Institutional Context of Science, Models, and Policy: The IIASA Energy Study', *Policy Sciences* (1984) 17: 277–320.

113 *Energy in a Finite World* quoted in Wynne, 'The Institutional Context', 284; see further 286–93.

114 William D. Nordhaus, 'Can We Control Carbon Dioxide?', working paper, International Institute of Applied Systems Analysis [later a pioneering IAM hub], Austria, 1975 (quotations from 24, 8, 39).

115 Cesare Marchetti, 'On Geoengineering and the CO_2 Problem', *Climatic Change* (1977) 1: 59–68.

116 Freeman J. Dyson, 'Can We Control the Carbon Dioxide in the Atmosphere?', *Energy* (1977) 2: 288, 291

117 Ibid., 288 (emphasis added).

118 Freeman J. Dyson, 'Warm-Blooded Plants and Freeze-Dried Fish', *Atlantic*, November 1977.

119 Avi Loeb, 'What Came First: The Astro-Chicken or the Egg?', *Medium*, 14 August 2022.

120 Freeman J. Dyson, 'Time without End: Physics and Biology in an Open Universe', *Reviews of Modern Physics* (1979) 51: 448.

121 On his denialism, see e.g. Nicholas Dawidoff, 'The Civil Heretic', *New York Times Magazine*, 25 March 2009; Steve Connor and Freeman Dyson, 'Letters to a Heretic: An Email Conversation with Climate Change Sceptic Professor Freeman Dyson', *Independent*, 25 February 2011.

122 Quotation from Simon Robertson, 'Transparency, Trust, and Integrated Assessment Models: An Ethical Consideration for the

Intergovernmental Panel on Climate Change', *WIREs Climate Change* (2021) 12: 4–5. See further 1–2; Masson-Delmotte et al., *Global Warming*, 121, 154 (here enunciating 'middle-of-the-road' development as the working assumption of IAMs); Ajay Gambhir, Isabela Butnar, Pei-Hao Li et al., 'A Review of Criticisms of Integrated Assessment Models and Proposed Approaches to Address These, through the Lens of BECCS', *Energies* (2019) 12: 5; Beck and Krueger, 'The Epistemic, Ethical', 628; Haikola et al., 'Map-makers and Navigators', 10; Carton, 'Carbon Unicorns', 37; Rivadineira and Carton, '(In)justice in Modelling', 6; Keppo et al., 'Exploring the Possibility', 14; Van Beek et al., 'Navigating the Political', 194.

123 Beck and Oomen, 'Imagining the Corridor', 170. Cf. Peter Newell, *Power Shift: The Global Political Economy of Energy Transitions* (Cambridge: Cambridge University Press, 2021), 80–1.

124 See e.g. Keen, 'The Appallingly Bad', 1150; Beck and Mahoney, 'The IPCC and the Politics'; Beck and Oomen, 'Imagining the Corridor'; Low and Schäfer, 'Is Bio-Energy', 7–8; Haikola et al., 'Map-makers and Navigators', 11; Rivadineira and Carton, '(In)justice in Modelling', 1.

125 Low and Schäfer, 'Is Bio-Energy', 6. The argument about performativity was prefigured in Wynne, 'The Institutional Context', 277, 282, 284.

126 Machines: Haikola et al., 'Map-makers and Navigators', 10; world-making: Beck and Mahoney, 'The Politics of Anticipation', 5; Beck and Oomen, 'Imagining the Corridor', 170.

127 Beck and Krueger, 'The Epistemic, Ethical', 639. Cf. Robertson, 'Transparency, Trust', 4.

128 Asefi-Najafabady et al., 'The Failure of Integrated', 1185. Note that we have here only provided a sample of the critique of IAMs; for a more comprehensive picture, the cited articles should be consulted.

129 For these properties of agency and the fallacies of Latourian theory, see further Malm, *The Progress*.

130 Theodor Adorno, *Philosophy and Sociology* (Cambridge: Polity, 2022 [1960]), 160.

131 Theodor Adorno, *Negative Dialectics* (New York: Bloomsbury, 2014 [1966]), 268. The famously substandard English translation of this magnum opus is here corrected: 'infrastructure' has been changed to 'base'.

132 Adorno, *Philosophy and Sociology*, 167.

133 Ellen Meiksins Wood, *Democracy against Capitalism: Renewing Historical Materialism* (Cambridge: Cambridge University Press 1995), e.g. 61, 68, 74.

134 Karl Marx and Friedrich Engels, *The German Ideology* (Amherst: Prometheus, 1998 [1845]), 67 (emphasis added).

135 On the choice between productive force determinism and constructivist Marxism, see e.g. Andreas Malm, *Fossil Capital: The Rise of Steam Power and the Roots of Global Warming* (London: Verso, 2016); Andreas Malm, 'Marx on Steam: From the Optimism of Progress to the Pessimism of Power', *Rethinking Marxism* (2018) 30: 166–85.

136 Louis Althusser, *Philosophy for Non-Philosophers* (London: Bloomsbury, 2017 [1977–78]), 113 (emphasis in original). 'It is not such-and-such an idea considered as an individual fantasy that counts, but *only the ideas endowed with a capacity for social action*.' Ibid., 112 (emphasis in original).

137 Theodor Adorno, *Minima Moralia: Reflections from Damaged Life* (London: Verso, 2005 [1951]), 208. For an excellent up-to-date study of this theme in Adorno, see Charles Andrew Prusik, *Adorno and Neoliberalism: The Critique of Exchange Society* (London: Bloomsbury, 2022).

138 Karl Marx, *Capital: A Critique of Political Economy. Volume I* (London: Penguin, 1990 [1867]), 169–70; Marx, *Grundrisse*, 83.

139 Asefi-Najafabady et al., 'The Failure of Integrated', 1183. 'She' in the original is here changed to 'he' in keeping with the gender of Robinson.

140 See e.g. Theodor Adorno, 'Marx and the Basic Concepts of Sociological Theory', *Historical Materialism* (2018) [1962] 26: 156–60; Adorno, *Negative*, e.g. 88, 94, 178, 354–5; Prusik, *Adorno and Neoliberalism*, 10–32.

141 Adorno, 'Marx and the Basic', 156, 161.

142 Ibid., 156. Cf. e.g. Theodor Adorno, *History and Freedom: Lectures 1964–5* (Cambridge: Polity, 2006), 135–6; Adorno, *Negative*, 355.

143 Cf. e.g. Beck and Oomen, 'Imagining the Corridor', 171; Asefi-Najafabady et al., 'The Failure of Integrated', 1178, 1182; Van Beek et al., 'Anticipating Futures', 10; Van Beek et al., 'Navigating the Political', 196; Rivadineira and Carton, '(In)justice in Modelling', 1, 6.

144 Marx, *Capital Volume I*, 169 (emphasis added).

145 Hickel, 'The Nobel Prize'. Whether this was more influential in shaping US climate policy than uninhibited, literal denial is perhaps a moot point. So much less sophisticated in its treatment of power than of biogeochemical matters, the IPCC even let itself be persuaded by Nordhaus' argument about the indoor economy being sheltered from climate change: Keen, 'The Appallingly Bad', 152–3. Given the complicity of bourgeois climate economics

in causing the delay, Keen names it 'the most significant and dangerous hoax in the history of science' – this, possibly, an exaggeration, but not necessarily by much. Ibid., 1151.

146 Theodor Adorno, 'Progress', in *Critical Models: Interventions and Catchwords* (New York: Columbia University Press, 2005 [1964]), 159.

147 Cf. Workman et al., 'Decision Making', 79.

148 We owe this point to Franciszek Korbanski.

149 Cf. Geden and Löschel, 'Define Limits'; Asayama et al., 'Why Setting', 571.

150 Adorno, *Negative*, 206–7 (emphasis added).

151 With 'bankruptcy' being the harshest word in the clerk's lexicon: immanent critique, or perhaps a term appropriately faithful to the accountings of the carbon budget.

152 John D. Sutter, Joshua Berlinger and Ralph Ellis, 'Obama: Climate Agreement "Best Chance We Have" to Save the Planet', *CNN*, 14 December 2015.

153 Brian Wingfield, 'U.S. Reverses Decades of Oil-Export Limits with Obama's Backing', *Bloomberg*, 18 December 2015.

154 Jie Jenny Zou, 'How Washington Unleashed Fossil-Fuel Exports and Sold Out on Climate', *Texas Tribune*, 16 October 2018; Alexandra Twin, 'The World's 10 Biggest Oil Exporters', *Investopedia*, 23 August 2022. In 2022, only Saudi Arabia and Russia were still ahead of the US.

155 And he himself continued the talk. In October 2016, when a string of nations had ratified the agreement, Obama repeated: 'today is a historic day in the fight to protect our planet for future generations. This gives us the best possible shot to save the one planet we got. With optimism and faith and hope, we are proving it is possible.' Oliver Milman, 'Paris Climate Deal a "Turning Point" in Global Warming Fight, Obama Says', *Guardian*, 5 October 2016. A few weeks later, Trump happened.

156 Christina Figueres, 'The Secret to Tackling Climate Change', *Nature* (2020) 577: 471.

157 Stefan C. Aykut, Edouard Morena and Jean Foyer, '"Incantatory" Governance: Global Climate Politics' Performative Turn and Its Wider Significance for Global Politics', *International Politics* (2021) 58: 31–2; cf. Aykut, 'Governing through Verbs', 33–4.

158 Aykut et al., '"Incantatory" Governance', 30–1.

159 Figueres, 'The Secret', 471.

160 'Stubborn Optimism Is a Deliberate Mindset', *Global Optimism*, n.d.

161 Adorno, *Minima Moralia*, 122.

162 Althusser, *Philosophy for Non-Philosophers*, 113–14. Cf. e.g.

Frieder Otto Wolf, 'The Problem of Reproduction: Probing the Lacunae of Althusser's Theoretical Investigations of Ideology and Ideological State Apparatuses', in Katja Diefenbach, Sara R. Farris, Gal Kirn and Peter D. Thomas (eds), *Encountering Althusser: Politics and Materialism in Contemporary Radical Thought* (New York: Bloomsbury, 2013), 249.

163 Étienne Balibar, 'Althusser's Dramaturgy and the Critique of Ideology', *Differences: A Journal of Feminist Cultural Studies* (2015) 26: 19 (emphasis in original). See further the comments on Balibar/Althusser in Warren Montag, 'Althusser's Authorless Theatre', *Differences: A Journal of Feminist Cultural Studies* (2015) 26: 50; Adi Ophir, 'On Linking Machinery and Show', *Differences: A Journal of Feminist Cultural Studies* (2015) 26: 59–65, 69.

164 Paul E. Little, 'Ritual, Power and Ethnography at the Rio Earth Summit', *Critique of Anthropology* (1995) 15: 276, 280; Carl Death, 'Summit Theatre: Exemplary Governmentality and Environmental Diplomacy in Johannesburg and Copenhagen', *Environmental Politics* (2011) 20: 2, 7, 10. As these early references suggest, the theatricality of climate summitry preceded Paris; but it made a qualitative leap there, as convincingly argued by Aykut et al.

165 Perhaps the endpoint was the blending of the two styles: 'when machinery and show become indistinguishable, individuals play their roles in the show and the order of power as if glued to them by force.' Ophir, 'On Linking', 69.

166 The incantatory, theatrical series of climate negotiations had an instructive counterpart in the so-called peace process, in which the state of Israel negotiated with the leadership of Fatah. These two diplomatic tracks were constructed in the same historical moment – the early 1990s – and proceeded in parallel for some time. Through strikingly similar mechanisms, both performed fictions: that the 'international community' was working to stabilise climate and give the Palestinians a state. But the 'peace process', of course, came to an ignominious end around 2005, when the state of Israel reconfigured its occupation of Gaza as the operation of a concentration camp. All that then remained was the never-ending *nakba*; and in this respect, as in so many others, the catastrophe of Palestine appeared to prefigure that of climate. For some more such respects, see Andreas Malm, 'The Walls of the Tank: On Palestinian Resistance', *Salvage* (2017) 4: 21–55; Andreas Malm, 'Warming', in John Parham (ed.), *The Cambridge Companion to Literature and the Anthropocene* (Cambridge: Cambridge University Press, 2021), 242–57.

167 Aykut et al., '"Incantatory" Governance', 31–3.
168 Ibid., 34; Stefan C. Aykut and Monica Castro, 'The End of Fossil Fuels? Understanding the Partial Climatisation of Global Energy Debates', in Aykut et al., *Globalising the Climate*, 182.
169 This point was already made in Tavoni and Socolow, 'Modeling Meets'.
170 Shue, *The Pivotal*, 17, 19–22.
171 'Global action is not going to stop climate change. The world needs to look harder at how to live with it.' *Economist*, 'Facing the Consequences'.
172 Martin Parry, Jason Lowe and Clair Hanson, 'Overshoot, Adapt and Recover', *Nature* 458 (2009): 1102–3.
173 T. M. L. Wigley, 'A Combined Mitigation/Geoengineering Approach to Climate Stabilization', *Science* (2006) 314: 453–4.
174 Nordhaus, 'To Slow', 928–9.
175 Reuters, 'Macron Announce un Forum Pour Conjurer les Périls Mondiaux', 4 January 2018.
176 Chloé Farand, 'As 1.5C Overshoot Looms, A High-Level Commission Will Ask: What Next?', *Climate Home News*, 22 April 2022.
177 Climate Overshoot Commission, 'The Commissioners', 2022.
178 Climate Overshoot Commission, 'Climate Overshoot Commission Launches, Meets for the First Time', n.d.
179 Climate Overshoot Commission, 'The Commission's Fourth Meeting Took Place in Jakarta (10 to 12 February)', n.d.
180 Climate Overshoot Commission, 'How Should the World Reduce the Risk of Temperature Overshoot?', 2022.
181 *Economist*, 'Goodbye 1.5°C'; see further *Economist*, 'The World Is'.
182 Gokul Iyer, Yang Ou, James Edmonds et al., 'Ratcheting of Climate Pledges Needed to Limit Peak Global Warming', *Nature Climate Change* (2022) 12: 1134.
183 *Economist*, 'The World Is'.
184 *Economist*, 'Goodbye 1.5°C' (emphasis added).

Part II: Fossil Capital Is a Demon

1 Kühne et al., '"Carbon Bombs"', 6.
2 Gabe Eckhouse, 'United States Hydraulic Fracturing's Short-Cycle Revolution and the Global Oil Industry's Uncertain Future', *Geoforum* (2021) 127: 250; Eckhouse, 'Covid-19', 1651. The exact dating of break even of course depends on the prices that happen to prevail.
3 Robert Brenner, 'The Economics of Global Turbulence', *New Left*

Review (1998) 229: 26. Needless to say, maintenance of the fixed capital, as well as labour-power, raw materials and other forms of circulating capital, would still have to be covered: but as such, the oilfield or equivalent is paid for. For the general mechanisms of the inertia of fixed capital, see David Harvey, *The Limits to Capital* (London: Verso, 1999 [1982]), e.g. 220–1, 394. On how this inertia works in the department of fossil fuels, see e.g. Vivien Fisch-Romito, Céline Guivarch, Felix Creutzig et al., 'Systematic Map of the Literature on Carbon Lock-In Induced by Long-Lived Capital', *Environmental Research Letters* (2021) 16: 2; Edenhofer et al., 'Reports of Coal's', 2, 8; Trout et al. 'Existing Fossil', 9. The problem has previously been discussed by the present authors in Wim Carton, 'Dancing to the Rhythms of the Fossil Fuel Landscape: Landscape Inertia and the Temporal Limits to Market-Based Climate Policy', *Antipode* (2017) 49: 47–9; Malm, *Fossil Capital*, 358–9; Andreas Malm, *How to Blow Up a Pipeline: Learning to Fight in a World on Fire* (London: Verso, 2021), 28–9.

4. The Political Economy of Asset Stranding

1 Heede, 'East Africa, 9; Augé, The Trans-Saharan', 10.
2 *Guyana Times*, 'ExxonMobil Taking Steps to Ensure Longevity of Oil Fields – Production Manager', 28 July 2022.
3 JA!/Friends of the Earth Mozambique, 'The Impacts', 7.
4 *Business Wire*, 'ConocoPhillips Reports Fourth-Quarter, Full-Year 2022 Results and 176% Preliminary Reserve Replacement Ratio; Announces 2023 Guidance and Planned Return of Capital of $11 Billion; Declares Quarterly Dividend and Variable Return of Cash Distribution', 2 February 2023.
5 Ole Reinert Omvik, 'Aker BP trenger 30.000 oljearbeidere', *NRK*, 5 February 2023.
6 Ryna Yiyun Cui, Nathan Hultman, Morgan R. Edwards et al., 'Quantifying Operational Lifetimes for Coal Power Plants under the Paris Goals', *Nature Communications* (2019) 10: 2–4; cf. Julie Rozenberg, Steven J. Davis, Ulf Narloch and Stephane Hallegatte, 'Climate Constraints on the Carbon Intensity of Economic Growth', *Environmental Research Letters* (2015) 10: 3; Alexander Pfeiffer, 'The "Decarbonisation Identity": Stranded Assets in the Power Generation Sector', in Ben Caldecott (ed.), *Stranded Assets and the Environment: Risk Resilience and Opportunity* (Abingdon: Routledge, 2018), 56.
7 Hauenstein, 'Stranded Assets', 2; Samantha Hepburn, 'Court Challenge Will Test Coal Mining's Climate Culpability', *Conversation*,

15 January 2015; Peter Hannam, '"Barbaric": Adani's Giant Coal Mine Granted Unlimited Water License for 60 Years', *Sydney Morning Herald*, 5 April 2017.

8 Reuters, 'Britain Approves First New Coal Mine in Decades Despite Climate Targets', 7 December 2022; Sandra Laville, 'Cumbria Coalmine Is Owned by Private Equity Firm with Caymans Base', *Guardian*, 8 December 2022. On the Elizabethan leap, see Malm, *Fossil Capital*, 320–6.

9 For the concept of primitive accumulation of fossil capital, see further Malm, *Fossil Capital*; Malm and the Zetkin Collective, *White Skin*.

10 This is the definition used in e.g. Shukla et al., *Climate Change*, 90; Atif Ansar, Ben Caldecott and James Tilbury, 'Stranded Assets and the Fossil Fuel Divestment Campaign: What Does Divestment Mean for the Valuation of Fossil Fuel Assets?', Stranded Asset Programme (Smith School of Enterprise and the Environment, and University of Oxford), October 2013, 9; Ben Caldecott, Alex Clark, Krister Koskelo et al., 'Stranded Assets: Environmental Drivers, Societal Challenges, and Supervisory Responses', *Annual Review of Environment and Resources* (2021) 46: 418; Nur Firdaus and Akihisa Mori, 'Stranded Assets and Sustainable Energy Transition: A Systematic and Critical Review of Incumbents' Response', *Energy for Sustainable Development* (2023) 73: 76. For several variations on the standard definition, see Ben Caldecott, 'Introduction: Stranded Assets and the Environment', in Caldecott, *Stranded Assets*, 4–5; Rob Aitken, 'Depletion Work: Climate Change and the Mediation of Stranded Assets', *Socio-Economic Review* (2023) 21: 273.

11 On the chronology of the discourse – a phenomenon of the second decade – see e.g. A. Shimbar, 'Environment-Related Stranded Assets: An Agenda for Research into Value Destruction within Carbon-Intensive Sectors in Response to Environmental Concerns', *Renewable and Sustainable Energy Reviews* (2021) 144: 1–12.

12 On the role of these movements in engendering the discourse, see e.g. Roland Benedikter, Kjell Kühne, Ariane Benedikter and Giovanni Atzeni, '"Keep It In the Ground." The Paris Agreement and the Renewal of the Energy Economy: Toward an Alternative Future for Globalized Resource Policy?', *Challenge* (2016) 59: 212–14; Yonatan Strauch, Truzaar Dordi and Angela Carter, 'Constraining Fossil Fuels Based on 2°C Carbon Budgets: The Rapid Adoption of a Transformative Concept in Politics and Finance', *Climatic Change* (2020) 160: 182, 191; Lorenzo Pellegrini and Murat Arsel, 'The Supply Side of Climate Policies: Keeping

Unburnable Fossil Fuels in the Ground', *Global Environmental Politics* (2022) 22: 3–4.

13 Georgia Piggot, 'The Influence of Social Movements on Policies that Constrain Fossil Fuel Supply', *Climate Policy* (2018) 18: 942–54; Benedikter et al., '"Keep It"', 10–11; Strauch et al., 'Constraining Fossil', 183, 191–2; and for these cycles of climate activism and further references, see Malm, *How to Blow*, 14–15.

14 Bill McKibben, 'Global Warming's Terrifying New Math', *Rolling Stone*, 19 July 2012. On the importance of his article and the US-based divestment movement, see e.g. Julie Ayling and Neil Gunningham, 'Non-State Governance and Climate Policy: The Fossil Fuel Divestment Movement', *Climate Policy* (2017) 17: 132, 134–6; Noam Bergman, 'Impacts of the Fossil Fuel Divestment Movement: Effects on Finance, Policy and Public Discourse', *Sustainability* (2018) 10: 2, 10–11; Sibylle Braungardt, Jeroen van den Bergh and Tessa Dunlop, 'Fossil Fuel Divestment and Climate Change: Reviewing Contested Arguments', *Energy Research and Social Science* (2019) 50: 191; Ansar et al., 'Stranded Assets', 19, 49; Shukla et al., *Climate Change*, 1744–5; Caldecott, 'Introduction', 6–7, 13; Strauch et al., 'Constraining Fossil', 188–91. Another much cited estimate from the same period suggested that the reserves were not five but three times larger than the 2°C budget, which did not, however, change the essence of the terrifying math: Malte Meinshausen, Nicolai Meinshausen, William Hare et al., 'Greenhouse-Gas Emission Targets for Limiting Global Warming to 2°C', *Nature* (2009) 458: 1158–62. This classic article would come to be considered 'the earliest assessment of stranded assets': Aitken, 'Depletion Work', 9; cf. e.g. Shimbar, 'Environment-Related Stranded', 5.

15 Nicholas Stern, 'A Profound Contradiction at the Heart of Climate Change Policy', *Financial Times*, 8 December 2011. Stern also wrote a preface to another key document of these years of stranded asset discourse formation: James Leaton, Nicola Ranger, Bob Ward et al., 'Unburnable Carbon 2013: Wasted Capital and Stranded Assets', Carbon Tracker Initiative and The Grantham Research Institute, 2013, 4, 13.

16 Al Gore and David Blood, 'The Coming Carbon Asset Bubble', *Wall Street Journal*, 29 October 2013. The two had broached the topic of stranded assets in a more upbeat manifesto for something they called 'sustainable capitalism' two years earlier, without causing the splash this later piece did (see further below): Al Gore and David Blood, 'A Manifesto for Sustainable Capitalism', *Wall Street Journal*, 14 December 2011.

17 David Roberts, 'Bernie Sanders and Jeff Merkley Have a New

Bill to Leave Fossil Fuels in the Ground', *Vox*, 4 November 2015; Suzanne Goldenberg, 'Bernie Sanders Backs New Climate Plan to Curb US Fossil Fuel Extraction', *Guardian*, 4 November 2015.

18 As chronicled by Strauch et al., 'Constraining Fossil', 193; and for a document from Paris, see Jeff McMahon, 'UN Climate Strategists Have a Powerful New Ally: Money', *Forbes*, 3 December 2015.

19 As pointed out in Strauch et al. 'Constraining Fossil', 186.

20 Ibid., 197. Cf. Ayling and Gunningham, 'Non-State Governance', 135; Bergman, 'Impacts of', 8, 14–15; Braungardt et al., 'Fossil Fuel', 193; Aitken, 'Depletion Work', 276.

21 Temper et al., 'Movements Shaping', 18. Cf. Strauch et al., 'Constraining Fossil', 192.

22 Dimitri Zenghelis, Roger Fouquet and Ralph Hippe, 'Stranded Assets: Then and Now', in Caldecott, *Stranded Assets*, 23–5, 29.

23 Caldecott, 'Introduction', 17.

24 Carlota Perez, *Technological Revolutions and Financial Capital: The Dynamics of Bubbles and Golden Ages* (Cheltenham: Edward Elgar, 2002), 36. Perez's theory is summed up in support of the normality of asset stranding in Caldecott, 'Introduction', 5–6 (although he, seemingly undecided, opens the possibility of such stranding being 'qualitatively and quantitively different from previous drivers of creative destruction': ibid., 6). For a critique of Perez and an alternative view, see Andreas Malm, 'Long Waves of Fossil Development: Periodizing Energy and Capital', in Brent Ryan Bellamy and Jeff Diamanti (eds), *Materialism and the Critique of Energy* (Chicago: M-C-M', 2018), 161–95. For some further considerations on Schumpeter and Marx in the context of asset stranding, see Sarah Knuth, 'Green Devaluation: Disruption, Divestment, and Decommodification for a Green Economy', *Capitalism Nature Socialism* (2017) 28: 101–4; Julia Dehm, 'Legally Constituting the Value of Nature: The Green Economy and Stranded Assets', in Isabel Feichtner and Geoff Gordon (eds), *Constitutions of Value: Law, Governance, and Political Ecology* (London: Routledge, 2023), 269–70.

25 Or, 'stranded assets are inherent to the process of disruptive innovation in competitive markets', with reference to Schumpeterian theory, filtered through the work of Caldecott: Kyra Bos and Joyeeta Gupta, 'Stranded Assets and Stranded Resources: Implications for Climate Change Mitigation and Global Sustainable Development', *Energy Research and Social Science* (2019) 56: 4; cf. 3, 10; Firdaus and Mori, 'Stranded Assets', 76.

26 Karl Marx, *Capital: A Critique of Political Economy. Volume II* (London: Penguin, 1992 [1885]), 308–9; see further e.g. 247–8; Marx, *Grundrisse*, e.g. 661, 679–85.

27 On this distinction, see further David Harvey, *A Companion to Marx's* Capital, Volume 2 (London: Verso, 2013), 113–16, 126; David Harvey, *A Companion to Marx's* Grundrisse (London: Verso, 2023), 301–8, 351–2; and cf. Shaikh, *Capitalism*, 208.

28 Marx, *Capital Volume II*, 242, 289.

29 Ibid., 243; see further e.g. 239–40, 246–7, 276, 296.

30 Ibid., 238, 240, 250, 298–9; cf. Harvey, *A Companion to* Volume 2, 130; Harvey, *A Companion to* Grundrisse, 307; Paul Burkett, *Marxism and Ecological Economics: Toward a Red and Green Political Economy* (Leiden: Brill, 2006), 197; and the emphasis on fixed capital formation as a 'metabolic' process in Michael Ekers and Scott Prudham, 'The Metabolism of Socioecological Fixes: Capital Switching, Spatial Fixes, and the Production of Nature', *Annals of the American Association of Geographers* (2017) 107: 1370–88; Michael Ekers and Scott Prudham, 'The Socioecological Fix: Fixed Capital, Metabolism, and Hegemony', *Annals of the American Association of Geographers* (2018) 108: 17–34.

31 Marx, *Capital Volume II*, 288; cf. Harvey, *A Companion to* Volume 2, 111; Carton, 'Dancing to'; Ekers and Prudham, 'The Metabolism of Socioecological', 1376.

32 Harvey, *A Companion to* Grundrisse, 359–60, 300; see further e.g. 313, 325, 339–40, 350–5, 386–7.

33 Marx, *Capital Volume II*, 250 (emphasis added); cf. Harvey, *A Companion to* Volume 2, 136.

34 Marx, *Capital Volume I*, 528; cf. 318, 509–10; Marx, *Capital Volume II*, 250; and further Harvey, *The Limits*, 197; Harvey, *A Companion to* Volume 2, 111, 117, 135–6, 139; Andrea Furnaro, 'The Role of Moral Devaluation in Phasing Out Fossil Fuels: Limits for a Socioecological Fix', *Antipode* (2021) 53: 1444–6.

35 Marx, *Capital Volume II*, 264.

36 Karl Marx, *Capital: A Critique of Political Economy. Volume III* (London: Penguin, 1991 [1894]), 522 (emphasis added); cf. e.g. 916.

37 Marx, *Capital Volume II*, 250 (emphasis added); cf. Harvey, *A Companion to* Volume 2, 136.

38 Karl Marx, *Theories of Surplus Value* (New York: Prometheus, 2000 [1862–3]), book II, 495 (emphasis in original).

39 Ibid., 495–6; cf. Marx, *Capital Volume III*, 362–3.

40 Paul M. Sweezy, *The Theory of Capitalist Development: Principles of Marxian Political Economy* (New York: Monthly Review Press), 155. On how crises preserve the capitalist mode of production and form an integral and necessary part of its operations, see further, in much more detail, Holgersen, *Against the Crisis*.

41 Marx, *Capital Volume II*, 264; cf. Harvey, *A Companion to* Volume 2, 137.

42 The Schumpeterian theory of long waves was, of course, inspired by Marx, via Kondratiev.

43 Marx, *Grundrisse*, 749–50. Although he, in the rest of the sentence, interprets this recurring sort of crisis as 'advice' to the capitalist mode of production to get out of the way and let something better take its place.

44 See e.g. ibid., 750; Marx, *Capital Volume III*, 357, 362; and further Mandel, *Late Capitalism*; Mandel, *Long Waves*.

45 E.g. Peter Linquity and Nathan Cogswell, 'The Carbon Ask: Effects of Climate Policy on the Value of Fossil Fuel Resources and the Implications for Technological Innovation', *Journal of Environmental Studies and Sciences* (2016) 6: 673; Gregor Semieniuk, Philip B. Holden, Jean-Francois Mercure et al., 'Stranded Fossil-Fuel Assets Translate to Major Losses for Investors in Advanced Economies', *Nature Climate Change* (2022) 12: 532; Dehm, 'Legally Constituting', 269–70; Ekers and Prudham, 'The Metabolism of Socioecological', 1382; and this difference was already noticed by McKibben in 'Global Warming's'. Dehm points out that this negatively reveals the political constitution of fossil fuel property in existence – i.e. it was always already political.

46 For a very rosy imagining of such a scenario, see Zenghelis et al., 'Stranded Assets', 33–4.

47 Lucas Bretschger and Susanne Soretz, 'Stranded Assets: How Policy Uncertainty Affects Capital, Growth, and the Environment', *Environmental and Resource Economics* (2022) 83: 265, 282 (see further 263–6). Several other proposed definitions of asset stranding align with this insight. 'We define stranded assets as high-carbon assets in the power sector that are still within design service life and have prematurely lost financial value or are devalued before the end of design service life *because of carbon allowance limits*.' Weirong Zhang, Yiou Zhou, Zhen Gong et al., 'Quantifying Stranded Assets of the Coal-Fired Power in China under the Paris Agreement Target', *Climate Policy* (2023) 23: 12 (emphasis added). Or, better yet, with the International Energy Agency: stranded assets are 'investments, which have already been made but which, at some time prior to the end of their economic life, are no longer able to earn an economic return as a result of changes in the market and regulatory environment *brought about by climate policy*'. Quoted in Thomas Auger, Johannes Trüby, Paul Balcombe and Iain Stafell, 'The Future of Coal Investment, Trade, and Stranded Assets', *Joule* (2021) 5: 1463 (emphasis added).

48 Mark Campanale, 'Investors Need to Look Carefully at Stranded Asset Risks', *Brink News*, 24 January 2023.

49 'When the government suddenly': Frederick van der Ploeg and Armon Rezai, 'Stranded Assets in the Transition to a Carbon-Free Economy', *Annual Review of Resource Economics* (2020) 12: 282 (cf. 291); 'a global extinction event': Christina Atanasova and Eduardo S. Schwartz, 'Stranded Fossil Fuel Reserves and Firm Value', *NBER Working Paper Series*, no. 26497, November 2019, 3.

50 Jason Channell, Elizabeth Cumri, Phuc Nguyen et al., 'Energy Darwinism: Why a Low Carbon Future Doesn't Have to Cost the Earth', Citi GPS: Global Perspectives and Solutions, August 2015, 94.

51 Here building on Malm, *Fossil Capital*.

52 See e.g. Brett Christophers, *Rentier Capitalism: Who Owns the Economy and Who Pays for It?* (London: Verso, 2020), xvi; Paul Langley, Gavin Bridge, Harriet Bulkeley and Bregje van Veelen, 'Decarbonizing Capital: Investment, Divestment and the Qualification of Carbon Assets', *Economy and Society* (2021) 50: 497–9, 510.

53 'Enforced prohibition': Jeff D. Colgan, Jessica F. Green and Thomas N. Hale, 'Asset Revaluation and the Existential Politics of Climate Change', *International Organization* (2021) 75: 604; cf. 588; Linquiti and Cogswell, 'The Carbon Ask', 674; Knuth, 'Green Devaluation', 109; Zenghelis et al., 'Stranded Assets', 37. A classic reference here, preceding the discourse on stranded assets, is Marc D. Davidson, 'Parallels in Reactionary Argumentation in the US Congressional Debates on the Abolition of Slavery and the Kyoto Protocol', *Climatic Change* (2008) 86: 67–82; and see further below.

54 Richard Pipes, *The Russian Revolution* (New York: Vintage, 1990), 672, 692.

55 For a clarifying discussion of slavery that dispels the misunderstanding by some scholars of slaves as fixed capital, see John Clegg, 'A Theory of Capitalist Slavery', *Journal of Historical Sociology* (2020) 33: 74–98. As Clegg shows, slaves belonged to the category of variable capital, because they produced surplus value.

56 It should be pointed out that nationalisation is identical to asset stranding from the standpoint of the capitalist; but if e.g. a railroad is nationalised – the ownership structure altered but operations not so – the means of production are not, of course, taken out of circulation. They are merely expropriated from their private owners to be run by the state. In the case of nationalisation of private fossil fuel companies, proposals have been floated for a somewhat similar repurposing of certain types of technologies and

skills in the pursuit of carbon dioxide removal: we shall return to these in the sequel to this book. The distinction between the destruction of value in the process of valorisation and that of means of production *sensu stricto* obviously has major implications for society. But here, the central point is precisely that *from the perspective of capital*, any asset stranding represents a woeful loss (even if the railroad remains in nationalised use).

57 On the necessity of value destruction, see e.g. James Goodman and James Anderson, 'From Climate Change to Economic Change? Reflections on "Feedback"', *Globalizations* (2021) 18: 1260; Shimbar, 'Environment-Related Stranded', 1.

58 Or, perhaps better, Leninian-Marxian.

59 Alan Livsey, 'The $900bn Cost of "Stranded Energy Assets"', *Financial Times*, 4 February 2020.

60 Thom Allen and Mike Coffin, 'Unburnable Carbon: Ten Years On', Carbon Tracker Initiative, June 2022, 5.

61 Marx, *Capital Volume III*, 772; see further e.g. 752, 754, 911.

62 Ibid., 909.

63 Upton Sinclair, *Oil!* (New York: Penguin, 2007 [1926]), 25.

64 On the status and contexts of this novel, see Stephanie LeMenager, 'The Aesthetics of Petroleum, after *Oil!*', *American Literary History* (2012) 24: 59–86. LeMenager unfortunately labels it a work of 'peak-oil fiction'. Ibid., 63. On the utter obsolescence of the term peak oil, see below. For an updated reading of Sinclair that more directly places his novel in the context of the climate crisis, see Michael Tondre, '*Oil!*: On the Petro-Novel', *Paris Review*, 1 March 2023.

65 Sinclair, *Oil!*, 96–7. The oil has been pushed to the surface by an earthquake. The son, Bunny, later defects to the socialist camp, under the influence of a friend converting to Bolshevism while serving in the American expedition in Siberia. This friend, Paul, sides with the political destruction of capital in general: 'our troops were in Siberia because American bankers and big businessmen had loaned enormous sums of money to the government of the Tsar, both before the war and during it; the Bolshevik government had repudiated these debts, and therefore our bankers and businessmen were determined to destroy it. It was not merely the amount of the money, but the precedent involved; if the government of any country could repudiate the obligations of a previous government, what would become of international loans? . . . The total amount of international loans was one or two hundred billions of dollars, and the creditor nations meant to make an example of Soviet Russia, and establish the rule that a government which repudiated its debts would be out of business': logic

of the original, pre-climatic, pan-capitalist asset stranding and the counter-revolution against it. Ibid., 250–1.

66 Ibid., 66.

67 Ibid., 60.

68 Abelrahman Munif, *Cities of Salt* (New York: Vintage, 1989 [1984]), 68.

69 The secondary literature on Munif and his entrance into the canon of ecocriticism begins with Amitav Gosh, 'Petrofiction: The Oil Encounter and the Novel', in *Incendiary Circumstances: A Chronicle of the Turmoil of Our Times* (Boston: Houghton Mifflin, 2005), 138–51. A complaint about Gosh ignoring Sinclair was filed in an important essay by Graeme Macdonald, 'Oil and World Literature', *American Book Review* (2012) 33: 7. A particularly compelling reading of Munif remains Rob Nixon, *Slow Violence and the Environmentalism of the Poor* (Cambridge, MA: Harvard University Press, 2011), 68–102. For a reading that puts him in the specific context of fossil fuel fiction, on which this paragraph draws, see Andreas Malm, '"This Is the Hell that I Have Heard of": Some Dialectical Images in Fossil Fuel Fiction', *Forum for Modern Language Studies* (2017) 53: 121–41.

70 Munif, *Cities of Salt*, 44.

71 Ibid., 71.

72 Hussein K. Abdel-Aal, *Economic Analysis of Oil and Gas Engineering Operations* (Boca Raton: CRC Press, 2021), 159; Saeid Asadzadeh, Wilson José de Oliveira and Carlos Roberto de Souza Filho, 'UAV-Based Remote Sensing for the Petroleum Industry and Environmental Monitoring: State-of-the-Art and Perspectives', *Journal of Petroleum Science and Engineering* (2022) 208: 10–11. These instruments first came into use in the post-war decades: see Homer Jensen, 'The Airborne Magnetometer', *Scientific American* (1961) 204: 151–65.

73 Asadzadeh et al., 'UAV-Based Remote', 11.

74 Marx, *Grundrisse*, 706.

75 See e.g. Sinclair, *Oil!*, 74; Munif, *Cities of Salt*, 30–1, 67–70, 83–4, 97–8. As for the former, it is remarkable that its explorationist's jargon remains unchanged to this day: the talk is still of 'wildcat wells' and 'dry wells'. See e.g. Abdel-Aal, *Economic Analysis*, 10, 14.

76 For some of the challenges and technologies of exploration in such waters, see Sidum Adumene, Faisal Khan, Sunday Adedigba et al., 'Offshore Oil and Gas Development in Remote Harsh Environments: Engineering Challenges and Research Opportunities', *Safety in Extreme Environments* (2022), online first, 1–17.

77 Equinor, 'Exploration for Oil and Gas', 2023. Equinor describes

its exploration activities in these places during the first years of the third decade in Equinor, 2021: *Annual Report and Form 20–F*, 40–5; Equinor, 2022: *Integrated Annual Report*, 105, 114–18.

78 Cf. e.g. Atanasova and Schwartz, 'Stranded Fossil', 3; Abdel-Aal, *Economic Analysis*, 157, 163. Thus the airborne magnetometer was presented as a crucial step, from discovering oil *on* the surface – as in Sinclair's crevice – to deep *under* it: Jensen, 'The Airborne Magnetometer', 151. Similar tendencies are, of course, at work in other extractive industries too. A very general theory about them is outlined in Stephen G. Bunker and Paul S. Ciccantell, *Globalization and the Race for Resources* (Baltimore: Johns Hopkins University Press, 2005). (Grandiloquently, Bunker and Ciccantell labelled their theory 'New Historical Materialism', a term that did not catch on).

79 Richard T. Kelly, *The Black Eden* (London: Faber and Faber, 2023). The tendency registers in one minor classic of fossil fuel fiction: *Greenvoe* by George Mackay Brown, likewise set in the North Sea around the time of the offshore breakthrough. Here the 'metal monsters' and 'engines of destruction' attacking the remote island village of Hellya have another order of technical enormity: a helicopter pad built for the explorationists, an extension to the pier making room for the larger boats, the faint whine of pneumatic drills in the bottom of the sea. George Mackay Brown, *Greenvoe* (Edinburgh: Polygon, 2004 [1972]), quotations from 210, 215. For contexts and a fine reading of *Greenvoe*, alongside the ever-present *Cities of Salt*, see Graeme Macdonald, '"Monstrous Transformer": Petrofiction and World Literature', *Journal of Postcolonial Writing* (2017) 53: 289–302. But the tendency is absent in the other major recent novel of petrofiction, possibly because it is set on the land: Imbolo Mbue, *How Beautiful We Were* (New York: Random House, 2021). In many other respects, the American oil companies of these two novels – Mbue naming hers Pexton, Kelly his Paxton – are similar in their destructivity. In another key work from the frontiers of African oil extraction, the level of technology is, likewise, not much more advanced than in Sinclair or Munif: Helon Habila, *Oil in Water* (London: Penguin, 2011).

80 Kelly, *The Black*, 52.

81 Ibid., 72, 98.

82 Ibid., 86.

83 Ibid., 152, 154.

84 Ibid., 148; cf. e.g. 391, 399, 407. The logic of investment could here also be read as a mirror of the form of the novel: once begun, the story must continue; it must not be broken off mid-sentence.

85 Ibid., 410; cf. 312–14.

86 Ibid., 103; cf. 110.

87 Sinclair, *Oil!*, 163–4.

88 Munif, *Cities of Salt*, 86. The emir personifying the reactionary dictatorships of the Arabian Peninsula, here defending the American explorationists as against the complaints from the inhabitants of the oasis.

89 Kelly, *The Black*, 97.

90 Equinor, *2020: Annual Report and Form 20–F*, 45. The buyer: Grayson Mill Energy.

91 Marx, *Capital Volume III*, 752. Some one hundred and forty pages later, however, Marx emphasises that 'extractive industry' should 'be clearly distinguished from agriculture', largely because fixed capital 'does play a major role in mining.' Ibid., 893–4.

92 Cf. Timothy Mitchell, *Carbon Democracy: Political Power in the Age of Oil* (London: Verso, 2011), 193.

93 Abdel-Aal, *Economic Analysis*, 158. Dhahran, it might be noted, is the Saudi city to which the former inhabitants of the destroyed oasis in *Cities of Salt* are herded, although Munif calls the place 'Harran', meaning 'the overheated'.

94 On this status, see Malm, *Fossil Capital*.

95 Cyrus Bina, 'Some Controversies in the Development of Rent Theory: The Nature of Oil Rent', *Capital and Class* (1989) 13: 89.

96 Cf. Abdel-Aal, *Economic Analysis*, 134, 174; Bård Mismund and Petter Osmundsen, 'Valuation of Proved vs. Provable Oil and Gas Reserves', *Cogent Economics and Finance* (2017) 5: 1.

97 Marx, *Capital Volume III*, 790.

98 Ibid., 756. Cf. *Capital Volume II*, 288. The concept of terre-capital has primarily been picked up by theorists of 'landesque capital', or agricultural land whose productivity has been enhanced by investment: see e.g. Mats Widgren, 'Precolonial Landesque Capital: A Global Perspective', in Alf Hornborg, J. R. McNeill and Joan Martinez-Alier (eds), *Rethinking Environmental History: World-System History and Global Environmental Change* (Lanham: AltaMira, 2007), 62–3; Eric Clark and Huei-Min Tsai, 'Islands: Ecologically Unequal Exchange and Landesque Capital', in Alf Hornborg, Brett Clark and Kenneth Hermele, *Ecology and Power: Struggles over Land and Material Resources in the Past, Present, and Future* (London: Routledge, 2012), 55–7.

99 In farming, terre-capital is the result of a change in the 'physical characteristics' and 'chemical properties' of the cultivable soil, not the appearance of the soil per se: Marx, *Capital Volume III*, 879. The latter would apply, on the other hand, to other extractive

industries as well; and one cannot analytically rule out a future scenario, however hypothetical, in which rare earth metals – or, indeed, uranium – take on a position commensurate to that of fossil fuels.

100 Abdel-Aal, *Economic Analysis*, 54. Such capital includes, of course, that wasted on dry holes.

101 Ibid., 174.

102 Equinor, 'Exploration for'.

103 Adumene et al., 'Offshore Oil', 2.

104 Bob Henderson, 'The Offshore Oil Business Is Gushing Again', *Wall Street Journal*, 21 January 2023. The *Journal* reported that 80 per cent of new capacity in Saudi Arabia would likewise come from offshore fields.

105 That is, bracketing out the reduction of the remaining carbon budget. But this could only happen, of course, insofar as the growth in reserves outpaces their depletion – of which nothing can be known, except that oil and gas companies in the early third decade seemed as determined as ever to make the most of their 'innovative ideas' and 'latest technologies'; see further below.

106 There is, in keeping with Marxian taxonomy, some ambiguity to the category of fossil terre–capital here. Insofar as it precedes commodity production – identified reserves whose extraction has not yet begun – there is a fictitious quality to it. Before value in motion has taken hold of material resources and transformed them into production, the capital is, at least when narrowly conceived, still fictitious and potential; but this makes little difference to the owners of the capital in question.

107 See e.g. Nils Johnson, Volker Krey, David L. McCollum et al., 'Stranded on a Low-Carbon Planet: Implications of Climate Policy for the Phase-Out of Coal-Based Power Plants', *Technological Forecasting and Social Change* (2015) 90: 90; Auger at al., 'The Future of Coal', 1463, 1472; Hauenstein, 'Stranded Assets', 2.

108 Cf. Marx, Bina, 'Some Controversies', 99–100; Marx, *Capital Volume III*, 759, 785. On this tiny analytical detail, we deviate from the analysis in *Rentier Capitalism*, in which rent is the catch-all category including the kind of gain hydrocarbon companies get from their reserves: Christophers, *Rentier Capitalism*, 99. Rent as a term for just owning things without doing or making anything – the thrust of Christophers's argument – fits poorly onto the activities of fossil fuel companies. The rent is rather pocketed by the owners of the land who let them use it, through one arrangement or another, in accordance with the

classical Marxian distinction between surplus value and ground rent (the latter, of course, derived from the former).

109 Paul Roberts, *The End of Oil: On the Edge of a Perilous New World* (New York: Houghton Mifflin, 2005); Andrew McKillop and Sheila Newman (eds), *The Final Energy Crisis* (London: Pluto, 2005); David Goodstein, *Out of Gas: The End of the Age of Oil* (New York: W. W. Norton, 2004); Jeremy Leggett, *Half Gone: Oil, Gas, Hot Air and the Global Energy Crisis* (London: Portobello, 2005). Embarrassingly, one of the present authors relayed this argument wholesale in a book fortunately published only in Swedish: Andreas Malm, *När kapitalet tar till vapen: Om imperialism i vår tid* (Stockholm: Agora, 2004); but some of it spilled into Andreas Malm and Shora Esmailian, *Iran on the Brink: Rising Workers and Threats of War* (London: Pluto, 2007).

110 Colin J. Campbell and Jean H. Laherrère, 'The End of Cheap Oil', *Scientific American* (1998) 278: 78. For summaries of the theory, see Mazen Labban, 'Oil in Parallax: Scarcity, Markets, and the Financialization of Accumulation', *Geoforum* (2010) 41: 542–3; Gavin Bridge and Andrew Wood, 'Less Is More: Spectres of Scarcity and the Politics of Resources Access in the Upstream Oil Sector', *Geoforum* (2010) 41: 566–7; Gavin Bridge, 'Past Peak Oil: Political Economy of Energy Crises', in Richard Peet, Paul Robbins and Michael Watts, *Global Political Ecology* (Abingdon: Routledge, 2011), 313–14; Ugo Bardi, 'Peak Oil, 20 Years Later: Failed Prediction or Useful Insight?', *Energy Research and Social Science* (2019) 48: 257–8.

111 Campbell and Laherrère, 'The End'.

112 Kenneth S. Deffeyes, *Beyond Oil: The View from Hubbert's Peak* (New York: Farrar, Straus and Giroux, 2005), 3.

113 Statista, 'Oil Production Worldwide from 1998 to 2022', n.d.; Enerdata, 'Crude Oil Production', *World Energy and Climate Statistics: Yearbook 2023*, n.d. The expected drop by 10 per cent: Deffeyes, *Beyond Oil*, 7. On the incorrect predictions of the peak date, cf. Bardi, 'Peak Oil', 259.

114 David Demin, 'M. King Hubbert and the Rise and Fall of Peak Oil Theory', *AAPG Bulletin* (2023) 107: 852; Matt Egan, 'America Is Now the World's Largest Oil Producer', *CNN*, 12 September 2018.

115 Nerijus Adomaitis, 'Norway Expects Jump in Oil Output and Gas Near Record Highs', Reuters, 9 January 2023.

116 On this depreciation, see Bardi, 'Peak Oil', 257–8. Ironically, interest in the theory formed a neat bell curve, with a conspicuous peak in 2005, as measured in e.g. the number of academic papers mentioning the term: ibid., 259–60. On scholars guilty of

peak oil illusions, see Bridge, 'Past Peak', 308–9, 315; and cf. note 550 above. But it is also worth remembering the extent to which mainstream policy and thinking in the Anglo-American imperial core was influenced by the peak oil theory in those years: see e.g. Bridge and Wood, 'Less Is', 566.

117 See e.g. Campbell and Laherrère, 'The End', 81–3; Deffeyes, *Beyond Oil*, 99–123. Particularly ridiculous was the argument that tar sands and shale would not be developed, because they would cause pollution: Campbell and Laherrère, 'The End', 82–3.

118 See e.g. Russell Gold, 'Why Peak-Oil Predictions Haven't Come True', *Wall Street Journal*, 28 September 2014.

119 Advertisement included in Bridge and Wood, 'Less Is', 570. On such enhanced recovery, cf. Labban, 'Oil in Parallax', 544.

120 Cf. Bridge and Wood, 'Less Is', 571–3.

121 Gold, 'Why Peak-Oil'.

122 For the former two, see e.g. Demin, 'M. King Hubbert'.

123 'How can you reduce [sic] carbon dioxide in the atmosphere? Run out of oil.' Deffeyes, *Beyond Oil*, xiv.

124 See e.g. Filip Johnsson, Jan Kjärstad and Johan Rootzén, 'The Threat to Climate Change Mitigation Posed by the Abundance of Fossil Fuels', *Climate Policy* (2019) 19: 258–74; Krista Halttunen, Raphael Slade and Iain Stafell, 'What if We Never Run Out of Oil? From Certainty of "Peak Oil" to "Peak Demand"', *Energy Research and Social Science* (2022) 85: 1–6; Ansari and Fareed, 'Stranded Assets', 1.

125 Van der Ploeg and Rezai, 'Stranded Assets', 287; Frederick van der Ploeg and Armon Rezai, 'The Risk of Policy Tipping and Stranded Carbon Assets', *Journal of Environmental Economic and Management* (2020) 100: 4.

126 Van der Ploeg and Rezai, 'The Risk of Policy', 12.

127 Nicholas Newman, 'The Rig of the Future', *Energy Focus*, 4 April 2019.

128 Louise Davis, 'Using an Offshore Platform beyond its Expected Lifespan', *Engineer Live*, 1 August 2018; cf. e.g. Eva Grey, 'Extending the Lifespan of Offshore Assets: Can We Live with It?', *Offshore Technology*, 25 May 2016; J. P. Casey, 'Asset Life Extension: Viable in the Long Term for Oil and Gas?', *Offshore Technology*, 6 May 2020.

129 E.g. *Offshore Technology*, 'PETRONAS Launches New Corrosion Protection Technology', 6 April 2023.

130 Adumene et al., 'Offshore Oil'. This later type of maintenance clearly renders established fixed capital not so 'costless', as in Brenner above; but it shows the gains to be expected from keeping costly structures intact.

131 Mehdi Hajinezhadian and Behrouz Behnam, 'A Probabilistic Approach to Lifetime Design of Offshore Platforms', *Nature Scientific Reports* (2023) 13: 1–24.

132 The average cost of a platform in the early 2020s was 650 million dollars. Trevor English, 'The Engineering and Construction of Offshore Oil Platforms', *Interesting Engineering*, 12 January 2020.

133 Cf. Abdel-Aal, *Economic Analysis*, 196.

134 Ibid., e.g. 216–19, 243; Rozenberg et al., 'Climate Constraints', 3.

135 Campanale, 'Investors Need'.

136 Amy Stillman, 'Mexico's Largest Refinery Is Now Open. It's Just Not Making Any Fuel', *Bloomberg*, 1 July 2020; Adriana Barrera, 'Mexico's Newest Oil Refinery Now Seen Working at Half Capacity in Mid-2023', Reuters, 24 December 2022; Tom Wilson, 'Saudi Aramco Strengthens China Ties with Two Refinery Deals', *Financial Times*, 27 March 2023.

137 Erwin Seba, 'Exclusive: Exxon Prepares to Start Up $2 bln Texas Oil Refinery Expansion', Reuters, 13 January 2023.

138 Abdel-Aal, *Economic Analysis*, 297. 'Pipelines power prosperity.' Ibid. See further 208–304.

139 Ibid., 314.

140 On these, see e.g. Angelika von Dulong, 'Concentration of Asset Owners Exposed to Power Sector Stranded Assets May Trigger Climate Policy Resistance', *Nature Communications* (2023) 14: 1–9.

141 Karen C. Seto, Steven J. Davis, Ronald B. Mitchell et al., 'Carbon Lock-In: Types, Causes, and Policy Implications', *Annual Review of Environment and Resources* (2016) 41: 426. The reference here is to 'the current global energy system', more than 80 per cent of which is powered by fossil fuels.

142 On power plants as part of this circuit, see Malm, *Fossil Capital*, 359.

143 Louison Cahen-Fourot, Emanuele Campiglio, Antoine Godin et al., 'Capital Stranding Cascades: The Impact of Decarbonisation on Productive Asset Utilisation', *Energy Economics* (2021) 103: 1–15; Caitlin Swalec, 'Pedal to the Metal: It's Not Too Late to Abate Emissions from the Global Iron and Steel Sector', GEM, June 2022; Lukas Hermwille, Stefan Lechtenböhmer, Max Åhman et al., 'A Climate Club to Decarbonize the Global Steel Industry', *Nature Climate Change* (2022) 12: 495.

144 A coal plant, it has been pointed out, can be turned into a museum, but this seems to be an inherently limited posthumous market. Van der Ploeg and Rezai, 'Stranded Assets', 289.

145 Cf. ibid., 282.

146 Technically, in Marxian terms, this would be the department producing means of production for the circuit of primitive accumulation of fossil capital.

147 Cf. Seto et al., 'Carbon Lock-In', 431–2.

148 Dawud Ansari and Ambria Fareed, 'Stranded Assets and Resource Rents: Between Flaws, Dependency, and Economic Diversification', *DIW Roundup*, 12 November 2019, 2.

149 Harvey, *The Limits*, 276–7; cf. Harvey, *A Companion to* Volume 2, 243–4, 249.

150 Marx, *Capital Volume III*, 598 (emphasis added).

151 See e.g. Mismund and Osmundsen, 'Valuation of Proved'; Atanasova and Schwartz, 'Stranded Fossil Fuel'; Langley et al., 'Decarbonizing Capital', 506; Christophers, *Rentier Capitalism*, 100–1; Jan Bebbington, Thomas Schneider, Lorna Stevenson and Alison Fox, 'Fossil Fuel Reserves and Resources Reporting and Unburnable Carbon: Investigating Conflicting Accounts', *Critical Perspectives on Accounting* (2020) 66: 2.

152 More precisely (or not so precisely), the 'probable' reserves are assigned a 50 per cent chance of having quantities larger than estimated, the 'possible' reserves a 10 per cent ditto and a 90 per cent chance of having smaller quantities – a system for assigning probabilities by petroleum engineers, fraught with uncertainty and indeed arbitrariness. See e.g. Shishir Khetan and Naveed Yahya, 'Valuation Methodologies in the Oil and Gas Industry', *Stout*, 1 March 2016; Chris Dumont, '5 Common Trading Multiples Used in Oil and Gas Valuation', *Investopedia*, 26 January 2022; Jason Fernando, 'Possible Reserves', *Investopedia*, 27 June 2022. The proved reserves dominate share valuation, with a smaller role for the probable and virtually none for possible: Misund and Osmundsen, 'Valuation of Proved', 2–3.

153 Bert Scholtens and Robert Wagenaar, 'Revisions of International Firms' Energy Reserves and the Reaction of the Stock Market', *Energy* (2011) 36: 3451–6.

154 Maria Lundin, *En 1900-talsresa: Från Odessa till Bromma* (Södertälje: Fingraf, 1993).

155 Robert Eriksson, *Adolf H. Lundin: Med olja i ådrorna och guld i blick* (Klippan: Sellin and Partner, 2003). For a brief biography in English, see Jonathan Kandell, 'The Life and Death of Adolf Lundin', *Institutional Investor*, 16 November 2006.

156 Namely Eva of the Wehtje family of the company Cementa: on its pro-Nazi activities during the war, see Joakim Berglund, *Quislingcentralen: Nazismen i Skåne på 30- och 40-talet* (Malmö: Weinco, 1994), 111–14. As of the early third decade,

the Cementa factory in Slite was the second largest point source of CO_2 emissions in Sweden. *SVT*, 'Lista: Sveriges fem största utsläppare', 29 September 2021.

157 Kerstin Lundell, *Affärer i blod och olja: Lundin Petroleum i Afrika* (Stockholm: Ordfront, 2010); and for the trial, e.g. Reuters, 'Sweden Charges Lundin Energy Executives with Complicity in Sudan War Crimes', 11 November 2021; Trial International, 'Lundin Energy – Alex Schneiter and Ian Lundin', 17 April 2023. For the broader context, see Malm and the Zetkin Collective, *White Skin*.

158 The discovery was made in 2010. See e.g. Lundin Petroleum, 'Johan Sverdrup Development: The Most Important Norwegian Industrial Project over the Next 80 Years', n.d., 2.

159 Misund and Osmundsen, 'Valuation of Proved', 2.

160 Simon Johnson and Christopher Jungstedt, 'Lundin's Avaldsnes Find Adds Extra Shine to Q3', Reuters, 2 November 2011.

161 Henrik Öhlin, 'Börs: Lundin Petroleum är Europas bästa aktie senaste 10 åren', *Aktiespararna*, 17 April 2012.

162 Nick Ferris, 'How One Field Has Transformed Norway's Oil Fortunes', *Energy Monitor*, 5 December 2022; Kari Lundgren, 'At North Sea Field a Scoreboard Tracks Norway's Rising Oil Clout', *Bloomberg*, 14 February 2023.

163 A fourth share was held by the Norwegian state-owned company Petoro. Equinor, 'Johan Sverdrup Phase 2 On Stream', 15 December 2022. Johan Sverdrup was the founder of Norwegian parliamentarism. On nationalism and oil and Norway, see Malm and the Zetkin Collective, *White Skin*, 118–32.

164 Lundin Energy, 'Press Release: Completion of the Combination of Lundin Energy's E&P Business with Aker BP', 30 June 2022.

165 Nerijus Adomaitis, 'Norway Ramps Up W. Europe's Largest Oilfield as Oil's Future Questioned', Reuters, 5 December 2019. Production began in October 2019, but the Norwegian government held the official ceremony on 7 January 2020; during the festivities, the minister of petroleum and energy, climate denialist Sylvi Listhaug declared that 'there is no reason to put a brake on oil.' Helene Halvorsen Rossholt, Maria Knoph Vigsnæs, Kjersti Hetland and Erlend Frafjord, 'Listhaug om Johan Sverdrup-åpningen: – Ingen grunn til oljebrems', *NRK*, 7 January 2020. On Listhaug, see further Malm and the Zetkin Collective, *White Skin*, 130. On some of the reasons for why Johan Sverdrup was such an example of why oil production should continue because it was now green, see further below.

166 Equinor, 'Johan Sverdrup'.

167 Lundgren, 'At North Sea'.

168 Halvorsen Rossholt et al., 'Listhaug om'; Nerijus Adomaitis, 'Norway Expects Jump in Oil Output and Gas Near Record Highs', Reuters, 9 January 2023.

169 Refinitiv Streetevents, 'Q4 2022 Equinor ASA Earnings Call and Capital Markets Update', 8 February 2023, 19.

170 We are not aware of any such study. For an old, hyper-empirical account of the events during said period, showing how investment in railways, steamships and coal mines drove much of the ascension, see Bishop C. Hunt, 'The Joint-Stock Company in England, 1830–1844', *Journal of Political Economy* (1935) 43: 331–64. Needless to say, the joint-stock company form had, just like coal mining and many other capitalist phenomena, its origins in the late medieval era: the early nineteenth century was a period of synthesisation of these elements.

171 Marx, *Capital Volume I*, 780; cf. Cédric Durand, *Fictitious Capital: How Finance Is Appropriating Our Future* (London: Verso, 2017 [2014]), 91–2.

172 Marx, *Capital Volume III*, 597, 608. Cf. e.g. 567, 1046; Harvey, *The Limits*, 276–7; Harvey, *A Companion to* Volume 2, 208, 231; Ekers and Prudham, 'The Metabolism of Socioecological', 1377.

173 On shares as simultaneously fictitious and real capital, see Durand, *Fictitious Capital*, 51–3; Harvey, *The Limits*, e.g. 277; Harvey, *A Companion to* Volume 2, 249.

174 Cf. e.g. Durand, *Fictitious Capital*, 63–4.

175 Marx, *Capital Volume III*, e.g. 464–71; Harvey, *The Limits*, 257–8; and on this type of capital, see further François Chesnais, *Finance Capital Today: Corporations and Banks in the Lasting Global Slump* (Chicago: Haymarket, 2016), 66–73.

176 Marx, *Capital Volume II*, 311–12. See further Harvey, *The Limits*, 271; Harvey, *A Companion to* Volume 2, 82, 213, 218, 235–6, 247–8, 265.

177 Marx, *Grundrisse*, 782; Harvey, *A Companion to* Grundrisse, 297, 358, 416–8, 420, 424; David Harvey, *Marx, Capital and the Madness of Economic Reason* (London: Profile, 2019), 147–8.

178 Kelly, *The Black*, 249.

179 Harvey, *The Limits*, 265, 269.

180 Marx, *Capital Volume III*, 490 (emphasis in original). But in the modern credit system, of course, it is the money of *all* classes, including the proletariat, that is sucked into the whirlpool: see e.g. Harvey, *The Limits*, 262–3. Hence the pension funds as targets for divestment campaigns.

181 Winta Beyene, Manthos Delis and Steven Ongena, 'Financial Institutions' Exposure to Fossil Fuel Assets: An Assessment of

Financial Stability Concerns in the Short Term and in the Long Run, and Possible Solutions', Study Requested by the ECON Committee, European Parliament, June 2022, 15. See further Irene Monasterolo, 'Climate Change and the Financial System', *Annual Review of Resource Economics* (2020) 12: 299–320. Mister Paxton reddens when someone proposes restraints on his industry in the national interest. 'You said *what?* Listen, I got creditors, patriot, and they ain't gonna wait. You Brits said you wanted the oil real fast.' Kelly, *The Black*, 377 (emphasis in original). Cf. 390. The creditors chivvy Paxton to conclude his part of the deal: to produce enough oil to repay the loan with interest.

182 Beyene et al., 'Financial Institutions', 16.

183 Monasterolo, 'Climate Change', 311–12; Zhang et al., 'Quantifying Stranded', 7–8.

184 Gregor Semieniuk, Emanuele Campiglio, Jean-Francois Mercure et al., 'Low-Carbon Transition Risks for Finance', *WIREs Climate Change* (2020) 12: 9.

185 Adam Hanieh, 'The Commodities Fetish? Financialisation and Finance Capital in the US Oil Industry', *Historical Materialism* (2021) 29: 70–113. On the role of asset managers, see further Lara Cuvelier, 'The Asset Managers Fueling Climate Chaos: 2022 Scorecard on Asset Managers, Fossil Fuels and Climate Change', Reclaim Finance, April 2022; Simon Mundy and Kaori Yoshida, 'Squaring Fossil Fuel Holdings with Climate Pledges', *Financial Times*, 21 April 2023; and the important study, confirming Hanieh's findings, by Joseph Baines and Sandy Brian Hager, 'From Passive Owners to Planet Savers? Asset Managers, Carbon Majors and the Limits of Sustainable Finance', *Competition and Change* (2023) 27: 449–71.

186 E.g. Marx, *Capital Volume III*, 379–93; Harvey, *A Companion to Volume II*, 147–65.

187 Blas and Farchy, *The World*. On commodity traders as contemporary instantiations of commercial or merchant capital, see Chesnais, *Finance Capital*, 115–17.

188 Blas and Farchy, *The World*, e.g. 16–17, 44–57, 293.

189 On oil and the commodity traders, see also Michael Watts, 'Trading Houses and Market Volatility: Global Commodity Traders and the Shifting Landscapes of Oil and Gas', *Environment and Planning A: Economy and Space* (2022) 54: 1653–7.

190 'Vitol 2022 Volumes and Review', Vitol, 20 March 2023; Fortune Global 500, 2023 (in which list Vitol was not included).

191 'Trafigura Group', Fortune Global 500, 2023.

192 Blas and Farchy, *The World*, 318–19.

193 Observatory of Economic Complexity, 'World', 2021.

194 UNCTAD, *Review of Maritime Transport 2023: Towards a Green and Just Transition* (Geneva: United Nations, 2023), 12. Note that, again, these were figures (metric tons) for 2021, growing in 2022.

195 Ibid., 30.

196 On 1.5°C requiring the abolition of the bulk of fossil fuel trade, with nothing to replace it, see K. Keramidas, F. Fosse, A. Diaz Rincon et al., *Global Energy and Climate Outlook 2022: Energy Trade in a Decarbonised World* (Luxemburg: Publications Office of the European Union, 2022).

197 Trafigura, 'Shipping and Marine Logistics', 2023; Blas and Farchy, *The World*, 118, 304.

198 Blas and Farchy, *The World*, 11, 13, 64, 171, 186–7, 260–4, 273, 292–3, 326.

199 Mark Carney, 'Breaking the Tragedy of the Horizon: Climate Change and Financial Stability', speech at Lloyd's of London, 29 September 2015, 9.

200 E.g. J.-F. Mercure, H. Pollitt, J. E. Viñuales et al., 'Macro-economic Impact of Stranded Fossil Fuel Assets', *Nature Climate Change* (2018) 8: 588; Brett Christophers, 'Climate Change and Financial Instability: Risk Disclosure and the Problematics of Neoliberal Governance', *Annals of the American Association of Geographers* (2017) 107: 1108, 1113; Hugh Miller and Simon Dikau, 'Preventing a "Climate Minsky Moment": Environmental Financial Risks and Prudential Exposure Limits', The Grantham Research Institute on Climate Change and the Environment, March 2022, 1, 7; Caldecott, 'Introduction', 7; David Comerford and Alessandro Spiganti, 'The Carbon Bubble: Climate Policy in a Fire-Sale Model of Deleveraging', *Scandinavian Journal of Economics* (2022), online first, 1–38; UNEP, *Emissions Gap*, 67–8; Beyene et al., 'Financial Institutions', 16–18; Van der Ploeg and Rezai, 'Stranded Assets', 293.

201 Hyman P. Minsky, 'The Financial Instability Hypothesis: An Interpretation of Keynes and an Alternative to "Standard" Theory', *Nebraska Journal of Economics and Business* (1977) 16: 10.

202 Ibid., 15 (emphasis added). For Marxist appraisals of Minsky, see Maria N. Ivanova, 'Marx, Minsky, and the Great Recession', *Review of Radical Political Economics* (2012) 45: 59–75; Jim Kincaid, 'Marx after Minsky: Capital Surplus and the Current Crisis', *Historical Materialism* (2016) 24: 105–46; Durand, *Fictitious Capital*, 29–30.

203 E.g. Anthony De Grandi and Christian Tutin, 'Marx and the

"Minsky Moment": Liquidity Crises and Reproduction Crises in *Das Kapital*', *The European Journal of the History of Economic Thought* (2020) 27: 853–80; Kincaid, 'Marx after', 110–14; Harvey, *The Limits*, 266–70; Harvey, *A Companion to* Volume 2, 177–8, 213–14, 251–2, 257–60; Chesnais, *Finance Capital*, 81–6; Durand, *Fictitious Capital*.

204 Harvey, *A Companion to* Volume 2, 203. Cf. e.g. 181, 214, 219, 389.

205 Ibid., 389. There are, of course, more twists to the relation between fictitious and fixed capital in Harvey's theories of crisis. The former has a tendency to 'flit away' from the latter, leaving it 'high and dry and subject to savage devaluation'. Ibid., 111. Here, fixed capital appears as a barrier to the mobility of capital, an 'increasingly sclerotic' landscape that prevents accumulation from fulfilling its potentials; but in the next moment, having been abandoned, it is a victim of devaluation – the urban wastelands of Baltimore or Detroit that fill the pages of Harvey. See e.g. ibid., 111–12, 138; David Harvey, *Spaces of Global Capitalism: Towards a Theory of Uneven Geographical Development* (London: Verso, 2006), 101–2; David Harvey, *The Enigma of Capital and the Crisis of Capitalism* (London: Profile, 2010), 191; Harvey, *Marx, Capital*, 150. On the other hand, fixed capital also offers the *solution* to the problem of surplus capital – too much money floating around on the various markets for its fictitious manifestations, finally released into profit-making channels when new urban landscapes arise. Fixed capital formation through Chinese urbanisation is here the paradigmatic example: see e.g. Harvey, *A Companion to* Grundrisse, 297, 360, 375, 415–19; Harvey, *Marx, Capital*, 150, 178–84. Neither of these scenarios quite correspond to the relation between fictitious and fixed capital in the run-up to asset stranding. The fixed is not a barrier to the fictitious; nor is it a solution for absorbing its surpluses. The two are rather more harmoniously integrated in a Gordian Knot that awaits some political sword.

206 Cf. Durand, *Fictitious Capital*, 31–4, 38–9; Ivanova, 'Marx, Minsky', 63.

207 Harvey, *A Companion to* Volume 2, 246; also in Harvey, *A Companion to* Grundrisse, 14, 428.

208 Harvey also speaks of 'disciplinary power' exercised in the moment of crisis, but this power is then located in the sphere of value production as such, which punishes fictitious capital with limitations and subordinates it to reality – something else, again. Harvey, *A Companion to* Volume 2, 181, 247.

209 Marx, *Capital Volume III*, 911 (emphases added).

210 Ibid. John Bellamy Foster quotes this sentence in *Marx's Ecology: Materialism and Nature* (New York: Monthly Review Press, 2000), 164; *Ecology Against Capitalism* (New York: Monthly Review Press, 2002), 161; with Brett Clark and Richard York, *The Ecological Rift: Capitalism's War on the Planet* (New York: Monthly Review Press, 2010), 442; with Paul Burkett, *Marx and the Earth: An Anti-Critique* (Leiden: Brill, 2016), 9; with Brett Clark, *Capitalism and the Ecological Rift: The Robbery of Nature* (New York: Monthly Review Press, 2020), 61; *Capitalism in the Anthropocene: Ecological Ruin or Ecological Revolution* (New York: Monthly Review Press, 2022), 71–2.

211 On the 'radical uncertainty' inherent in this exercise, making it almost futile, see Hugues Chenet, Josh Ryan-Collins and Frank van Lerven, 'Finance, Climate-Change and Radical Uncertainty: Towards a Precautionary Approach to Financial Policy', *Ecological Economics* (2021) 83: 4.

212 T. A. Hansen, 'Stranded Assets and Reduced Profits: Analyzing the Economic Underpinnings of the Fossil Fuel Industry's Resistance to Climate Stabilization', *Renewable and Sustainable Energy Reviews* (2022) 158: 1–14.

213 *Macrotrends*, 'World GDP 1960–2023', 2023.

214 Four trillion: Mercure et al. 'Macroeconomic Impact'; 185 trillion: Linquity and Cogswell, 'The Carbon Ask'. (The scenario of the former had a stated 75 per cent likelihood of staying at 2°C.) Citigroup arrived at 100 trillion in unburnable reserves for 2°C. Citigroup, 'Energy Darwinism', 82–4.

215 Livsey, 'The $900bn Cost'. Scenarios that smack of such underestimation extrapolated into 6°C are reported in Thomä, 'The Stranding of Upstream', 121–2.

216 BMI Industry Research quoted in Aitken, 'Depletion Work', 284. A carbon budget for 1.5°C would leave space for the exploitation of *some* existing reserves before the limit is reached, meaning some value would be recoverable – the part lost would be more than a third, not necessarily 100 per cent; but it would seem, especially in light of the fast depletion of said budget, closer to the latter share than the former.

217 *Fortune*, 'Global 500', 2020.

218 Swalec, 'Pedal to the Metal'. Relaxed assumptions: the net-zero plans of various countries.

219 Van der Ploeg and Rezai, 'Stranded Assets', 293.

220 Allen and Coffin, 'Unburnable Carbon', e.g. 31–33.

221 Stefano Battiston, Antoine Mandel, Irene Monasterolo et al., 'A Climate Stress-Test of the Financial System', *Nature Climate Change* (2017) 7: 283–8.

222 Cuvelier, 'The Asset Managers', 10.

223 Moody's as reported in *Bloomberg*, 'The World Needs a Plan for Stranded Assets', 20 October 2021; the Dutch system: Robert Vermeulen, Edo Schets, Melanie Lohuis et al., 'The Heat Is On: A Framework for Measuring Financial Stress under Disruptive Energy Transition Scenarios', *Ecological Economics* (2021) 190: 1–11; the Mexican: Alan Roncoroni, Stefano Battiston, Luis O. L. Escobar-Farfán and Serafin Martinez-Jaramillo, 'Climate Risk and Financial Stability in the Network of Banks and Investment Funds', *Journal of Financial Stability* (2021) 54: 1–27. For other approximations, see Semieniuk et al., 'Low-Carbon Transitions', 11.

224 Campanale, 'Investors Need'.

225 Ian Simm, CEO of Impax Asset Management, whose business profile was fossil free, quoted in Sheryl Tian Tong Lee and Alastair Marsh, 'Biden's Climate Law Exposes Mispriced Assets Due for Writedowns', *Bloomberg*, 11 April 2023; Livsey, 'The $900bn Cost'.

226 Cf. Mercure et al., 'Macroeconomic Impact', 592.

227 See e.g. Dawud Ansaro and Franziska Holz, 'Between Stranded Assets and Green Transformation: Fossil-Fuel-Producing Developing Countries towards 2055', *World Development* (2020) 130: 10.

228 Welsby et al., 'Unextractable Fossil', 233; cf. e.g. Simone Tagliapietra, 'The Impact of the Global Energy Transition on MENA Oil and Gas Producers', *Energy Strategy Reviews* (2019) 26: 1–6.

229 Trout et al., 'Existing Fossil', 5.

230 Hauenstein, 'Stranded Assets', 5–7.

231 Arthur Rempel, 'An Unsettled "Stranded Asset Debt?": Proposing a Supply-Side Counterpart to the "Climate Debt" in a Bid to Guide a Just Transition from Fossil Fuels in South Africa and Beyond', *Antipode* (2023) 55: 243–67.

232 Semieniuk et al., 'Stranded Fossil-Fuel', 537 ('private persons': 534). Cf. the similar findings in Von Dulong, 'Concentration of Asset'. On the enduring imperialist patterns of fossil fuel ownership, cf. Christopers, *Rentier Capitalism*, 108–10. The assets indirectly held by working-class people with savings in pensions funds were almost too small to register, further demonstrating how concentrated the losses would be to the top of the pyramid: Gregor Semieniuk, Lucas Chancel, Eulalie Saisset et al., 'Potential Pension Fund Losses Should Not Deter High-Income Countries from Bold Climate Action', *Joule* (2023) 7: 1383–93.

5. How to Kill a Spectre

1 Dawud Ansari and Franziska Holz, 'Between Stranded Assets and Green Transformation: Fossil-Fuel-Producing Developing Countries towards 2055', *World Development* (2020) 130: 1. The same invocation appeared in Ansari and Fareed, 'Stranded Assets', 1.

2 Mark Fisher, *Ghosts of My Life: Writings on Depression, Hauntology and Lost Futures* (Winchester: Zero, 2014), 18 (emphases in original). With thanks to Franciszek Korbanski for his Fisherian insights.

3 See e.g. Shimbar, 'Environment-Related Stranded', 6; Chenet et al., 'Finance, Climate-Change', 6; Semieniuk et al., 'Low-Carbon Transitions', 13; Monasterolo 'Climate Change', 313.

4 Carney, 'Breaking the Tragedy', 8.

5 Patricia Loria and Matthew B. H. Bright, 'Lessons Captured from 50 Years of CCS projects', *The Electricity Journal* (2021) 34: 1–6; Heleen de Coninck and Sally M. Benson, 'Carbon Dioxide Capture and Storage: Issues and Prospects', *Annual Review of Environment and Resources* (2014) 39: 243–270; Jinfeng Ma, Lin Li, Haofan Wang et al., 'Carbon Capture and Storage: History and the Road Ahead', *Engineering* (2022) 14: 33–43; D. M. Reiner, 'Learning through a Portfolio of Carbon Capture and Storage Demonstration Projects', *Nature Energy* (2016) 1: 1–7.

6 On this history, see Malm and the Zetkin Collective, *White Skin*.

7 Andreas Tjernshaugen, 'Technological Power as a Strategic Dilemma: CO_2 Capture and Storage In the International Oil and Gas Industry', *Global Environmental Politics* (2012) 12: 19–22; Jennie C. Stephens, 'Technology Leader, Policy Laggard: CCS Development for Climate Mitigation in the US Political Context', in James Meadowcroft and Oluf Langhelle (eds), *Caching the Carbon: The Politics and Policy of Carbon Capture and Storage* (Northampton, MA: Edward Elgar, 2009), 36; Howard Herzog, Baldur Eliasson and Olav Kaarstad, 'Capturing Greenhouse Gases', *Scientific American* (2000) 282: 72–9. With thanks to Ulrika Winter for research on the oil and gas industry and its strategies regarding CCS.

8 IEA, *Prospects for CO_2 Capture and Storage*, 2004, 3.

9 Ibid.

10 Ibid., 113.

11 Ibid., 18.

12 Ibid., 113.

13 Ibid, 19, 116.

14 Daiju Narita, 'Managing Uncertainties: The Making of the IPCC's Special Report on Carbon Dioxide Capture and Storage', *Public Understanding of Science* (2012) 21: 89.

15 Ibid.

16 Bert Metz quoted in ibid. He was also co-chair of the IPCC's Working Group on mitigation at this time.

17 Bert Metz, Ogunlade Davidson, Heleen de Coninck et al. (eds), *IPCC Special Report on Carbon Dioxide Capture and Storage* (Cambridge: Cambridge University Press, 2005), 12, 44.

18 Heleen de Coninck and Karin Bäckstrand, 'An International Relations Perspective on the Global Politics of Carbon Dioxide Capture and Storage', *Global Environmental Change* (2011) 21: 369; Reiner, 'Learning through', 1.

19 De Coninck and Benson, 'Carbon Dioxide', 257; Sam Morgan, 'Post-Mortem: Auditors Analyse EU's Failed Carbon Capture Projects', *Euractiv*, 24 October 2018.

20 Brigitte Nerlich and Rusi Jaspal, 'UK Media Representations of Carbon Capture and Storage: Actors, Frames and Metaphors', *Metaphor and the Social World* (2013) 3: 35–53; on positive representations of carbon capture beyond the UK context, see also Andreas M. Feldpausch-Parker, Morey Burnham, Maryna Melnik et al., 'News Media Analysis of Carbon Capture and Storage and Biomass: Perceptions and Possibilities', *Energies* (2015) 8: 3058–74; Shinichiro Asayama and Atsushi Ishii, 'Selling Stories of Techno-Optimism? The Role of Narratives on Discursive Construction of Carbon Capture and Storage in the Japanese Media', *Energy Research and Social Science* (2017) 31: 50–9.

21 De Coninck and Benson, 'Carbon Dioxide'.

22 Reiner, 'Learning through', 4; Timmo Krüger, 'Conflicts over Carbon Capture and Storage in International Climate Governance', *Energy Policy* (2017) 100: 65; Vivian Scott, Stuart Gilfillan, Nils Markusson et al., 'Last Chance for Carbon Capture and Storage', *Nature Climate Change* (2013) 3: 105.

23 Damian Carrington, 'UK Cancels Pioneering £1bn Carbon Capture and Storage Competition', *Guardian*, 25 November 2015.

24 European Court of Auditors, 'Special Report: Demonstrating Carbon Capture and Storage and Innovative Renewables at Commercial Scale in the EU: Intended Progress not Achieved in the Past Decade', 2018, 8.

25 Ibid., 20.

26 De Coninck and Benson, 'Carbon Dioxide', 245.

27 Ibid., 10. For a forensic analysis of the ideology of disclosure, see Christophers, 'Climate Change'; see further Nadia Ameli, Paul Drummond, Alexander Bisaro et al., 'Climate Finance and Disclosure for Institutional Investors: Why Transparency Is Not Enough', *Climatic Change* (2020) 160: 565–89; and cf. e.g. Beyene

et al., 'Financial Institutions', 22; Chenet et al., 'Finance, Climate-Change', 2–3.

28 Shukla et al., *Climate Change*, 1583; see further e.g. *Economist*, 'How to Deal with Worries about Stranded Assets', 24 November 2016; Sandra Batten, Rhiannon Sowerbutts and Misa Tanaka, 'Climate Change: What Implications for Central Banks and Financial Regulators?', in Caldecott, *Stranded Assets*, 265–7; Ameli et al., 'Climate Finance', 566–7; Christophers, 'Climate Change', 1113–15; Monasterolo, 'Climate Change', 301–2.

29 Miller and Dikau, 'Preventing a "Climate"', 15.

30 E.g. Christophers, 'Climate Change', 1114–15; Beyene et al, 'Financial Institutions', 12–13.

31 Chenet et al., 'Finance, Climate-Change', 3; cf. Miller and Dikau, 'Preventing a "Climate"', 15–16.

32 E.g. Mark Carney, François Villeroy de Galhau [governor of Banque de France] and Frank Elderson, 'Open Letter on Climate-Related Financial Risks', Bank of England, 17 April 2019.

33 Citigroup, 'Energy Darwinism'; and see e.g. Kate Aronoff, 'Bank of America, Citigroup, and Wells Fargo Vote to Keep Financing Fossil Fuels', *New Republic*, 26 April 2022; Attracta Mooney and Aime Williams, 'Nuns Urge Citigroup to Rethink Financing of Fossil Fuel Projects', *Financial Times*, 10 April 2023; Frances Schwartzkopff and Alastair Marsh, 'Citi, BofA Lead Wall Street Banks Funding Fossil-Fuel Expansion', *Bloomberg*, 17 January 2023; Rothfeder and Maag, 'How Wall Street's'.

34 David Mackie and Jessica Murray, 'Risky Business: The Climate and the Macroeconomy', JPMorgan, 14 January 2020, 1.

35 Ibid., 4. 'Nature is the lifeblood of the planet and critical for human existence, providing food, energy and medicine and also playing a fundamental role in communities and cultures.' Ibid., 15.

36 Ibid., 18–19.

37 Ibid., 20. Or, 'no global solution is in sight.' Ibid., 1. One of the world's sharpest climate journalists summed up the gist of the document: Kate Aronoff, 'The Planet Is Screwed, Says Bank that Screwed the Planet', *New Republic*, 25 February 2020.

38 Rothfeder and Maag, 'How Wall Street's'; 'Coast GasLink Pipeline', GEM, 6 January 2023. In 2022, the Royal Bank of Canada eclipsed JPMorgan; but it was unclear if this was a momentary feat. Kevin Orland, 'RBC Becomes World's Biggest Fossil-Fuel Bank, Topping JPMorgan', *Bloomberg*, 13 April 2023.

39 Larry Fink, 'A Fundamental Reshaping of Finance', BlackRock, 16 January 2020.

40 Joanna Partridge, 'World's Biggest Fund Manager Vows to Divest

from Thermal Coal', *Guardian*, 14 January 2020; Jasper Jolly, 'BlackRock Holds $85bn in Coal Despite Pledge to Sell Fossil Fuel Shares', *Guardian*, 13 January 2021; 'One Year after Net-Zeto Commitments, BlackRock and Vanguard Are Still Largest Investors in Coal', *BlackRocks Big Problem*, 22 February 2022; Buller, *The Value*, 160–3. Not even the Big Three's ESG funds – acronym for Environmental, Social and Governance – behaved in any way differently, instead consistently supporting the most climate-destructive trajectories of the companies they invested in: see Baines and Hager, 'From Passive'.

41 Shaji Mathew, 'Saudi Aramco Closes $15.5BN BlackRock-Led Gas Pipeline Deal', *Al Jazeera*, 23 February 2022; Yousef Saba, 'BlackRock-Led Investors in Aramco Pipelines to Get $4.5 bln in Bond Sale', Reuters, 9 February 2023.

42 Brendan McDermid, 'BlackRock Tells UK "No" to Halting Investment in Coal, Oil and Gas', Reuters, 18 October 2022; Jessye Waxman, 'Myth-Busting Larry Fink's Annual Letter: BlackRock CEO Drags His Feet on Climate Change', *Sierra Club*, 11 April 2023; Rothfeder and Maag, 'How Wall Street's'; Baines and Hager, 'From Passive', 450.

43 'Larry Fink's Annual Chairman's Letter to Investors', BlackRock, 17 March 2023.

44 'Summary of the 56th Session of the Intergovernmental Panel on Climate Change and the 14th Session of Working Group III: 21 March–4 April 2022', *Earth Negotiations Bulletin* (2022) 12: 14.

45 Adam Morton, 'Australian Government Refuses to Join 40 Nations Phasing Out Coal, Saying It Won't "Wipe Out Industries"', *Guardian*, 5 November 2021.

46 Earth Negotiations Bulletin, 'Glasgow Climate Change Conference: 31 October – 13', *IISD* (2021) 12: 16–17.

47 Earth Negotiations Bulletin, 'Summary of the Sharm El-Sheikh Climate Change Conference: 6–20 November 2022', *IISD* (2022) 12: 4.

48 Benjamin Zycher, 'No – Carbon Taxes and Green Policies Harm Economic Growth and Jobs', *Financial Times*, 7 May 2020.

49 Quoted in Davidson, 'Parallels in Reactionary', 75.

50 Jean Eaglesham and Vipal Monga, 'Trillions in Assets May Be Left Stranded as Companies Address Climate Change', *Wall Street Journal*, 20 November 2021.

51 Holger Dieter, 'New York Needs to Brace for Climate Change, Regulator Says', *Wall Street Journal*, 29 October 2020.

52 E.g. *Economist*, 'A Green Light', 29 March 2014; *Economist*, 'The Elephant in the Atmosphere', 19 July 2014; Mike Scott, 'Investors Set to Pressure Oil Industry over $1.1 Trillion Exposure to

High-Cost Projects', *Forbes*, 15 May 2014; Liam Denning, 'Big Oil's Disruptive Climate Change', *Wall Street Journal*, 6 May 2015; Citigroup, 'Energy Darwinism', 82, 95–6; Ansar et al., 'Stranded Assets', 17; Bergman, 'Impacts of', 9.

53 On 2014 as an eventful year in the battle with the spectre, see Ayling and Gunningham, 'Non-State Governance', 138–9.

54 BHP Billiton, *Value through Performance: Annual Report 2014*, 12.

55 Ibid., 51.

56 ExxonMobil, *Energy and Carbon: Managing the Risks*, March 2014, 1. The report was, more specifically, written in response to a shareholder resolution from the fund manager Arjuna Capital that demanded 'explanations and actions on environmental threats to the firm'. *Economist*, 'A Green Light'.

57 ExxonMobil, *Energy and Carbon*, 12.

58 Ibid., 13–14.

59 ExxonMobil, *Advancing Climate Solutions: 2023 Progress Report*, December 2022, 61.

60 Paul Takahashi, 'Exxon Warns Investors of Climate Risk to its Oil and Gas Assets', *Houston Chronicle*, 3 November 2021.

61 None of this implied, it should be noted, transparent disclosure: see e.g. Akshat Rathi and Kevin Crowley, 'Exxon, Oil Rivals Shield their Carbon Forecasts from Investors', *Bloomberg*, 4 October 2020.

62 Michael Szabo, 'Shell Says Fossil Fuel Reserves Won't Be "Stranded" by Climate Regulation', Reuters, 19 May 2014; *Carbon Brief*, 'Shell Says its Assets Won't Get Stranded by a Carbon Bubble. With Climate Change on the International Agenda, Is it Right?', 20 May 2014. The wordings of this letter (no longer publicly available at the time of this writing) were virtually identical to ExxonMobil's: 'Shell does not believe that any of its proven reserves will become "stranded" as a result of current or reasonably foreseeable future legislation concerning carbon.' Furthermore, we 'do not see governments taking the steps now that are consistent with the two degree scenario' – or, in other words, Shell explicitly planned for a world hotter than 2°C. Quotations from *Carbon Brief*, 'Shell Says'.

63 Repetitions: Dmitry Zjhdannikov, 'Shell Sees No Risk of "Stranded Assets" as Reserve Life Shrinks', Reuters, 12 April 2018; Ron Bousso, 'Shell Plays Down Risk of Stranded Oil and Gas Reserves', Reuters, 15 April 2021; addressing the threat (not to be confused with full disclosure): Shell plc, *Powering Progress: Annual Reports and Accounts for the Year Ended December 31, 2022*, 84, 88–9, 223–4.

64 See e.g. 'Shell Scenarios: Sky – Meeting the Goals of the Paris Agreement', Shell, 2018, 32–3, 60–1; Andrei Sokolov, Sergey Paltsev, Angelo Gurgel et al., 'Temperature Implications of the 2023 Shell Energy Security Scenarios: Sky 2050 and Archipelagos', *MIT Joint Program on the Science and Policy of Global Change*, March 2023.

65 'Shell Energy Transition Report', Shell, 2018, 8.

66 Bob Dudley in Manus Cranny, 'BP Rejects Concern over Stranded Assets Amid "Slow" Energy Shift', *Bloomberg*, 20 February 2018 (emphases added).

67 Jeff McMahon, '"What Carbon Bubble?" Says Oil Company Economist', *Forbes*, 17 March 2016; ConocoPhillips, 'Financial Planning', 2023.

68 Chevron, *Managing Climate Change Risks: A Perspective for Investors*, March 2017, 1, 12.

69 Chevron, *Climate Change Resilience: Advancing a Lower Carbon Future*, October 2021, 14. The existential threat: 10.

70 Ibid., 21.

71 Ian Taylor in Blas and Farchy, *The World*, 318.

72 Glencore, *Climate Change Considerations for Our Business*, 2016, 23.

73 CEO Ivan Glasenberg in Jesse Riseborough, 'Biggest Coal Exporter Says Climate Change Won't Strand Assets', *Bloomberg*, 28 April 2015; and see further *Bloomberg NEF*, 'Glencore Doubles Down on Coal as Mining Rivals Head for Exit', 13 June 2017; Neil Hume, 'Glencore's Coal Business in the Spotlight', *Financial Times*, 16 February 2020; Glencore, *Climate Change*, 16, 19, 22–3.

74 Amin H. Nasser, 'Lifting the Hood on the Real Future Facing the Petroleum Industry', Aramco, 6 March 2018.

75 Allesandra Galloni and Dmitry Zhdanikov, 'Aramco CEO Warns of Global Oil Crunch due to Lack of Investment', Reuters, 23 May 2022. Hauntology not, of course, precluding some excessive whining also going on here.

76 Or, the anti-revolution against an imagined revolution that lives a full life of its own in the consciousness of the dominant classes, to follow Budraitskis.

77 Ayling and Gunningham, 'Non-State Governance' 135; Strauch et al., 'Constraining Fossil', 185.

78 'Foreword by Lord Stern', in Leaton et al., *Unburnable Carbon*, 7. Cf. 4, 16; *Economist*, 'How to Deal'; McMahon, 'UN Climate'; Gore and Blood above.

79 As observed by many, e.g. Ayling and Guggenheim, 'Non-State Governance', 135–6; Bruangardt et al., 'Fossil Fuel', 194; Semieniuk

et al., 'Stranded Fossil-Fuel', 536–7; Beyene et al., 'Financial Institutions', 25; Shukla et al., *Climate Change*, 1744.

80 There is some evidence to suggest that stock markets lost some of their appreciation for declarations of growth in the reserves of oil companies after Paris, reported in Atasanova and Schwartz, 'Stranded Fossil'; and that divestment pledges that go viral on social media cause a reduction in the valuation of 'carbon-intensive companies' for a few days, reported in Attracta Mooney, 'Fossil Fuel Groups Hit Extra Hard by Divestment Pledges that Go Viral', *Financial Times*, 10 April 2023. But none of this had, as of this writing, done anything to stem the aggregate flow of capital into and through the primitive circuit.

81 Cf. Goodman and Anderson, 'From Climate', 1263–4.

82 Sigmund Freud, 'Negation', in *The Standard Edition of the Complete Psychological Works of Sigmund Freud, Volume XIX* (London: Vintage, 2001 [1925]), 235 (emphases in original).

83 Ibid., 236.

84 Livsey, 'The $900bn'; cf. Bergman, 'Impacts of', 10.

85 Cf. Mark Buchanan, 'High Fossil Fuel Valuations Are a Political Weapon', *Bloomberg*, 21 July 2022.

86 Jude Clemente, '"Stranded Asset" Argument against Coal, Oil and Natural Gas Isn't Real', *Forbes*, 2 March 2018.

87 ExxonMobil, *Energy and Carbon*, 1. Almost identical phrasing appeared in Clemente, '"Stranded Asset"'; and cf. also Benjamin Zycher, 'No – Carbon'.

88 Chevron, *Managing Climate*, 5 (cf. Chevron, *Climate Change*, 13); further Riseborough, 'Biggest Coal'; BHP Billiton, *Value through*, 51; BP, *Sustainability Report 2014*, 16; TotalEnergies, 'Strategy, Sustainability and Climate: Transcript', 3; Equinor, 'Exploration for'. On the prevalence of this argument, cf. Bebbington et al., 'Fossil Fuel', 13.

89 On the game, cf. Ansari and Fareed, 'Stranded Assets', 2.

90 Sigmund Freud, 'Inhibitions, Symptoms and Anxiety', in *The Standard Edition of the Complete Psychological Works of Sigmund Freud: Volume XX* (London: Vintage, 2001 [1926]), 128; Sigmund Freud, 'New Introductory Lectures on Psychoanalysis', in *The Standard Edition of the Complete Psychological Works of Sigmund Freud: Volume XXII* (London: Vintage, 2001 [1933]), 81–2. See further e.g. Sigmund Freud, 'Introductory Lectures on Psychoanalysis', in *The Standard Edition of the Complete Psychological Works of Sigmund Freud: Volume XVI* (London: Vintage, 2001 [1917]), 393–4.

91 Freud, 'Inhibitions, Symptoms', 137.

92 Freud, 'New Introductory', 82.

93 For a brief, excellent survey, see Joseph Dodds, 'The Psychology of Climate Anxiety', *BJPsych Bulletin* (2021) 45: 222–6.

94 Ashima Sharma, 'Wildfires in Canada Force Shutdown of Oil and Gas Production', *Offshore Technology*, 9 May 2023. On physical risks, see e.g. George C. Unruh, 'The Real Stranded Assets of Carbon Lock-In', *One Earth* (2019) 1: 399–401; Caldecott et al., 'Stranded Assets', 419, 422; Shukla et al., *Climate Change*, 1580–1.

95 Chevron, *Managing Climate*, 10; cf. Chevron, *Climate Change*, 9.

96 Marx, *Capital Volume II*, 249.

97 Brett Christophers, 'Environmental Beta or How Institutional Investors Think about Climate Change and Fossil Fuel Risk', *Annals of the American Association of Geographers* (2019) 109: 763. The interviews were conducted in 2017 and 2018. Cf. Christophers, *Rentier Capitalism*, 132; and Goodman and Anderson, 'From Climate'.

98 Cf. Christophers, 'Environmental Beta', 764; Chenet et al., 'Finance, Climate-Change', 3.

99 Comerford and Spiganti, 'The Carbon Bubble', 2. The authors later acknowledge that this assumption is not exactly backed up by real world events, but then clarify that their scenario 'imagines something that shifts global climate policy from its current trajectory, towards strictly following the scientific advice' – a scenario which does not necessarily rely on rationalism: see further below. Ibid., 30.

100 Letter by Rael Jean Isaac in the collection of letters headlined 'Al Gore and David Blood are Mistaken about "The Coming Carbon Asset Bubble"', *Wall Street Journal*, 3 November 2013. Isaac was the author of the book *Roosters of the Apocalypse: How the Junk Science of Global Warming is Bankrupting the Western World*, as well as various Zionist screeds denying the ethnic cleansing of Palestine, e.g. 'Whose Palestine?', *Commentary*, July 1986. Another letter was even more contemptuous in tone. 'Thank you Al Gore and David Blood for pointing out there is still time to hit it big in green investing. Unfortunately, I ignored Mr. Gore's similar advice some years ago and missed the chance to get in on the ground floor of green energy gold mines such as Solyndra, Abound Solar, A123 Systems, Beacon Power, Nevada Geothermal Power, Range Fuels, Nordic Windpower, etc.' This was David Kreutzer of the denialist Heritage Foundation, who would later go on to work with Donald Trump in demolishing Obamaesque climate policy: see *DeSmog*, David Kreutzer. But what must be conceded is that Isaac and Kreutzer were essentially right.

101 Clemente, '"Stranded Asset"'.

102 Ameli et al., 'Climate Finance', 578.

103 Cf. Buchanan, 'High Fossil'; Zenghelis et al., 'Stranded Assets', 32; Braungardt et al., 'Fossil Fuel', 194; Chenet, 'Finance, Climate-Change', 8; Van der Ploeg and Rezai, 'Stranded Assets', 291.

104 Ameli et al., 'Climate Finance', 575.

105 Christophers, 'Environmental Beta', 766–7; Ameli et al., 'Climate Finance', 577. This was Mark Carney's original 'tragedy of the horizon'.

106 Christophers, 'Environmental Beta', 759; and cf. Ameli et al., 'Climate Finance', 577–8. Cf. note 683. This might also explain the lacklustre embrace of CCS: because mitigation policy was totally ineffectual, it was also totally redundant to invest in something as expensive as carbon capture. It was more economically rational to do nothing at all besides business as usual, with impunity granted by governments.

107 JPMorgan, *Risky Business*, 1, 20; Shukla et al., *Climate Change*, 90, 355, 697–8, 1583.

108 Marx, *Capital Volume III*, 914. This central point is repeated on end in the literature: see e.g. Alexander Pfeiffer, Cameron Hepburn, Adrien Vogt-Schilb and Ben Caldecott, 'Committed Emissions from Existing and Planned Power Plants and Asset Stranding Required to Meet the Paris Agreement', *Environmental Research Letters* (2018) 13: 4–6; Johnson et al., 'Stranded on a', 90, 93, 100–1; Rozenberg et al., 'Carbon Constraints', 5, 7; Seto et al., 'Carbon Lock-In', 429, 432; Strauch et al., 'Constraining Fossil', 186–7; Chenet et al., 'Finance, Climate-Change', 9; Roncoroni et al., 'Climate Risk', 15; Mercure et al., 'Macroeconomic Impact', 588; Beyene et al., 'Financial Institutions', 21, 27; Semieniuk et al., 'Low-Carbon Transition', 15; Auger et al., 'The Future of Coal', 1463; Zhang, 'Quantifying Stranded', 7; Hauenstein, 'Stranded Assets', 8.

109 Mark Campanale, '$1 Trillion of Oil and Gas Assets Risk Being Stranded by Climate Change', *Brink News*, 22 January 2023.

110 E.g. Shukla et al., *Climate Change*, 698.

111 See e.g. Carney, 'Breaking the Tragedy', 4, 9; Battiston et al., 'A Climate Stress-Test', 287; Batten et al., 'Climate Change', 256, 268; Monasterolo, 'Climate Change', 304; Beyene et al., 'Financial Institutions', 18.

112 Semieniuk et al., 'Low-Carbon Transition', 13.

113 Beyene et al., 'Financial Institutions', 26; Miller and Dikau, 'Preventing a "Climate"', 1, 9–10 cf. Semieniuk et al., 'Low-Carbon Transition', 12.

114 Adam Tooze, 'Central Banks', *Foreign Policy* (2019) no. 233: 21.

115 Chenet et al., 'Finance, Climate-Change', 9.

116 Benoît Lallemand of Finance Watch quoted in Frances Schwartz-kopff, 'Europe's Biggest Pension Fund Issues ESG Warning to Banks', *Bloomberg*, 23 January 2023; on the stoking, see Kevin Crowley, David Wethe and Mitchell Ferman, 'Oil Investors Get $128 Billion Handout as Doubts Grow About Fossil Fuels', *Bloomberg*, 4 March 2023. For hauntology in 2021, see e.g. Collin Eaton, 'Houston's Big Oil Conference Goes Green as Energy Transition Accelerates', *Wall Street Journal*, 2 March 2021; Saijel Kishan, 'Wall Street Banks Face New Pressure to Cut Fossil-Fuel Financing', *Bloomberg*, 16 December 2021.

117 Alastair Marsh, 'Exxon Faces Investor Demand for More Disclosure of Climate Risks', *Bloomberg*, 17 April 2023.

118 As argued forcefully in Colgan et al., 'Asset Revaluation'; cf. e.g. Hansen, 'Stranded Assets', 1.

119 E.g. Rozenberg, 'Climate Constraints', 2; Linquiti and Cogswell, 'The Carbon Ask', 662–6, 672–3.

120 CREA and GEM, 'China Permits', 3; and cf. e.g. Buchanan, 'High Fossil'; 'Seto et al., 'Carbon Lock-In', 434; Pfeiffer et al., 'Committed Emissions', 9; Trout et al. 'Existing Fossil', 9.

121 Weber, 'Big Oil's', 156.

122 Marx, *Grundrisse*, 703 (emphasis in original); Adorno, *History and Freedom*, 171.

123 For two recent, eminent, reconstructions of the concept of 'passive revolution', see Peter D. Thomas, 'Gramsci's Revolutions: Passive and Permanent', *Modern Intellectual History* (2020) 17: 117–46; Panagiotis Sotiris, 'Revisiting the Passive Revolution', *Historical Materialism* (2022) 3: 3–45.

124 On these two presumed categories of capital, see further below.

6. We Are Going to Be Driven by Value

1 Svante Arrhenius, *Kemien och det moderna livet* (Stockholm: Hugo Gebers, 1919), 171. The classic paper is Svante Arrhenius, 'On the Influence of Carbonic Acid in the Air upon the Temperature of the Ground', *London, Edinburgh, and Dublin Philosophical Magazine and Journal of Science* (1896) 41: 237–76.

2 Arrhenius, *Kemien*, 201–5, 321.

3 Ibid., 321, 329; Svante Arrhenius, *Världarnas utveckling* (Stockholm: Hugo Gebers, 1917), 57–62.

4 Arrhenius, *Kemien*, 170–1, 196, 200, 205.

5 Daniela Russ, '"Socialism Is Not Just Built for a Hundred Years":

Renewable Energy and Planetary Thought in the Early Soviet Union (1917–1945)', *Contemporary European History* (2022) 31: 491–508.

6 Bent Sørensen, 'Energy and Resources', *Science* (1975) 189: 255–60. On this paper as the origins of the modern research agenda, see Kenneth Hansen, Christian Breyer and Henrik Lund, 'Status and Perspectives on 100% Renewable Energy Systems', *Energy* (2019) 175: 472; Christian Breyer, Siavash Khalili, Dmitrii Bogdanov et al., 'On the History of Future of 100% Renewable Energy Systems Research', *IEEE Access* (2022) 10: 78178–9.

7 Sørensen, 'Energy and Resources', 260.

8 For this definition, see e.g. Rainer Hinrichs-Rahlwes, David Renné, Monica Oliphant et al., 'Towards 100% Renewable Energy: Utilities in Transition', IRENA [International Renewable Energy Agency] Coalition for Action, 2020, 7.

9 For some of these developments, see e.g. Robert A. Marotto, 'Subtexts of Solar: Community and Conservation in the Solar Capital', *Capitalism Nature Socialism* (1990) 1: 97–118; John Wihbey, 'Jimmy Carter's Solar Panels: A Lost History that Haunts Today', *Yale Climate Connections*, 11 November 2008; Geoffrey Jones and Loubna Bouamane, '"Power from Sunshine": A Business History of Solar Energy', Harvard Business School working paper, no. 12–105, May 2012; Jordan Michael Scavo, *False Dawn of a Solar Age: A History of Solar Heating and Power during the Energy Crisis, 1973–1986*, PhD dissertation, University of California Davis, 2015.

10 Mark Z. Jacobson and Mark A. Delucchi, 'A Path to Sustainable Energy by 2030', *Scientific American* (2009) 301: 58–65; Mark Z. Jacobson and Mark A. Delucchi, 'Providing All Global Energy with Wind, Water and Solar Power, Part I: Technologies, Energy Resources, Quantities and Areas of Infrastructure, and Materials', *Energy Policy* (2011) 39: 1154–69; Mark Z. Jacobson and Mark A. Delucchi, 'Providing All Global Energy with Wind, Water and Solar Power, Part II: Reliability, System and Transmission Costs, and Policies', *Energy Policy* (2011) 39: 1170–90. An important follow-up for the US was Mark Z. Jacobson, Mark A. Delucchi, Mary A. Cameron and Bethany A. Frew, 'Low-Cost Solution to the Grid Reliability Problem with 100% Penetration of Intermittent Wind, Water, and Solar for All Purposes', *PNAS* (2015) 112: 15060–5.

11 For the history of the research, see Hansen et al., 'Status and Perspectives'; Breyer et al., 'On the History'.

12 The two significant papers from this camp were B. P. Heard, B. W. Brook, T. M. L. Wigley and C. J. A. Bradshaw, 'Burden of Proof:

A Comprehensive Review of the Feasibility of 100% Renewable-Electricity Systems', *Renewable and Sustainable Energy Reviews* (2017) 76: 1122–33; Christopher T. M. Clack, Staffan A. Qvist, Jay Apt et al., 'Evaluation of a Proposal for Reliable Low-Cost Grid Power with 100% Wind, Water, and Solar', *PNAS* (2017) 114: 6722–7. For the more disinhibited ecomodernists training their nuclear-fuelled anger on Jacobson et al., see Ted Nordhaus, 'On Anti-Nuclear Bullshit', *Breakthrough Institute*, 10 February 2021; Ted Nordhaus, 'The Death of Anti-Nuclearism', *Breakthrough Institute*, 10 August 2023. Jacobson advances four compelling arguments for privileging renewables and excluding nuclear: the latter is far more expensive than wind and solar, takes far longer time to construct, has risks associated with waste and accidents not shared by renewables – wind turbines cannot undergo meltdowns; solar panels produce no radioactive material – and stands exposed to the impacts of global warming itself, since nuclear plants require cooling from rivers (which can run dry, as in France in 2022) or seas (which can rise and surge perilously). Mark Z. Jacobson, *No Miracles Needed: How Today's Technology Can Save Our Climate and Clean Our Air* (Cambridge: Cambridge University Press, 2023), 156–73, 284. Less impressively, Jacobson grossly inflates the casualty numbers from Fukushima, generally overstates the risk of accidents and exaggerates the CO_2 emissions intensity of nuclear power: ibid., 164, 171. The most prudent position on nuclear appears to be neutrality, or perhaps agnosticism: as of the 2020s, this form of energy is neither the problem nor the solution. Fossil fuels are the problem, solar and wind and the rest of the flow the solution.

13 For a representative anarchist rant against renewables, see Alexander Dunlap, 'Does Renewable Energy Exist? Fossil Fuel+ Technologies and the Search for Renewable Energy', in Susana Batel and David Rudolph (eds), *A Critical Approach to the Social Acceptance of Renewable Energy Infrastructures: Going Beyond Green Growth and Sustainability* (Cham: Palgrave Macmillan, 2021), 83–102; and cf. e.g. Alexander Dunlap and Diego Marin, 'Comparing Coal and "Transition Materials"? Overlooking Complexity, Flattening Reality and Ignoring Capitalism', *Energy Research and Social Science* (2022) 89: 1–9. A major, inveterate critic of renewables in general and Jacobson in particular was the Australian anarchist Ted Trainer: see 'A Critique of Jacobson and Delucchi's Proposals for a World Renewable Energy Supply', *Energy Policy* (2012) 44: 476–81; '100% Renewable Supply? Comments on the Reply by Jacobson and Delucchi to the Critique by Trainer', *Energy Policy* (2013) 57: 634–40. His anarchism

is put on unflattering display in Ted Trainer, 'An Anarchism for Today: The Simpler Way', *Capitalism Nature Socialism* (2019) 30: 87–103. For a brief but effective takedown of his hostility to renewables, see David Schwartzman, 'Ted Trainer and the Simpler Way: A Somewhat Less Sympathetic Critique', *Capitalism Nature Socialism* (2014) 25: 112–17.

14 Trainer, 'Anarchism'; Seibert and Rees, 'Through the Eye'. 'Frugal lifestyle': Trainer, 'A Critique', 481. 'One billion or so': Seibert and Rees, 'Through the Eye', 13. The latter two authors proposed an enforced, universal ban on siblings as a measure for removing 7 billion humans from the globe, only to declare that it would be 'insufficient' for the purpose; how it could be implemented, they did not tell. Ibid. Unlike Trainer, Seibert had no academic credentials. On the website of her organisation, she described herself as the daughter of a military family, pursuing a career in the defence industry before reverting to the influence she felt from her 'environmentally oriented relatives and scholarly German heritage.' Furthermore, a 'regular yoga practice that began out of college, later catalyzed by exposure to Eastern philosophies in graduate school, led her into the world of shamanism, animism, astrology, channeling, and teacher plants. After a decade-long journey of truth seeking that culminated in a spiritual awakening, she's stepped into an entirely new way of being.' 'People: Megan Seibert, Executive Director', The REAL Green New Deal Project, n.d. Anarchism should be acquitted of association with this individual, who rather seemed to represent a character type still all too familiar from the fringes of the environmental movement in the global North. Ironically and incoherently, of course, all these detractors had no other option but to acknowledge that their alternative worlds would also require renewable energy, insofar as humanity (even if only 'one billion or so') would need some form of energy to live: Trainer, 'Anarchism', 95; Dunlap, 'Does Renewable', 97; Seibert and Rees, 'Through the Eye', 13.

15 Rebutting Heard et al., 'Burden of Proof': T. W. Brown, T. Bischof-Niemz, K. Blok et al., 'Response to "Burden of Proof: A Comprehensive Review of the Feasibility of 100% Renewable-Electricity Systems"', *Renewable and Sustainable Energy Reviews* (2018) 92: 834–47. Rebutting Clack et al., 'Evaluation of a Proposal': Mark Z. Jacobson, Mark A. Delucchi, Mary A. Cameron and Bethany A. Frew, 'The United States Can Keep the Grid Stable at Low Cost with 100% Clean, Renewable Energy in All Sectors Despite Inaccurate Claims', *PNAS* (2017) 114: E5021–3. Rebutting Trainer: Mark A. Delucchi and Mark Z. Jacobson, 'Response to "A Critique of Jacobson and Delucchi's Proposals for a World

Renewable Energy Supply" by Ted Trainer', *Energy Policy* (2012) 44: 482–4; Mark Z. Jacobson and Mark A. Delucchi, 'Response to Trainer's Second Commentary on a Plan to Power the World with Wind, Water, and Solar Power', *Energy Policy* (2013) 57: 641–3. Rebutting Seibert and Rees, 'Through the Eye': Mark Diesendorf, 'Comment on Seibert, M. K.; Rees, W. E. Through the Eye of a Needle: An Eco-Heterodox Perspective on the Renewable Energy Transition. *Energies* 2021, 14, 4508', *Energies* (2022) 15: 1–5.

16 See results of a comprehensive literature study in Hansen et al., 'Status and Perspectives', 474; and cf. Breyer et al., 'On the History', 78187.

17 Mark Z. Jacobson, Mark A. Delucchi, Zack A. F. Bauer et al., '100% Clean and Renewable Wind, Water, and Sunlight All-Sector Energy Roadmaps for 139 Countries of the World', *Joule* (2017) 1: 108–21; Mark Z. Jacobson, Anna-Katharina von Kraul-and, Stephen J. Coughlin et al., 'Low-Cost Solutions to Global Warming, Air Pollution, and Energy Insecurity for 145 Countries', *Energy and Environmental Science* (2022) 15: 3343–59 (quotation from 3343). See further e.g. Mark Z. Jacobson, Mark A. Delucchi, Mary A. Cameron et al., 'Impacts of Green New Deal Energy Plans on Grid Stability, Costs, Jobs, Health, and Climate in 143 Countries', *One Earth* (2019) 1: 449–63; Dmitrii Bogdanov [!], Manish Ram, Arman Aghahosseini et al., 'Low-Cost Renewable Electricity as the Key Driver of the Global Energy Transition towards Sustainability', *Energy* (2021) 227: 1–12; David S. Renné, 'Progress, Opportunities and Challenges of Achieving Net-Zero Emissions and 100% Renewables', *Solar Compass* (2022) 1: 1–11.

18 Jacobson, *No Miracles*.

19 Ibid., 101. On these widely envisioned contours, see e.g. ibid., 20–2, 288–9; Jacobson et al. 'Low-Cost Solutions', 3346; Bogdanov et al., 'Low-Cost Renewable', 4–5; Renné, 'Progress, Opportunities', 6–7; Saul Griffith, *Electrify: An Optimist's Playbook for Our Clean Energy Future* (Cambridge, MA: MIT Press, 2021); Rupert Way, Matthew C. Ives, Penny Mealy and J. Doyne Farmer, 'Empirically Grounded Technology Forecasts and the Energy Transition', *Joule* (2022) 6: 2067; Gunnar Luderer, Silvia Madeddu, Leon Merfort et al., 'Impact of Declining Renewable Energy Costs on Electrification in Low Emission Scenarios', *Nature Energy* (2022) 7: 36–7.

20 On the dominance of solar, see e.g. Jacobson et al. 'Low-Cost Solutions', 3350; Bogdanov et al., 'Low-Cost Renewable', 4; Renné, 'Progress, Opportunities', 7; Marta Victoria, Nancy Haegel, Ian Marius Peters et al., 'Solar Photovoltaics is Ready to Power a Sustainable Future', *Joule* (2021) 19: 1041–56.

21 On the definition of the flow, see further Malm, *Fossil Capital*, 38–40. The shares of geothermal that stem from the transfer of energy from the Earth's core, volcanos, and the decay of radioactive elements – rather than the downward conduction of sunlight – fall outside of this definition. Hence the vision was for perhaps a 99 per cent rather than a 100 per cent flow economy. (Geothermal makes up 1.22 per cent of total energy in the plan for 2050 outlined in Jacobson et al. 'Low-Cost Solutions', 3350.) This conceptual compromise mirrors that of Jacobson himself, who, somewhat incongruously, includes geothermal in the residual category 'water', together with tidal, wave and (non-expanded) hydro (hence his formula 'WWS' or 'water, wind, solar' for 100 per cent renewables).

22 Arrhenius, *Kemien*, 207.

23 Russ, '"Socialism Is"', 503–4.

24 E.g. Griffith, *Electrify*, 90–1; Jacobson, *No Miracles*, 289–90, 294, 301; Jacobson et al., 'Low-Cost Solutions', 3343–4, 3354–4; D. P. Schlachtberger, T. Brown, S. Schramm and M. Greiner, 'The Benefits of Cooperation in a Highly Renewable European Electricity Network', *Energy* (2017) 134: 469–81; Michael Child, Claudia Kemfert, Dmitrii Bogdanov and Christian Beyer, 'Flexible Electricity Generation, Grid Exchange and Storage for the Transition to a 100% Renewable Energy System in Europe', *Renewable Energy* (2019) 139: 80–101; Mark Z. Jacobson, 'The Cost of Grid Stability with 100% Clean, Renewable Energy for All Purposes when Countries Are Isolated versus Interconnected', *Renewable Energy* (2021) 179: 1065–75.

25 *Designboom*, 'Soleolico Unveils Rotating Solar Panels on Wind Turbine that Can Generate Green Energy 24/7', 9 October 2023; see also soleolico.com.

26 See e.g. Jacek Kapica, Fausto A. Canales and Jakub Jurasz, 'Global Atlas of Solar and Wind Resources Temporal Complementarity', *Energy Conversion and Management* (2021) 246: 1–13; Sonia Jerez, David Barriopedro, Alejandro Garcia-López et al., 'An Action-Oriented Approach to Make Most of the Wind and Solar Power Complementarity', *Earth's Future* (2023) 11: 1–16.

27 E.g. Griffith, *Electrify*, 83–6; Jacobson, *No Miracles*, 36–46, 289; Jacobson et al., 'Low-Cost Solutions', 3348–9; Richard Perez, Karl R. Rábago, Mike Trahan et al., 'Achieving Very High PV Penetration: The Need for an Effective Electricity Remuneration Framework and a Central Role for Grid Operators', *Energy Policy* (2016) 96: 3.

28 Matthew Stocks, Ryan Stocks, Bin Lu et al., 'Global Atlas of Closed-Loop Pumped Hydro Energy Storage', *Joule* (2021) 5: 270–84.

Concerns over the environmental impact of such reservoirs are dealt with, and the differences with traditional dams clarified, in ibid., 278–9.

29 Jacobson, *No Miracles*, 294–5, 300; Marc Perez, Richard Perez, Karl R. Rábago and Morgan Putnam, 'Overbuilding and Curtailment: The Cost-Effective Enablers of Firm PV Generation', *Solar Energy* (2019) 180: 412–22.

30 Griffith, *Electrify*, 86–90; Jacobson, *No Miracles*, 291–2, 295. Or, charging the electric vehicles by night with solar electricity stored in batteries: one study claimed to show that the entire car fleet of Spain could be thus safely powered by photovoltaics. Tobias Boström, Bilal Babar, Jonas Berg Hansen and Clara Good, 'The Pure PV-EV Energy System: A Conceptual Study of a Nationwide Energy System Based on Photovoltaics and Electric Vehicles', *Smart Energy* (2021) 1: 1–11.

31 On this argument from both camps, see Clack et al., 'Evaluation of a Proposal', 6722–3; Heard et al., 'Burden of Proof', 1124, 1129; Trainer, 'A Critique of Jacobson', 476–7; on the battery of solutions, e.g. T. Brown, D. Schlachtberger, A. Kies et al., 'Synergies of Sector Coupling and Transmission Reinforcement in a Cost-Optimised, Highly Renewable European Energy System', *Energy* (2018) 160: 720–39; Rupert Way, Matthew C. Ives, Penny Mealy and J. Doyne Farmer, 'Supplemental Information: Empirically Grounded Technology Forecasts and the Energy Transition', *Joule* (2022) 6: 34–6; Diesendorf and Elliston, 'The Feasibility of 100%', 321, 324; Breyer et al., 'On the History', 78189, 78196.

32 Kevin Palmer-Wilson, James Donald, Bryson Robertson et al., 'Impact of Land Requirements on Electricity System Decarbonisation Pathways', *Energy Policy* (2019) 129: 193; John van Zalk and Paul Behrens, 'The Spatial Extent of Renewable and Non-Renewable Power Generation: A Review and Meta-Analysis of Power Densities and their Application in the U.S.', *Energy Policy* (2018) 123: 83.

33 This argument is made from an avowedly Marxist standpoint in Matthew T. Huber and James McCarthy, 'Beyond the Subterranean Energy Regime? Fuel, Land Use and the Production of Space', *Transactions of the Institute of British Geographers* (2017) 42: 655–68. It is based not on Marxist theory, however, but on Ricardian-Malthusian theory as laid out by E. A. Wrigley and, in particular, Rolf Peter Sieferle – the latter being so reactionary that he killed himself in despair over the influx of refugees into Germany in the year before this article appeared. His theory of renewable energy occupying too much *Raum* was of a piece with his worry over non-whites doing the same. For an excellent

account of Ricardian-Malthusian theory and the especially perni-
cious thought of Sieferle, see Thomas Turnbull, 'Energy, History,
and the Humanities: Against a New Determinism', *History
and Technology* (2021) 37: 247–92; and for further critique of
Ricardianism-Malthusianism, Malm, *Fossil Capital*. In the case
of Huber, his flawed view of the land demands of renewables
might have contributed to him taking the path down the most
extreme form of ecomodernism on offer from the left, in a duo
with Leigh Phillips; the ideological alliance with the Breakthrough
Institute was clinched by uninhibited advocacy of nuclear power.
See e.g. Matt Huber, 'Mish-Mash Ecologism', *Sidecar*, 18 August
2022; 'The Problem with Degrowth', *Jacobin*, 16 July 2023.

34 Jacobson, *No Miracles*, 315; Mark Z. Jacobson, Anna-Katharina
von Krauland, Stephen J. Coughlin et al., 'Supplementary Information: Low-Cost Solutions to Global Warming, Air Pollution, and
Energy Insecurity for 145 Countries', *Energy and Environmental
Science* (2022), 66. Note that this figure included footprint but
not spacing of wind turbines; if the latter was included, it rose
to 0.84 (on this distinction, see further below). For a discussion
of findings that point to a similar emancipation for the US, see
Bryn Huxley-Reicher, 'How Much Land Will a Renewable Energy
System Use?', *Frontier Group*, 21 November 2022.

35 Damon Turney and Vasilis Fthenakis, 'Environmental Impacts
from the Installation and Operation of Large-Scale Solar Power
Plants', *Renewable and Sustainable Energy Reviews* (2011) 15:
3261–70.

36 The lower in Jacobson et al., 'Low-Cost Solutions'; the higher figure
in Jacobson et al., '100% Clean', 113 (again, spacing excluded;
when included, 0.17 rose to 0.36). Cf. Jacobson, *No Miracles*,
314–15. Already the 1975 *Science* article put the likely share at
'less than 1 percent of the total land area'. Sørensen, 'Energy and
Resources', 260.

37 One study that found a great increase in land use in the US under
80 per cent renewables was consequently based on very significant
components of bioenergy and hydro, while recognising that solar
outshone both in 'density': Van Zalk and Behrens, 'The Spatial
Extent'. On the biodiversity impacts of the various renewables – a
proxy for their land hunger – see Alexandros Gasparatos, Christopher N. H. Doll, Miguel Esteban et al., 'Renewable Energy and
Biodiversity: Implications for Transitioning to a Green Economy',
Renewable and Sustainable Energy Reviews (2017) 70: 161–84;
Jose E. Rehbein, James E. M. Watson, Joe L. Lane et al., 'Renew-
able Energy Development Threatens Many Globally Important
Biodiversity Areas', *Global Change Biology* (2020) 3040–51.

Another reason for emitting biomass and hydro was, of course, their emissions intensity, of CO_2 and CH_4 respectively.

38 Huber and James McCarthy, 'Beyond the Subterranean', 657.

39 Jacobson, *No Miracles*, 250–1, 324. Huber and McCarthy make the following argument for the land voracity of wind. 'Such wind farms and associated infrastructure would necessarily be very spatially extensive: modern turbines weigh well over a thousand tons apiece when the base is included (mainly concrete and steel), towers 100 metres or taller, have rotor diameters of over 100 metres and are typically spaced anywhere from three to ten rotor diameters apart from each other.' Huber and McCarthy, 'Beyond the Subterranean', 664. Every single factor here is irrelevant to the matter. We learn nothing about how much land the Eiffel Tower occupies by counting its weight, or indeed its height. Rotor diameters do not occupy land, but air. Spacing of the bases is, by definition, land available for other uses. The confusion of the vertical and the horizontal here appears complete.

40 See Malm and the Zetkin Collective, *White Skin*, e.g. 86–9, 108, 147, 274–9, 299–300.

41 See e.g. Weronika Strzyzynska, 'Sámi Reindeer Herders File Lawsuits Against Norway Windfarm', *Guardian*, 18 January 2021; Dorothée Cambou, Per Sandström, Anna Skarin and Emma Borg, 'Reindeer Husbandry vs. Wind Energy', in Monica Tennberg, Else Grete Broderstand and Hans-Kristian Hernes (eds), *Indigenous Peoples, Natural Resources and Governance: Agencies and Interactions* (London: Routledge: 2022), 39–58. But there were also observations of reindeer accustoming themselves to wind farms and continuing to migrate as before, suggesting compatibility between the two activities: Diress Tsegaye, Jonathan E. Colman, Sindre Eftestøl et al., 'Reindeer Spatial Use Before, During and After Construction of a Wind Farm', *Applied Animal Behaviour Science* (2017) 195: 103–11. For a measured assessment of the impacts of both onshore and offshore wind on biodiversity, see Gasparatos et al., 'Renewable Energy', 164, 169. Alexander Dunlap charges wind power with 'mass killing avian life' – a charge also made by Donald Trump and most other opponents of wind – but in the global North as of the 2010s, predation by domestic cats killed more birds than turbines did by two orders of magnitude. Dunlap and Marin, 'Comparing Coal', 7; Scott R. Loss, Tom Will and Peter P. Marra, 'Direct Mortality of Birds from Anthropogenic Causes', *Annual Review of Ecology, Evolution, and Systematics* (2015) 46: 99–120. If concern for avian life should dictate policy, the single most effective measure would thus be the abolition of the trade in cats, and moves in

that direction, for dogs as well, would have several other benefi-
cial environmental impacts, including emissions reductions: Pim
Martens, Bingtao Su and Samantha Deblomme, 'The Ecological
Paw Print of Companion Dogs and Cats', *BioScience* (2019) 69:
476–74; and further Troy Vettese, 'Want to Truly Have Empathy
for Animals? Stop Owning Pets', *Guardian*, 4 February 2023.
A superb account of struggles over wind is David McDermott
Hughes, *Who Owns the Wind? Climate Crisis and the Hope of
Renewable Energy* (London: Verso, 2021).

42   One study concluding that a shift to renewables would take up
more land made two cardinal errors: it counted the spacing area
of wind rather than the footprint of the base; and it forgot to
count the decommissioned fossil fuel infrastructure freeing up
land – ironically, in the Canadian province of Alberta, the case
study for this paper, home of the tar sands, the world's largest
oil reservoirs, whose scars of death on the landscape are visible
from space: Palmer-Wilson et al., 'Impact of Land'. Still, it only
reached the figure 0.92 per cent of Alberta's total land area for a
decarbonised electricity system. Ibid., 202.

43   Jacobson et al., 'Low-Cost Solutions', 3352.

44   Dirk-Jan van den Ven, Iñigo Capellan-Peréz, Iñaki Arto et al.,
'The Potential Land Requirements and Related Land Use Change
Emissions of Solar Energy', *Nature Scientific Reports* (2021) 11:
1–12.

45   On this difference, see e.g. Gasparatos et al., 'Renewable Energy',
162.

46   Ali Sawafta and Nidal al-Mughrabi, 'Palestinians Turn to the Sun
to Reduce their Power Shortfall', Reuters, 9 August 2018; Bahira
Amin, 'The Woman Bringing Solar Energy to Gaza and Rebuilding
it from the Rubble', *Scene Arabia*, 27 February 2019; UNRWA,
'Palestinian Refugee Women Pioneer the Use of Solar Panels in
Women's Programme Center in Askar Refugee Camp', 21 March
2022.

47   E.g. 'Solar Photovoltaics', 1047.

48   C. Dupraz, H. Marrou, G. Talbot et al., 'Combining Solar Pho-
tovoltaic Panels and Food Crops for Optimising Land Use:
Towards New Agrovoltaic Schemes', *Renewable Energy* (2011)
36: 2725–32; Stefano Ammaduci, Xinyou Yin and Michele
Colauzzi, 'Agrivoltaic Systems to Optimise Land Use for Electric
Energy Production', *Applied Energy* (2018) 220: 545–61; Greg
A. Barron-Gafford, Mitchell A. Pavao-Zuckerman, Rebecca L.
Minor et al., 'Agrivoltaics Provide Mutual Benefits across the
Food-Energy-Water Nexus in Drylands', *Nature Sustainability*
(2019) 2: 848–55; Takashi Sekiyama and Akira Nagashima, 'Solar

Sharing for Both Food and Clean Energy Production: Performance of Agrivoltaic Systems for Corn, A Typical Shade-Intolerant Crop', *Environments* (2019) 6: 1–12; Manoch Kumpanalaisatit, Worajit Setthapun, Hathaithip Sintunya et al., 'Current Status of Agrivoltaic Systems and their Benefits to Energy, Food, Environment, Economy, and Society', *Sustainable Production and Consumption* (2022) 33: 952–63; Mohammad Abdullah Al Mamun, Paul Dargusch, David Wadley et al., 'A Review of Research on Agrivoltaic Systems', *Renewable and Sustainable Energy Reviews* (2022) 16: 1–16. An exception so far, rice is reported to risk a 20 per cent drop in yields under agrivoltaic systems – still recommended as highly promising – in Ruth Anne Gonocruz, Ren Nakamura, Kota Yoshino et al., 'Analysis of the Rice Yield under an Agrivoltaic System: A Case Study in Japan', *Environments* (2021) 8: 1–18.

49 Alex Sandro Campos Maia, Eric de Andrade Culhari, Vinícius de França Carvalho Fonsêca et al., 'Photovoltaic Panels as Shading Resources for Livestock', *Journal of Cleaner Production* (2020) 258: 1–9; Robert Handler and Joshua M. Pearce: 'Greener Sheep: Life Cycle Analysis of Integrated Sheep Agrivoltaic Systems', *Cleaner Energy Systems* (2022) 3: 1–9.

50 Al Mamun et al., 'A Review', 9; see further the splendid survey Rebecca R. Hernandez, Alona Armstrong, Jennifer Burney et al., 'Techno-Ecological Synergies of Solar Energy for Global Sustainability', *Nature Sustainability* (2019) 2: 566.

51 Hernandez et al., 'Techno-Ecological Synergies', 562, 564–6; Kumpanalaisatit et al., 'Current Status', 953–7. Some of these options are also recognised in Van den Ven et al, 'The Potential Land', 7.

52 Figure from Al Mamun et al., 'A Review', 9. Note that Jacobson et al. did not include any agrivoltaics when reaching the figures 0.17 and 0.22 of global land area for 100 per cent 'WWS' energy production.

53 Nathan Lee, Ursula Grunwald, Evan Rosenlieb et al., 'Hybrid Floating Solar Photovoltaics-Hydropower Systems: Benefits and Global Assessment of Technical Potential', *Renewable Energy* (2020) 162: 1415–27; Raniero Cazzaniga and Marco Rosa-Clot, 'The Booming of Floating PV', *Solar Energy* (2021) 219: 3–10; Hernandez et al., 'Techno-Ecological Synergies', 565; Carly Wanna, 'How Floating Solar Panels Are Being Used to Power Electrical Grids', *Bloomberg*, 7 March 2023.

54 Jonathan Tirone, 'Floating Solar Panels Turn Old Industrial Sites into Green Energy Goldmines', *Bloomberg*, 3 August 2023.

55 Rafael M. Almeida, Rafael Schmitt, Steven M. Grodsky et al.,

'Floating Solar Power: Evaluate Trade-Offs', *Nature* (2022) 606: 247.

56 Yubin Jin, Shijie Hu, Alan D. Ziegler et al., 'Energy Production and Water Savings from Floating Solar Photovoltaics on Global Reservoirs', *Nature Sustainability* (2023) 6: 867; Almeida et al., 'Floating Solar', 248. Cf. Lee et al., 'Hybrid Floating', 1423.

57 Lee et al., 'Hybrid Floating', 1418–19; Almelda et al., 'Floating Solar', 246; Cazzaniga and Rosa-Clot, 'The Booming', 5; on Lake Kariba: Godfrey Marawanyika, 'World's Largest Man-Made Dam Weighs Using Floating Solar Panels', *Bloomberg*, 16 February 2023. Covering one third of a dam is estimated to cut evaporation by almost half; covering all to bring it to zero. Jin et al., 'Energy Production', 868. Likewise, floatovoltaics counteracts algae blooms, eliminating them at full coverage; but at that point, the aquatic ecosystem will be deprived of all oxygen and direct sunlight and collapse. From an ecological perspective, the installations should be neither small nor so large as to fill out the water body, but of moderate size. J. Haas, J. Khalighi, A. de la Fuente et al., 'Floating Photovoltaic Plants: Ecological Impacts versus Hydropower Operation Flexibility', *Energy Conservation and Management* (2020) 206: 1–8.

58 Dan-Bi Um, 'Exploring the Operational Potential of the Forest-Photovoltaic Utilizing the Simulated Solar Tree', *Nature Scientific Reports* (2022) 12: 1–12.

59 Even if offshore installations are discounted.

60 The final frontier for the verticality and abundance of solar power is space itself. Here the panels would be attached to satellites swimming in constant sunlight and beaming the energy down to antennas on Earth. This would have the potential to fully emancipate solar power from any spatial or temporal constraints, as the waves of energy could be dispatched to any place on Earth at any point in time. In the early 2020s, the technology registered several advances, but several obstacles remained on the path to development at scale: see e.g. Paul Jaffe, 'Space Solar', in Trevor M. Letcher (ed.), *Future Energy: Improved, Sustainable and Clean Options for our Planet* (Amsterdam: Elsevier, 2020), 519–42; Bloomberg News, 'Solar Panels Floating in Space Could One Day Power Your Home', *Bloomberg*, 16 September 2022; Stuart Clark, 'Beam Me Down: Can Solar Power from Space Help Solve Our Energy Needs?', *Guardian*, 9 October 2022; Peggy Hollinger, 'How to Make Space-Based Solar Power a Reality', *Financial Times*, 17 October 2023. Similarly, but within far easier reach, more powerful winds could be harnessed at higher altitudes by means of tethered, sky-high kite-like devices, rather than

fixed turbines: see e.g. Philip Bechtle, Mark Schelbergen, Roland Schmehl et al., 'Airborne Wind Energy Resource Analysis', *Renewable Energy* (2019) 14: 1103–16; Nicola Jones, 'After a Shaky Start, Airborne Wind Energy Is Slowly Taking Off', *Yale Environment 360*, 23 February 2022; Tim Schauenberg, 'Sky-High Kites Aim to Tap Unused Wind Power', *DW*, 27 March 2023.

61 Dunlap, 'Does Renewable', 89.

62 Anmar Frangoul, '"World's First Fossil-Free Steel" Produced in Sweden and Delivered to Volvo', *CNBC*, 19 August 2021.

63 Jinsoo Kim, Benjamin K. Sovacool, Morgan Bazilian et al., 'Decarbonizing the Iron and Steel Industry: A Systematic Review of Sociotechnical Systems, Technological Innovations, and Policy Options', *Energy Research and Social Science* (2022) 89: 1–32. Cf. Jacobson, *No Miracles*, 111–18; Swalec, 'Pedal to the Metal'; Hermwille et al., 'A Climate Club'.

64 This has been patiently pointed out in e.g. Jim Krane and Robert Idel, 'On the Reduced Supply Chain Risks and Mining Involved in the Transition from Coal to Wind', *Energy Research and Social Science* (2022) 89: 2; Diesendorf, 'Comment on Seibert', 2; Breyer et al., 'On the History', 78189.

65 This argument draws on the splendid Jim Krane and Rob Idle, 'More Transitions, Less Risk: How Renewable Energy Reduces Risk from Mining, Trade and Political Dependence', *Energy Research and Social Science* (2021) 82: 1–11. Cf. Jacobson, *No Miracles*, 64. Another qualitative difference is that commodities out of the stock cannot be recycled – once oil is burnt, it is irredeemably lost – whereas the transitional materials can. Even when the ladders and sticks have expired and must be renewed, the rates of extraction would therefore be on another scale entirely. Indra Overland, 'The Geopolitics of Renewable Energy: Debunking Four Energy Myths', *Energy Research and Social Science* (2019) 49: 37; Krane and Idle, 'More Transitions', 5.

66 Volumes measured in tonnes: Krane and Idle, 'More Transitions', 6.

67 See further ibid.

68 See e.g. Laura J. Sonter, Marie C. Dade, James E. M. Watson and Rick K. Valenta, 'Renewable Energy Production Will Exacerbate Mining Threats to Biodiversity', *Nature Communications* (2020) 11: 1–6; Benjamin K. Sovacool, Saleem H. Ali, Morgan Bazilian et al., 'Sustainable Minerals and Metals for a Low-Carbon Future', *Science* (2020) 367: 30; Benjamin K. Sovacool, 'Who Are the Victims of Low-Carbon Transitions? Towards a Political Ecology of Climate Change Mitigation', *Energy Research and Social Science* (2021) 73: 1–16.

69   As pointed out in e.g. Overland, 'The Geopolitics', 37.

70   Casey Crownhart, 'This Abundant Material Could Unlock Cheaper Batteries for EVs', *MIT Technology Review*, 9 May 2023; Casey Crownhart, 'How Sodium Could Change the Game for Batteries', *MIT Technology Review*, 11 May 2023; cf. Jacobson, *No Miracles*, 64–5.

71   P. J. Verlinden, 'Future Challenges for Photovoltaic Manufacturing at the Terawatt Level', *Journal of Renewable and Sustainable Energy* (2020) 12: 1–6; Matt Ferrell, 'The Solar Industry Has a Major Flaw: Its Reliance on Silver. Although there Is an Alternative, Most Manufacturers Are Hesitant to Make the Switch – Why?', *PV Buzz*, 21 February 2023; see further e.g. Felix Creutzig, Peter Agoston, Jan Christoph Goldschmidt et al., 'The Underestimated Potential of Solar Energy to Mitigate Climate Change', *Nature Energy* (2017) 2: 4; Sovacool et al., 'Sustainable Minerals', 31–2; Victoria et al., 'Solar Photovoltaics', 1046; Breyer et al., 'On the History', 78191.

72   Damien Giurco, Elsa Dominish, Nick Florin et al., 'Requirements for Minerals and Metals for 100% Renewable Scenarios', in Sven Teske (ed.), *Achieving the Paris Climate Agreement Goals: Global and Regional 100% Renewable Energy Scenarios with Non-energy GHG Pathways for +1.5°C and +2°C* (Cham: Springer, 2019), 437–57.

73   See e.g. Vitalii Lundaev, A. A. Solomon, Tien Le et al., 'Review of Critical Materials for the Energy Transition, An Analysis of Global Resources and Production Databases and the State of Material Circularity', *Minerals Engineering* (2023) 23: 2, 14.

74   Beyer et al., 'On the History', 78191; cf. e.g. Diesendorf, 'Comment on Seibert', 2; Griffith, *Electrify*, 186.

75   Thea Riofrancos, Alissa Kendall, Kristi K. Dayemo et al., 'Achieving Zero Emissions with More Mobility and Less Mining', Climate + Community Project, University of California, Davis, January 2023.

76   Griffith, *Electrify*, 61. (Cf., however, his averred preference for gorillas over electrical vehicles in ibid., 186). As for red meat, it deserves to be repeated that we are here leaving out this entire enormous question.

77   Scott M. Katalenich and Mark Z. Jacobson, 'Toward Battery Electric and Hydrogen Fuel Cell Military Vehicles for Land, Air, and Sea', *Energy* (2022) 254: 1–13. This is also a constantly recurring theme in Jacobson, *No Miracles*.

78   Katalenich and Jacobson, 'Toward Battery', 1.

79   For a report on state-of-the-art fusion research, see Philip Ball, 'Star Power', *Scientific American* (2023) 328: 28–35; and for

the difference from solar and wind, Brown et al., 'Response to "Burden"', 840; Jacobson, *No Miracles*, 160–1.

80 As pointed out in e.g. Brown et al., 'Response to "Burden"', 840; Jacobson et al., '100% Clean', 110; Jacobson et al., 'Low-Cost Solutions', 3343, 3346; and throughout Jacobson, *No Miracles*.

81 Management of the materials problem did, however, as we have seen, rest on ongoing and anticipated innovations.

82 Andreas Schäfer, Steven R. H. Barrett, Khan Doyme et al., 'Technological, Economic and Environmental Prospects of All-Electric Aircraft', *Nature Energy* (2019) 4: 160–66; Venkatasubramanian Viswanathan and B. Matthew Knapp, 'Potential for Electric Aircraft', *Nature Sustainability* (2019) 2: 88–9; Patrick Wheeler, Thusara Samith Sirimanna, Serhiy Bozhko and Kiruba S. Haran, 'Electric/Hybrid-Electric Aircraft Propulsion Systems', *Proceedings of the IEEE* (2021) 109: 1115–27. This is where the density of fossil (jet) fuels comes fully into play. Jacobson and colleagues placed the transition of global aviation to the flow close to the mid-century: Jacobson et al., '100% Clean', 110; Jacobson et al., 'Low-Cost Solutions', 3346. This had consequences, of course, for the idea that US fighter jets could be transitioned: see Katalenich and Jacobson, 'Toward Battery', 9.

83 Judith Magyar, 'Sleek, Silent Aircraft Are Electrifying Aviation; It Will Never Be the Same', *Forbes*, 3 August 2023; and for other recent advances, Jacobson, *No Miracles*, 61–3.

84 Amy Schwab, Anna Thomas, Jesse Bennett et al., 'Electrification of Aircraft: Challenges, Barriers, and Potential Impacts', National Renewable Energy Laboratory, U.S. Department of Energy, October 2021; Gökçin Çınar, 'Electric Planes Are Coming: Short-Hop Regional Flights Could Be Running on Batteries in a Few Years', *Conversation*, 19 September 2022.

85 Anthony Cuthbertson, 'NASA Invents "Incredible" Battery for Electric Planes', *Independent*, 10 October 2022; Schwab et al., *Electrification of Aircraft*, 23.

86 George Mallouppas and Elias Ar. Yfantis, 'Decarbonization in Shipping Industry: A Review of Research, Technology Development, and Innovation Proposals', *Journal of Marine Science and Engineering* (2021) 9: 1–40; Jerzy Herdzik, 'Decarbonization of Marine Fuels: The Future of Shipping', *Energies* (2021) 14: 1–10; Peyman Ghaforian Masodzadeh, Aykut I. Ölçer, Fabio Ballini and Anastasia Christodoulou, 'A Review on Barriers to and Solutions for Shipping Decarbonization: What Could Be the Best Policy Approach for Shipping Decarbonization?', *Marine Pollution Bulletin* (2022) 184: 1–8. Liquid hydrogen might be an option for

aviation as well (first used in planes in the Soviet Union in 1988): Jacobson, *No Miracles*, 68, 328. For one recent revival of sail on cargo ships, see Jasmina Ovcina Mandra, 'Oceanbird Takes Flight: Full-Scale Land-Based Prototype Set to be Assembled in Sweden', *Offshore Energy*, 30 March 2023.

87 Judith van Leeuwen and Jason Monios, 'Decarbonisation of the Shipping Sector: Time to Ban Fossil Fuels?', *Marine Policy* (2022) 146: 1.

88 Elizabeth J. Biddinger and Miguel A. Modestino, 'Electro-Organic Syntheses for Green Chemical Manufacturing', *Electrochemical Society Interface* (2020) 29: 43; and see, in particular, Adam Hanieh, 'Petrochemical Empire: The Geo-Politics of Fossil-Fuelled Production', *New Left Review* (2021) 130: 25–52.

89 Peter G. Levi and Jonathan M. Cullen, 'Mapping Global Flows of Chemicals: From Fossil Fuel Feedstocks to Chemical Products', *Environmental Science and Technology* (2018) 52: 1725–34; Elizabeth J. Biddinger and Miguel A. Modestino, 'Electro-Organic Syntheses for Green Chemical Manufacturing', *Electrochemical Society Interface* (2020) 29: 43–7; Ibrahim Eryazici, Naryan Ramesh and Carlos Villa, 'Electrification of the Chemical Industry: Materials Innovations for a Lower Carbon Future', *MRS Bulletin* (2021) 46: 1197–1204; Gabriele Centi and Siglinda Perathoner, 'Status and Gaps Toward Fossil-Free Sustainable Chemical Production', *Green Chemistry* (2022) 24: 7305–31; Dharik S. Mallapragada, Yury Dvorkin, Miguel A. Modestino et al., 'Decarbonization of the Chemical Industry through Electrification: Barriers and Opportunities', *Joule* (2023) 7: 23–41. The situation, in other words, was not quite as simple and bleak as there being 'no viable alternative to petroleum as a material feedstock'. Hanieh, 'Petrochemical Empire', 28.

90 Cf. e.g. Luderer et al., 'Impact of Declining', 37–8. On the problem of concrete, see below.

91 As argued in Van Leeuwen and Monios, 'Decarbonisation of the Shipping', the authority in question being the International Maritime Organization. Cf. e.g. Malloupas and Yfantis, 'Decarbonization in Shipping', 33.

92 Potentially, this could apply to something like nuclear fusion as well: 'Soviet fusion visionary Lev Artsimovich, the "father of the tokamak," once said that the world will have nuclear fusion when it decides it needs it. "When we realize what climate change will do as an existential threat, the delivery of fusion will accelerate enormously," Chapman says, drawing an analogy to the quick development of COVID-19 vaccines.' Ball, 'Star Power', 35.

93 In terms of CO_2 emissions from fossil fuel combustion: aviation and shipping each accounted for around 2 per cent of those in 2022. IEA, 'Aviation'; 'International Shipping', 2022. As for the petrochemical industry, it held 5.5 per cent. Levi and Cullen, 'Mapping Global', 1725; Biddinger and Modestino, 'Electro-Organic Syntheses', 43. Slightly more than half of the emissions in this industry stemmed from the use of fossil-fuelled electricity and heat, both eminently substitutable by the early 2020s, the rest from the processing of feedstock. Eryazici et al., 'Electrification of the Chemical', 1198, 1200. The latter did not, however, of course, include the emissions at the end of the lives of these products, e.g. the incineration of plastics. Somewhere between 30 and 40 per cent of the fossil fuels absorbed by the petrochemical industry made its way into the final chemicals and materials leaving its premises as commodities. Centi and Perathoner, 'Status and Gaps', 7309. If we then count feedstocks as non-substitutable, we would reach a very approximative figure of 4.7 per cent of total emissions (this being sum of emissions from processing feedstocks in the industry itself plus the emissions eventually realised from the fossil fuels embedded in its goods). $2 + 2 + 4.7 = 8.7$.

94 Jacobson et al., '100% Clean', 108; Jacobson et al., 'Low-Cost Solutions', 3343–4, 3347; Jacobson, *No Miracles*, xv, 318–22; Breyer et al., 'On the History', 78176–7, 78184; Bogdanov et al., 'Low-Cost Renewable', 1, 9; Renné, 'Progress, Opportunities', 1, 7. Some more radical modelling by Jacobson et al. had the whole transition completed by 2035. Note that 2050 as completion date was already suggested in Sørensen, 'Energy and Resources'.

95 Brown et al., 'Synergies of Sector', 737; cf. e.g. Diesendorf and Elliston, 'The Feasibility of 100%', 326–7; Way et al., 'Empirically Grounded', 2072; Kim et al., 'Decarbonizing the Iron', 17; Griffith, *Electrify*, 73; Jacobson, *No Miracles*, e.g. xii, 318.

96 Sonter et al., 'Renewable Energy', 4; cf. e.g. Sovacool et al., 'Sustainable Minerals', 32–3; Rehbein et al., 'Renewable Energy', 3040, 3047.

97 As argued, with full recognition of the problems caused by some large-scale renewables projects around the world, in Breyer et al., 'On the History', 78192.

98 IEA, *World Energy Outlook 2020*, 202, 214.

99 Victoria Masterson, 'Renewables Were the World's Cheapest Source of Energy in 2020, New Report Shows', *World Economic Forum*, 5 July 2021.

100 Will Mathis, 'Building New Renewables Is Cheaper than Burning Fossil Fuels', *Bloomberg*, 23 June 2021; Nat Bullard, 'Renewable

Energy Provides Relief from Rising Power Prices', *Bloomberg*, 14 July 2022; IEA, 'Renewable Energy Market Update', May 2022; IEA, *World Energy Investment*, 10–12, 20–1, 36–7.

101 Way et al., 'Empirically Grounded', 2057–8; Way et al., 'Supplemental Information', 72.

102 Philipp Beiter, Aubryn Cooperman, Eric Lantz et al., 'Wind Power Costs Driven by Innovation and Experience with Further Reductions on the Horizon', *WIREs Energy and Environment* (2021) 10: 14; Ryan Wiser, Joseph Rand, Joachim Seel et al., 'Expert Elicitation Survey Predicts 37% to 49% Declines in Wind Energy Costs by 2050', *Nature Energy* (2021) 6: 555–65.

103 Brett Christophers, *The Price Is Wrong: Why Capitalism Won't Save the Planet* (London: Verso, 2024), 124; cf. Beiter et al., 'Wind Power', 3; Victoria et al., 'Solar Photovoltaics', 1044–5.

104 E.g., 'experts predict that both solar and wind will be generally cheaper than fossil fuels sometime before 2025.' Malm, *Fossil Capital*, 369.

105 For some contingent circumstances surrounding this burst, see Martin A. Green, 'How Did Solar Cells Get So Cheap?', *Joule* (2019) 3: 631–40.

106 Yifeng Chen, Pietro P. Altermatt, Daming Chen et al., 'From Laboratory to Production: Learning Models of Efficiency and Manufacturing Cost of Industrial Crystalline Silicon and Thin-Film Photovoltaic Technologies', *IEEE Journal of Photovoltaics* (2018) 8: 1531–8.

107 C. Wilson, A. Grubler, N. Bento et al., 'Granular Technologies to Accelerate Decarbonization', *Science* (2020) 368: 36–9; Felix Creutzig, Jérôme Hilaire, Gregory Nemet et al., 'Technological Innovation Enables Low Cost Climate Change Mitigation', *Energy Research and Social Science* (2023) 105: 5.

108 Beiter et al., 'Wind Power', e.g. 2–3, 5–6, 11, 14; Wiser et al., 'Expert Elicitation'; Creutzig et al., 'Technological Innovation', 5.

109 Beiter et al., 'Wind Power', 1, 14; Victoria et al., 'Solar Photovoltaics', 1045. Indeed, the scaling up of renewables in and of itself will depress their prices further: Griffith, *Electrify*, 108–9.

110 Richard York and Shannon Elizabeth Bell, 'Energy Transitions or Additions? Why a Transition from Fossil Fuels Requires More than the Growth of Renewable Energy', *Energy Research and Social Science* (2019) 51: 40–3; and see also Richard York, 'Do Alternative Energy Sources Displace Fossil Fuels?', *Nature Climate Change* (2012) 2: 441–3.

111 CREA and GEM, 'China Permits', 2, 8.

112 Cf. Jacobson, *No Miracles*, 328.

113 York and Bell, 'Energy Transitions', 43.

114 Marc Jaxa-Rosen and Evelina Trutnevyte, 'Sources of Uncertainty in Long-Term Global Scenarios of Solar Photovoltaic Technology', *Nature Climate Change* (2021) 11: 266; Beiter et al., 'Wind Power', 2; Renné, 'Progress, Opportunities', 2; Haohui Liu, Vijay Krishna, Jason Lun Leung et al., 'Field Experience and Performance Analysis of Floating PV Technologies in the Tropics', *Progress in Photovoltaics: Research and Applications* (2018) 26: 958; Lee et al., 'Hybrid Floating', 1415; Almelda et al., 'Floating Solar', 247; IEA, 'Renewables', 11 July 2023; and see further e.g. Creutzig et al., 'The Understimated Potential', 1; Victoria et al., 'Solar Photovoltaics', 1050; Dieter Holger, 'After a Bumpy Year, Renewable Energy Looks Poised for Boom Times', *Wall Street Journal*, 29 December 2022. 5.5: hydro included, biofuels not.

115 York and Bell, 'Energy Transitions', 41.

116 Christophers, *The Price*, 7, 343–4.

117 Jillian Ambrose, 'Greenhouse Gas Emissions from Global Energy Industry Still Rising – Report', *Guardian*, 26 June 2023.

118 Michelle Grayson, 'Energy Transitions', *Nature* (2017) 551: 133.

119 Masson-Delmotte, *Global Warming*, 41, 315; on the function: York and Bell, 'Energy Transitions', 41.

120 Two examples shall suffice here. 'In today's energy context, with an ongoing energy transition towards renewables', begins one paragraph in Kühne et al., '"Carbon Bombs"', 6. 'The passage of the Inflation Reduction and CHIPS Acts suggests that the US state is tilting in the direction, not just of fiscal stimulus, but of industrial policy and a green transition', begins Aaron Benanav, 'A Dissipating Glut?', *New Left Review* 140/141: 53 (although he demonstrates awareness of some of the problems later in the same essay: ibid., 62). The problem here is the unthinking conflation of transition with support for renewables, forgetting about the suppressive element and misunderstanding the category in question.

121 'The world has not yet really even attempted the energy system transformation that will be required.' Matthews et al., 'Focus on Cumulative', 7.

122 Jacobson et al., 'Low-Cost Solutions', 3344. This is a constant theme in Jacobson's writings: see e.g. *No Miracles*, 1–4.

123 But residual causes of air pollution, such as certain agricultural activities and deforestation, would not thereby be removed.

124 Luderer et al., 'Impact of Declining', 32, 35; Way et al., 'Empirically Grounded', 2067; Jacobson et al., 'Low-Cost Solutions', 3343, 3347; Breyer et al., 'On the History', 78177, 78189; Bogdanov et al., 'Low-Cost Renewable', 1, 14; Jacobson et al.,

'100% Clean', 108, 111; Jacobson, *No Miracles*, 102–4, 266, 270, 274, 279; Griffith, *Electrify*, 55–8.

125 Cf. e.g. Michael Child, Claudia Kemfert, Dmitrii Bogdanov and Christian Breyer, 'Flexible Electricity Generation, Grid Exchange and Storage for the Transition to a 100% Renewable Energy System in Europe', *Renewable Energy* (2019) 139: 80–101; Renné, 'Progress, Opportunities', 8; Jacobson et al., '100% Clean', 115; Jacobson et al., 'Low-Cost', 3350–2; Jacobson, *No Miracles*, 274–5, 311–14.

126 Way et al., 'Empirically Grounded', e.g. 2063, 2074. Note that any costs of climate change (or air pollution, for that matter) were entirely omitted from this calculation.

127 Jacobson et al., 'Low-Cost Solutions', 3352; Bogdanov et al., 'Low-Cost Renewables', 7. And because the world would have to pay for about half as much energy as consumed in a fossil economy: Jacobson et al., 'Low-Cost Solutions', 3352.

128 Hiroto Shiraki and Masahiro Sugiyama, 'Back to the Basics: Toward Improvement of Technoeconomic Representation in Integrated Assessment Models', *Climatic Change* (2020) 162: 15; Mengzhu Xiao, Tobias Junne, Jannik Haas and Martin Klein, 'Plummeting Costs of Renewables: Are Energy Scenarios Lagging?', *Energy Strategy Reviews* (2021) 35: 5, 7. For the underestimation of wind, cf. Wiser et al., 'Expert Elicitation'.

129 Xiao et al., 'Plummeting Costs', e.g. 8.

130 Way et al., 'Empirically Grounded', 2062–3.

131 Creutzig et al., 'The Underestimated Potential', 1–2; Xiao et al., 'Plummeting Costs', 1.

132 Gerry Carrington and Janet Stephenson, 'The Politics of Energy Scenarios: Are International Energy Agency and Other Conservative Projections Hampering the Renewable Energy Transition?', *Energy Research and Social Science* (2018) 46: 110; Victoria et al., 'Solar Photovoltaics', 1042–3; Creutzig et al., 'The Underestimated', 1, 6.

133 See further e.g. Jaxa-Rosen and Trutnevyte, 'Sources of Uncertainty'; Luderer et al., 'Impact of Declining'; Way et al., 'Empirically Grounded', 2062–3; Way et al., 'Supplemental Information', 34–5; Breyer et al., 'On the History', 78193–4; Gambhir et al., 'A Review', 5–6.

134 Femke J. M. M. Nijsse, Jean-Francois Mercure, Nadia Ameli et al., 'The Momentum of the Solar Energy Transition', *Nature Communications* (2023) 14: 2; Way et al., 'Empirically Grounded', 2063.

135 Xiao et al., 'Plummeting Costs', 10.

136 Creutzig et al., 'Technological Innovation', 1.

137 Cf. Carrington and Stephenson, 'The Politics of Energy'; Jaxa-Rozen and Trutnevyte, 'Sources of Uncertainty', 266.

138 Xiao et al., 'Plummeting Costs', 1.

139 Nijsse et al., 'The Momentum of', 2; Carrington and Stephenson, 'The Politics of Energy', 106; Shiraki and Sugiyama, 'Back to the Basic', 15, 17–18; Breyer et al., 'On the History', 78193–4; Victoria et al., 'Solar Photovoltaics', 1041–3; Creutzig et al., 'Technological Innovation'.

140 Cf. Shiraki and Sugiyama, 'Back to the Basic', 14.

141 Carrington and Stephenson, 'The Politics of Energy', 110; Xiao et al., 'Plummeting Costs', 10–11; Breyer et al., 'On the History', 78194. It might be noted, as an aside, that one of the authors of an ecomodernist critique of Jacobson et al. was Tom Wigley, progenitor of the overshoot idea: Heard et al., 'Burden of Proof'. BECCS was also promoted as a better option than 100 per cent renewables in Clack et al., 'Evaluation of a Proposal', 2.

142 Victoria et al., 'Solar Photovoltaics', 1042, 1051; Victoria et al., 'Supplemental Information: Solar Photovoltaics Is Ready to Power a Sustainable Future', *Joule* (2022) 5: 5; and cf. Creutzig et al., 'The Underestimated Potential', 1.

143 Creutzig et al., 'Technological Innovation', 3.

144 Ibid., 2.

145 Luderer et al., 'Impact of Declining'. More precisely, the overshoot modelled here was the minimal 0.1°C – i.e. average temperature would hit 1.6°C by mid-century, after which this one tenth of a centigrade would be cleaned up. Ibid., 33. The Oxford study showing how a transition would save trillions of dollars marked a more radical methodological break with the IAMs, in that it relied strictly on *empirically* grounded forecasting, rather than on theoretical assumptions: see Evelina Trutnevyte, Nik Zielkonka and Xin Wen, 'Crystal Ball to Foresee Energy Technology Progress?', *Joule* (2022) 6: 1969–70.

146 'For individual technologies (in particular solar energy), IAM projections have been conservative regarding deployment rates and cost reductions . . . IAMs may underreport the potential for supply-side technological change assumed in 1.5°C-consistent pathways, but may be more optimistic about the systemic ability to realize incremental changes.' Masson-Delmotte et al., *Global Warming*, 322.

147 By acquiring the solar company Lucas Energy Systems, one branch of the Lucas industrial empire of Lucas Plan fame: Jonathan Pinkse and Daniel van den Buuse, 'The Development and Commercialization of Solar PV Technology in the Oil Industry', *Energy Policy* (2012) 40: 15.

148 Ibid., 15; Damian Miller, 'Why the Oil Companies Lost Solar', *Energy Policy* (2013) 60: 53–4.

149 Terry Macalister and Eleanor Cross, 'BP Rebrands on a Global Scale', *Guardian*, 25 July 2000.

150 Terry Macalister, 'BP Axes Solar Business', *Guardian*, 21 December 2011; Pinkse and Van den Buuse, 'The Development and Commercialization', 16; Miller, 'Why the Oil', 52, 57–8.

151 Scott Carpenter, 'After Abandoned "Beyond Petroleum" Re-Brand, BP's New Renewables Push Has Teeth', *Forbes*, 4 August 2020.

152 BP, 'BP Sets Ambition for Net Zero by 2050, Fundamentally Changing Organisation to Deliver', 12 February 2020.

153 Matt McGrath, 'COP27: BP Chief Listed as Delegate for Mauritania', *BBC*, 10 November 2022.

154 Interview with Nicolai Tangen of Norges Bank Investment Management in February 2022, quoted in Jenny Strasburg, 'BP's CEO Plays Down Renewables Push as Returns Lag', *Wall Street Journal*, 1 February 2023.

155 BP, 'BP (BP) Q4 2021'.

156 Strasburg, 'BP's CEO'; Jenny Strasburg, 'BP Slows Transition to Renewable Energy as Oil Bonanza Continues', *Wall Street Journal*, 7 February 2023; Stuart Braun, 'Shell, BP Boost Profit, Sink Investment in Renewable Energy', *DW*, 10 February 2023. On how unique the ambition was, see Allen and Coffin, 'Paris Maligned', 23.

157 Strasburg, 'BP's CEO'; Ron Bousso, Shadia Nasralla and Sarah Mcfarlane, 'Inside BP's Plan to Reset Renewables as Oil and Gas Boom', Reuters, 7 Marsh 2023 (Anja-Isabel Dotzenrath quoted in the latter).

158 Quoted in Kate Aronoff, 'Oil Companies Are Finally Being Honest about their Feelings on Renewable Energy', *New Republic*, 8 February 2023. On BP, see further Brett Christophers, 'Fossilised Capital: Price and Profit in the Energy Transition', *New Political Economy* (2022) 27: 146–59.

159 Or, the renewable energy adventures of fossil capital faced an intermittency problem of their own. They turned with the political winds, and when those winds were down, it was back to the combustion line.

160 M. Absi Halabi, A. Al-Qattan and A. Al-Otaibi, 'Application of Solar Energy in the Oil Industry: Current Status and Future Prospects', *Renewable and Sustainable Energy Reviews* (2015) 43: 307–8; Pinkse and Van den Buuse, 'The Development and Commercialization', 15.

161 Halabi and Al-Otaibi, 'Application of Solar', 301–2.

162 Quoted in Lylla Younes, 'Exxon Reports Record Profits, Doubles Down on Fossil Fuels', *Grist*, 1 February 2023; cf. Sam Meredith, 'Big Oil poised to Smash Annual Profit Records – Sparking Outcry from Campaigners and Activists', *CNBC*, 27 January 2023; David Blackmon, 'Big Profits for Big Oil from the High Price Environment of 2022', *Forbes*, 31 January 2023; Partridge, 'ExxonMobil's Record-Breaking'.

163 Gregory Meyer, 'Texas: How the Home of US Oil and Gas Fell in Love with Solar Power', *Financial Times*, 7 April 2020.

164 Halabi et al., 'Application of Solar', 307–8.

165 Chevron, 'ChevronTexaco Installs California's First Solar Project to Power Oil Production', 5 June 2003; largest plant of flexible, amorphous-silicon solar panels: Halabi et al., 'Application of Solar', 302.

166 Quoted in Billy Gridley, 'Chevron's Radical Gamble on Climate Change', 26 May 2022.

167 Robert Tuttle, 'Chevron Is Installing Solar Panels – To Produce Oil More Cheaply', *Bloomberg*, 22 July 2020.

168 Edward Hirs of the University of Houston quoted in Meyer, 'Texas: How'.

169 CEO Vicki Hollub in ibid. Cf. *NS Energy*, 'Occidental Opens 16MW Solar Facility to Power Permian Basin Operations', 4 October 2019.

170 Occidental Petroleum should not be confused with Occidental Power, a small solar business in California.

171 Solarcraft, 'Solar Powered Cathodic Protection: Benefits of Using CPX Controller Compared to Conventional Methods', n.d.; Ali O. M. Maka, Salem Salem and Mubbashar Mehmood, 'Solar Photovoltaic (PV) Applications in Libya: Challenges, Potential, Opportunities and Future Perspectives', *Cleaner Engineering and Technology* (2021) 5: 2, 7–8; and cf. Halabi et al., 'Application of Solar', 302. The technology is called cathodic protection.

172 Bernd Radowitz, 'GE Wind Turbines in Finland to Power Lundin Oil Fields off Norway', *Recharge News*, 31 January 2020.

173 Equinor, 'Hywind Tampen', n.d.; Jeremy Beckman, 'Floating Wind Turbines to Power North Sea Gullfaks, Snorre Platforms', *Offshore*, 1 February 2021.

174 See e.g. *Offshore Engineer*, 'Norway's Giant Offshore Oil Field Fully Developed', 15 December 2022; Adomaitis, 'Norway Ramps Up'; Equinor, 'Johan Sverdrup Phase 2'; Lundgren, 'At North Sea'.

175 Equinor, 'Equinor Fourth Quarter 2022 and Year End Results', 8 February 2023.

176 David Tong and Kelly Trout, 'Big Oil Reality Check 2023:

Equinor', Oil Change International, May 2023, 4. Figures for 2022.

177 Tuuka Mäkitie, Håkon E. Normann, Taran M. Thune and Jakoba Sraml Gonzalez, 'The Green Flings: Norwegian Oil and Gas Industry's Engagement in Offshore Wind Power', *Energy Policy* (2019) 127: 269–79.

178 ConocoPhillips, 'Plan for the Net-Zero Energy Transition', 4 May 2022, 4; Nermina Kulovic, 'ConocoPhillips and CNOOC to Harness Wind Energy to Power Chinese Oilfield', *Offshore Energy*, 7 November 2022. 'Investments in technology development will be disciplined and commensurate with the likely returns, market size, timing of development and technology risk inherent in renewable energy projects . . . Renewable energy opportunities that complement our existing processes will be prioritized.' ConocoPhillips, 'Renewable Energy Position', 2023.

179 Carol Ryan, 'Big Oil Gushes Cash because It Doesn't Know Where to Invest', *Wall Street Journal*, 9 March 2023.

180 Flavian Picchonat, 'The Capitalist Dilemma in the Energy Sector. Profit and Renewable Energy Investments of Major Oil Companies: The Case of TotalEnergies SE', master's thesis, University of Genève, 2023, 57–8 (the measure of profit here being 'operating margin'). Total also had an earlier fling with solar in the 2000s: see Pinkse and Van den Buuse, 'The Development and Commercialization', 17. On this company, see further Pierre-Louise Choquet, 'Piercing the Corporate Veil: Towards a Better Assessment of the Position of Transnational Oil and Gas Companies in the Global Carbon Budget', *Anthropocene Review* (2019) 6: 243–62; Christophers, 'Fossilised Capital'.

181 Patrick Pouyanné in TotalEnergies, 'Strategy, Sustainability and Climate: Transcript', 19.

182 Tong and Trout, 'Big Oil Reality Check 2023: TotalEnergies', 4.

183 Pinkse and Van den Buuse, 'The Development and Commercialization', 16–19; Miller, 'Why the Oil', 53–7.

184 Julian Ambrose, 'Shell Unveils Plans to Become Net-Zero Carbon Company by 2050', *Guardian*, 16 April 2020.

185 *Seeking Alpha*, 'Shell PLC (SHEL) Q4 2022 Earnings Call Transcript', 2 February 2023.

186 Ron Bousso and Shadia Nasralla, 'Shell Boosts Dividend, Steadies Oil Output under New CEO's Plan', Reuters, 14 January 2023. Cf. Jasper Jolly, 'Shell Drops Target to Cut Oil Production as CEO Aims for Higher Profits', *Guardian*, 14 June 2023; 'Shell, BP'; Lottie Limb, 'Shell Joins BP and Total in U-Turning on Climate Pledges "to Reward Shareholders"', *Euronews*, 15

June 2023. See further Christophers, 'Fossilised Capital'; Christophers, *The Price*, 216.

187 Halabi et al., 'Application of Solar', 305.

188 Dario Kenner and Richard Heede, 'White Knights, or Horsemen of the Apocalypse? Prospects for Big Oil to Align Emissions with a 1.5°C Pathway', *Energy Research and Social Science* (2021) 79: 4; Li et al., 'The Clean Energy', 13, 17–18. Cf. Christophers, 'Fossilised Capital', 155.

189 Li et al., 'The Clean Energy', 20; Tyler A. Hansen, *Three Essays on the Political Economy of Global Inaction on Climate Change*, PhD dissertation, University of Massachusetts Amherst, 2021, 105.

190 Nakhle, 'Oil and Gas'.

191 Ca 0.72 per cent as against ca 0.26: ibid.

192 Christophers, *The Price*.

193 Hansen, 'Stranded Assets', 9. See also the figures in Halttunen et al., 'What If', 2.

194 Hansen, *Three Essays*, 118. See further Christophers, 'Fossilised', 151–2; Christophers, *The Price*, 211.

195 Quoted in Kate Aronoff, 'BP Is not Woke. It's an Imperialist Success Story', *New Republic*, 6 June 2020.

196 'BP (BP) Q4 2021'.

197 Tom Wilson and Emma Dunkley, 'BP Slows Oil and Gas Retreat after Record $28bn Profit', *Financial Times*, 7 February 2023.

198 See ibid.; Strasburg, 'BP's CEO'.

199 Justin Worland, 'BP's Green U-Turn Shows Exactly Why the Energy Transition Is So Hard', *Time*, 10 February 2023.

200 Rothfeder and Maag, 'How Wall Street's'.

201 Beyene et al., 'Financial Institutions', 24. Cf. IEA, *World Energy Investment*, 203.

202 See e.g. Nijsse et al., 'The Momentum of', 4–5.

203 Christophers, *The Price*, xxvii–xxix, 216–7.

204 Quoted in Christophers, 'Environmental Beta', 9.

205 Steven Mufson and Douglas MacMillan, 'BlackRock's Larry Fink Tells Fellow CEOs that Businesses Are Not "Climate Police"', *Washington Post*, 18 January 2022; McDermid, 'BlackRock Tells'; Rothfeder and Maag, 'How Wall Street's'.

206 Way et al., 'Empirically Grounded', 2057–60; Way et al., 'Supplemental Information', 24, 72, 79–83; and for an earlier study finding similar trends for the US, see Robert S. Pindyck, 'The Long-Run Evolution of Energy Prices', *Energy Journal* (1999) 20: 1–27. See also Nasdaq, 'Crude Oil Prices from 1861', 22 December 2022.

207 With the two blades meeting around the year 2010: see in particular the diagram in Way et al., 'Empirically Grounded', 2058.

208 This also has consequences for the 'energy return on investment', or EROI, a measure that has become something of a hobbyhorse for critics of renewable energy, whether they be of a fossil fuel-chauvinist, ecomodernist or anarcho-primitivist bent. The contention here is that renewables have lower EROI than fossil fuels. In fact, however, fossil fuels have declining EROI, because more energy must be invested in the extraction of any given piece of fuel, whereas the EROI of renewables is heading upwards due to technological advancements – a physical counterpart to the scissors' movement in prices, already obliterating any advantages the former might have once had. But the EROI of fossil fuels has historically also been vastly exaggerated, because it has failed to account for conversion losses and measured only the *primary* energy returned after the moment of extraction (rather than final, actually useful energy in, say, an automobile). A transition would thus tend to improve the overall EROI in the world economy, just as it would offer savings in expenses on energy. On these points, see the illuminating study by Paul E. Brockway, Anne Owen, Lina K. Brand-Correa and Lukas Hardt, 'Estimation of Global Final-Stage Energy-Return-on-Investment for Fossil Fuels with Comparison to Renewable Energy Sources', *Nature Energy* (2019) 4: 612–21; and further e.g. M. Diesendorf and T. Wiedmann, 'Implications of Trends in Energy Return on Energy Invested (EROI) for Transitioning to Renewable Electricity', *Ecological Economics* (2020) 176: 1–8.

209 Cf. Way et al., 'Empirically Grounded', 2059; Way et al., 'Supplemental Information', 72.

210 The debate on the value of the labour theory of value for understanding the ecological crisis rumbles on: for an excellent demonstration of how high that value is, see Elke Pirgmaier, 'The Value of Value Theory for Ecological Economics', *Ecological Economics* (2021) 179: 1–10. For a superb general exposition, see Andrew Brown, 'A Materialist Development of Some Recent Contributions to the Labour Theory of Value', *Cambridge Journal of Economics* (2008) 32: 125–46.

211 Marx, *Capital Volume III*, 280. For an impressive recent demonstration of how the law of value regulates prices, see Güney Isıkara and Patrick Mokre, 'Price-Value Deviations and the Labour Theory of Value: Evidence from 42 Countries, 2000–2017', *Review of Political Economy* (2022) 34: 165–80. At this level of abstraction, we need not detain ourselves in further

investigations of the relations between market prices, prices of production and direct prices.

212 Shaikh, *Capitalism*, 232; cf. e.g. 212–13, 237.

213 Cf. Marx, *Capital Volume I*, 435–6, 645; Shaikh, *Capitalism*, e.g. 259–62, 266, 287–8, 333, 338. This first source of super-profit, to stick with Marxian categories, is relative surplus value. Companies can, of course, also use their advantage to set its price below the common price, so as to undersell their competitors and seize larger market shares; or, ideally, place its price between these two levels, so as to win *both* a super-profit and a larger share.

214 Cf. Marx, *Capital Volume III*, 752–3, 780, 784–6; Shaikh, *Capitalism*, 266, 295, 337; Isıkara and Mokre, 'Price-Value Deviations', 174–5. This second source of super-profit is ground rent; and as Marx makes clear in his discussion, this remains the case whether the company owns the superior patch of land or must pay rent to a landowner, who thereby pockets some of the ground rent. Marx, *Capital Volume III*, 784–5. On relative surplus value and ground rent as sources of super-profit in oil extraction, cf. Chesnais, *Finance Capital*, 121–2.

215 See e.g. Manuel Monge, María Fátima Romero Rojo and Luis Alberiko Gil-Alana, 'The Impact of Geopolitical Risk on the Behavior of Oil Prices and Freight Rates', *Energy* (2023) 269: 1–9; Jeff D. Colgan, Alexander S. Gard-Murray and Miriam Hinthorn, 'Quantifying the Value of Energy Security: How Russia's Invasion of Ukraine Exploded Europe's Fossil Fuel Costs', *Energy Research and Social Science* (2023) 103: 1–2, 8; Krane and Idel, 'More Transitions', 1–2; *Macrotrends*, 'Crude Oil Prices: 70 Year Historical Chart', 2023.

216 Weber, 'Big Oil's', 152–3; Ian Palmer, 'Oil and Gas Profits Very High Once Again: What this Feels Like to Energy Consumers', *Forbes*, 4 November 2022. Or, in the highly general terms used by CEO Darren Woods to explain the superb performance of the ExxonMobil gas fields in Papua New Guinea (gas and oil essentially behaving in the same manner during this period): 'one of the changes that we've made over time here is we brought a lot of the optimization skills and techniques that we've historically used in our downstream businesses and our refineries, where, you know, you're eking out pennies on the barrel to be successful, and taking that technology and approach to some of our upstream assets, and sharing that technology and working with the upstream teams to improve the productivity of the assets on the ground. That's paying off. Really, all around the world, we're applying that technique.' ExxonMobil, 'ExxonMobil (XOM) Q4 2021'.

217 See e.g. Norwegian Petroleum Directorate [for all practical purposes, Equinor], *Resource Report 2022*, 33; TotalEnergies, 'Strategy, Sustainability and Climate: Transcript', 4, 7, 28; Jolly, 'Shell Drops'; Strasburg, 'BP's CEO'.

218 Marx, *Capital Volume III*, 954; Adorno, 'Marx and the Basic', 161. Or, more elaborately, value is *'the social labour we do for others as organised through commodity exchanges in competitive price-fixing markets.'* Harvey, *Marx, Capital*, 4 (emphasis in original).

219 Marx, *Capital Volume III*, 787. See further 781–6; Malm, *Fossil Capital*, 311–12.

220 Marx, *Capital Volume III*, 785; and for some excellent renditions of these contrasts, see further McDermott Hughes, *Who Owns*, 20–6, 79.

221 See e.g. Miller, 'Why the Oil', 53; Pinkse and Van den Buuse, 'The Development and Commercialization', 16; Li et al., 'The Clean Energy', 17.

222 Strasburg, 'BP's CEO'; Bousso et al., 'Inside BP's'; Halabi et al., 'Application of Solar', 308; Amanda Leigh Mascarelli, 'Gold Rush for Algae', *Nature* (2009) 461: 460–1.

223 Ben Elgin and Kevin Crowley, 'Exxon Retreats from Major Climate Effort to Make Biofuels from Algae', *Bloomberg*, 10 February 2023.

224 Spatarshi Das, Eric Hittinger and Eric Williams, 'Learning Is not Enough: Diminishing Marginal Revenues and Increasing Abatement Costs of Wind and Solar', *Renewable Energy* (2020) 156: 634–44; Dev Millstein, Ryan Wiser, Andrew D. Mills et al., 'Solar and Wind Grid System Value in the United States: The Effect of Transmission Congestion, Generation Profiles, and Curtailment', *Joule* (2021) 5: 1749–75. Cf. e.g. James Temple, 'The Lurking Threat to Solar Power's Growth', *MIT Technology Review*, 14 July 2021; Christophers, *Price*, 224.

225 Das, 'Learning Is'; Millstein et al., 'Solar and Wind'.

226 Millstein et al., 'Solar and Wind', 1763.

227 Modelling suggests that storage can counteract revenue decline up to 50 per cent flow penetration: Millstein et al., 'Solar and Wind', 1766.

228 Millstein et al., 'Solar and Wind', 1765. All of this is, of course, the nail in the coffin of Jason W. Moore's theory of 'cheap nature'.

229 Das et al., 'Learning Is', 637–8 (cf. 642); Temple, 'The Lurking Threat'.

230 One expression of this is that 'value decline' will tend to run faster and outpace the diminution of costs for producing renewables – the exact inverse of fossil fuels, where the average socially

necessary labour time amenable to undercutting is reliably stable or rising. Millstein et al., 'Solar and Wind', 1749, 1765; Das et al., 'Learning Is', 642.

231 On this cost profile, see Christophers, *The Price*, 75–7, 91–2.

232 Jacobson, *No Miracles*, 17–18, 325–6; Jacobson, 'Low-Cost Solutions', 3344; Colgan et al., 'Quantifying the Value', 1, 8; Krane and Idel, 'More Transitions', 1–2; Overland, 'The Geopolitics', 37–8; and cf. Keramidas et al., 'Global Energy'.

233 As pointed out in Colgan et al., 'Quantifying the Value', 1. As for inter-regional supergrids, a mutual dependency between countries would emerge, in principle vulnerable to geopolitical crises; however, such ties would probably be of a more reciprocal nature – a region rich in sunlight connected to one swept in wind – and certainly less commodified than the relations between importers and exporters of oil. An exporter of solar-powered electricity could never put it in tankers for subsequent use or postpone extracting the fields until later. Overland, 'The Geopolitics', 37–8.

234 Alexander Villegas and Ernest Scheyder, 'Chile Plans to Nationalize its Vast Lithium Industry', Reuters, 21 April 2023. Cf. e.g. Natasha Turak, 'Iran Says It's Discovered What Could Be the World's Second-Largest Lithium Deposit', *CNBC*, 6 March 2023; Lisa O'Carroll, 'The East German Town at the Centre of the New "Gold Rush" . . . For Lithium', *Guardian*, 23 September 2023.

235 As demonstrated in Krane and Idel, 'More Transitions'; and see further Overland, 'The Geopolitics'.

236 In the oil business, again, the equivalent situation would be one where the oil itself remained uninterrupted, while the steel for building new platforms and pipes became costlier due to some embargo: Overland, 'The Geopolitics', 2, 4.

237 There is at least one other possible such factor. It has often been argued that mitigation would reduce demand for fossil fuels and therefore cause their prices to fall. If such a situation were to transpire, it would, ironically, again demonstrate exactly what it is about the stock that makes it so profitable. On the assumption that prices for the fuels do collapse with demand, in a world where the noose tightens around fossil capital, producers with the lowest costs would be able to stick it out the longest. And in the early third decade, nowhere in the world was there still so much oil so cheap to produce as in the Middle East: Saudi Aramco, for instance, extracted oil at the piddling average cost of 3 dollars per barrel, from reserves three times larger than those of the five biggest competitors combined. Lex, 'Saudi Aramco'.

A company like this would be best placed to squeeze out the last drops of profit, for the reasons examined above. Cf. e.g. Thomä, 'The Stranding of Upstream', 117; Semieniuk et al., 'Stranded Fossil-Fuel', 533–4; Shukla et al. *Climate Change*, 1584. Eventually mitigation would have to kill all profit-making in fossil fuels; but on the way there, even a lowered price would allow the best-positioned companies to accumulate capital. It should be noted that both these factors with a potential to close the scissors – a fast transition driving up the prices of critical materials; a ditto sapping demand for fossil fuels – *require a break with business as usual*; in and of itself, it seems incapable of generating any such factor (and hence its extreme tenacity).

238 Beatriz Santos, 'Top PV Module Manufacturers by Shipment Volume in 2022', *PV Magazine*, 13 April 2023; LONGi, 'LONGi Ranks 177th on the Fortune China 500 List in 2023', 27 July 2023.

239 A rival of Longi, Tongwei, did write history when it became the first photovoltaic company to enter Fortune Global 500 in 2023, ranking 476. (Even a minor actor in the oil industry such as Canadian National Resources – operations limited to the Alberta tar sands – ranked higher, at 471.) Fortune Global 500, 2023; *Energy Trend*, 'From Green "Intelligent" Pioneer Manufacturing to Fortune Global 500 Listing, TW (Tongwei) Solar Leads the Industry's Upgrading and Transformation', 30 August 2023.

240 SpaceX and The Boring Company respectively. Alejandro De La Garza, 'How Elon Musk Built His Private Fortune – And Became the Richest Private Citizen in the World', *Time*, 13 December 2021; Aimee Picchi, 'The World's Wealthiest Person: How Did Elon Musk Get So Rich?', *CBS News*, 30 September 2022.

241 Christophers, *The Price*, 144; and see further 145–6, 202. Cf. e.g. Hansen, *Three Essays*, 103.

242 On this fallacy and other flaws of the monopoly capitalism school, see Shaikh, *Capitalism*, 355–6, 367–8, 378–9.

243 This is the argument in Jack Copley, 'Decarbonizing the Downturn: Addressing Climate Change in an Age of Stagnation', *Competition and Change* (2023) 27: 440. But the thrust of this article, based on Robert Brenner's theory of the long downturn, is precisely that overcapacity was the *universal* predicament of capitalism after the 1970s; if so, it cannot explain the particular inability of the renewables industry to generate profits.

244 Marx, *Grundrisse*, 552 (emphasis in original).

245 Ibid., 414, 552, 649–52, 657–7; Marx, *Capital Volume I*, 433; Marx, *Capital Volume III*, 311; Paresh Chattophadhyay, 'Competition', in Ben Fine and Alfredo Saad-Filho (eds), *The Elgar*

Companion to Marxist Economics (Cheltenham: Edward Elgar, 2012), 72–7; and further Shaikh, *Capitalism*.

246 Paul A. Samuelson, 'The Pure Theory of Public Expenditure', *Review of Economics and Statistics* (1954) 36: 387.

247 Simon Marginson, 'Higher Education and Public Good', *Higher Education Quarterly* (2011) 65: 415 (emphasis added). The classical example of such a non-rivalrous good is, of course, knowledge (hence higher education).

248 John A. Tatom, 'Should Government Spending on Capital Goods Be Raised?', *Federal Reserve Bank of St. Louis Review* (1991) 73: 5.

249 However, any given patch of *land* used for solar and wind power generation is, of course, a rivalrous good: if Z puts up solar panels on a field, Y cannot also do so; it has been taken. The situation here would be that supra-surface supplies are non-rivalrous, while surface supplies are rivalrous. This remains qualitatively different from the situation in fossil fuels, where subsurface and surface supplies are *both* rivalrous. It follows from this that *land for renewable power generation* can be an eminently private good, bought and sold on the market, an object of investment and speculation – unlike the fuel itself, that is. A 100 per cent flow economy would then still allow for ground rent in energy production, even if there is no surplus value in it (and the sources for this ground rent would then have to be surplus value produced in other parts of the economy). Cf. McDermott Hughes, *Who Owns*, 185–6. Likewise, the instruments for capturing the fuels – panels, turbines – are rivalrous, which is why, as stated above, profits in the production of these commodities are possible. As for the second criterion of public goods, non-excludability, this is not an inherent feature of solar and wind power *qua* fuels; but then few public goods fulfil both.

250 There is also an important factor holding consumers back from demanding solar and wind even if they are cheaper than fossil fuels (or on the verge of becoming so), namely, the fixed capital sunk in the consumption of the latter. The prime example would perhaps again be the steel industry, the largest industrial consumer of coal.

251 Shaikh, *Capitalism*, 615.

252 Ibid., 7; see further e.g. xxxv–xxxvi, 6, 616. This is also the point of departure for Christophers, *The Price*, xxi–xxv, 136.

253 Shaikh, *Capitalism*, 66, 260, 264, 334.

254 And this, it is important to note, may well include state-owned enterprises that are just as profit-oriented as any other capitalist actors.

255 Cf. Luderer et al., 'Impact of Declining', 35–6; Nijsse et al., 'The Momentum of', 1–2.

256 Marx, *Capital Volume I*, 503; Marx, *Grundrisse*, 470 (emphasis in original), 831; Marx, *Theories of Surplus*, book III, 456.

257 David McNally, *Monsters of the Market: Zombies, Vampires and Global Capitalism* (Leiden: Brill, 2011), e.g. 13, 115, 118.

258 Harvey, *Marx, Capital*, 5. There's nothing mystical about this, any more than about democracy or national identity being invisible and immaterial qualities: ibid.

259 For Harvey, with his focus on urbanisation, the emblematic manifestations of this power are 'the impressive and daunting skylines of contemporary world cities': Harvey, *A Companion to* Grundrisse, 345. But we may as well extend it along the lines of all fossil fuel infrastructure.

260 The compulsion to produce exchange value emanates, of course, from the separation between the means of production and the direct producers and their reunification via the universal equivalent. 'The exchange-relation is, in reality, performed by class relations: that there is an unequal control of the means of production: that is the heart of the theory.' Adorno, 'Marx and the Basic', 158.

261 McNally, *Monsters*, 244. See further e.g. 7, 115, 121–32, 199, 250. In this contemporary classic, McNally does not deal with the ecological crisis, except for mentioning it once in the briefest passing: 16.

262 'Profit drives capitalism. If profit fails, the firm goes into shock and its capital begins to atrophy. Economic theory and business sentiment are in complete agreement on this point.' Shaikh, *Capitalism*, 206. See further e.g. 259, 672, 726. Growth is a *result* of profit-driven accumulation of capital, an epiphenomenal not primary phenomenon. Moreover, some research suggests that the rate of profit correlates far more closely with CO_2 emissions levels than GDP growth does: Matthew Soener, 'Profiting in a Warming World: Investigating the Link between Greenhouse Gas Emissions and Capitalist Profitability in OECD States', *Sociological Forum* (2019) 34: 974–98; and cf. Matthew Soener, 'Between Green Growth and De-growth: Locating the Roots of Climate Change in Capitalist Accumulation' in Marlène Benquet and Théo Bourgeron (eds), *Accumulating Capital Today: Contemporary Strategies of Profit and Dispossessive Politics* (London: Routledge, 2021), 47–60.

263 McNally, *Monsters of the Market*, 130.

264 Adorno, *History and Freedom*, 50.

265 Theodor W. Adorno, 'Sociology and Empirical Research', in

Theodor W. Adorno, Hans Albert, Ralf Dahrendorf et al., *The Positivist Dispute in German Sociology* (London: Heinemann, 1977 [1969]), 80 (emphases added). One could of course turn this around and say that the reality of e.g. fixed capital dominates the illusion of e.g. green capitalism. We shall offer a more systematic discussion of rationality and irrationality in the sequel to this book.

266 R. Daniel Bressler, 'The Mortality Cost of Carbon', *Nature Communications* 12 (2021): 1–12. For an earlier study in a similar vein, see John Nolt, 'How Harmful Are the Average American's Greenhouse Gas Emissions?', *Ethics, Policy and the Environment* (2011) 14: 3–10.

267 Oliver Millman, 'Three Americans Create Enough Carbon Emissions to Kill One Person, Study Finds', *Guardian*, 29 July 2021.

268 Details on the calculation: one barrel of crude oil generates on average 0.43 metric tons of CO_2. United States Environmental Protection Agency, 'Greenhouse Gas Equivalencies Calculator: Calculations and References', epa.gov, n.d. (data from 2019 and 2020). EACOP will transport 216,000 barrels of crude oil per day: see above. 216,000 × 0.43 = 92,880 metric tons of CO_2 from EACOP per day. 92,880 × 365 = 33,901,200 metric tons of CO_2 from EACOP per year (i.e. nearly 34 million metric tons). One million metric ton of CO_2 in 2020 causes 226 deaths, according to Bressler, 'The Mortality'. 33.9 × 226 = 7,661 deaths caused by EACOP in an average year of its operations.

269 E.g. Nicolae Buldumac, 'Top 25 Public Companies in France by Revenue in 2022', *Global Database*, 23 August 2023; Ed Reed, 'Macron Says "Count on Me" for EACOP Support', *Energy Voice*, 4 May 2021.

270 See e.g. Li et al., 'The Clean Energy', 12, 19. Looney: 'At the end of the day, we're responding to what society wants.' Strasburg, 'BP Slows'.

271 Cf. Calverley and Anderson, *Phaseout Pathways*, 11, 27.

272 TotalEnergies, 'Key Figures', 2021.

273 Calculation performed as above.

274 For an extended argument for classifying the extraction and marketing of fossil fuels as crimes, see Ronald C. Kramer, *Carbon Criminals, Climate Crimes* (New Brunswick: Rutgers University Press, 2020); on the question of intention and knowledge, see particularly 60–2.

275 Victoria Bekiempis, 'Derek Chauvin Trial: Jury Begins Deliberations over Killing of George Floyd – As It Happened', *Guardian*, 20 April 2021; and cf. McKinnon, *Climate Change*, 96, 111, 116–18.

276 Christophe Bonneuil, Pierre-Louise Choquet and Benjamin Franta, 'Early Warnings and Emerging Accountability: Total's Responses to Global Warming, 1971–2021', *Global Environmental Change* (2021) 71: 1–10.

277 Patrick Pouyanné in TotalEnergies, 'Strategy, Sustainability and Climate: Transcript', 18 (emphasis added). Elaborating on offshore resources in Brazilian waters, he continued: 'that's the way we work: giant deepwater fields, that's good: let's go. It's fitting our criteria. When you have this type of opportunities, you cannot avoid them.' Ibid.

278 Marx, *Grundrisse*, 91. An influential study of consumption as driver of environmental degradation (which does, however, pay tribute to Marxist theories of capital accumulation) is Wiedmann et al., 'Scientists' Warming'.

279 Benedikt Bruckner, Klaus Hubacek, Yuli Shan et al., 'Impacts of Poverty Alleviation on National and Global Carbon Emissions', *Nature Sustainability* (2022) 5: 311–20; Lukas Chancel, 'Global Carbon Inequality over 1990–2019', *Nature Sustainability* (2022) 5: 931–38; Lucas Chancel, Philipp Bothe and Tancrède Voituriez, 'Climate Inequality Report 2023: Fair Taxes for a Sustainable Future in the Global South', World Inequality Lab, 2023.

280 Bruckner et al., 'Impacts of Poverty', 313.

281 Tim Gore, 'Confronting Carbon Inequality: Putting Climate Justice at the Heart of the COVID-19 Recovery', Oxfam, 21 September 2020. This 'more than twice' figure was in line with the findings reported in Ilona M. Otto, Kyoung Mi Kim, Nika Dubrovsky and Wolfgang Lucht, 'Shift the Focus from the Super-Poor to the Super-Rich', *Nature Climate Change* (2019) 9: 82.

282 Chancel, 'Global Carbon'; Chancel et al., *Climate Inequality*, 28–9.

283 Sivan Kartha, Eric Kemp-Benedict, Emily Ghosh and Anisha Nazareth, 'The Carbon Inequality Era: An Assessment of the Global Distribution of Consumption Emissions among Individuals from 1990 to 2015 and Beyond', Oxfam and SEI, September 2020, 8.

284 Gore, 'Confronting Carbon', 2 (emphasis added); cf. e.g. Kartha et al. 'The Carbon Inequality', 27.

285 Gore, 'Confronting Carbon', 3.

286 Chancel, 'Global Carbon', 932. After 2050, this allotment would be zero.

287 Bruckner et al., 'Impacts of Poverty', 312–13.

288 Beatriz Barros and Richard Wilk, 'The Outsized Carbon Footprints of the Super-Rich', *Sustainability: Science, Practice and*

Policy (2021) 17: 318. By 2022, however, Musk had reached a footprint of 2,000 tons *from his own private jet solely*, on which he took 171 flights that year, shuttling from one company HQ to another. This made him the most frequent private flier among the rich men of the world. Marie Patino, Leonardo Nicoletti and Sophie Alexander, 'Elon Musk Is So Busy His Private Jet Is Taking 13-Minute Flights', *Bloomberg*, 2 March 2023.

289 Jared Starr, Craig Nilson, Michael Ash et al., 'Assessing U.S. Consumers' Carbon Footprints Reveals Outsized Impact of the Top 1%', *Ecological Economics* (2023) 205: 1–27. This study, again, counted exclusively consumption; adding investment would inflate the figures by several orders of magnitude. If emissions at the point of production were attributed to the billionaires investing in it, those individuals would attain levels not thousands of times the 1.5°C allotment but *more than one million* times. Alex Maitland, Max Lawson, Hilde Stroot et al., 'Carbon Billionaires: The Investment Emissions of the World's Richest People', Oxfam, November 2022. But these figures were *also* based on conservative assumptions, as they counted only so-called Scope 1 and 2 emissions, not Scope 3 – that is, not the emissions generated from the consumption of the commodities produced (i.e. a billionaire investing in ExxonMobil would only be charged with the emissions generated by that company in the act of extraction). On the attribution of investment emissions to rich individuals, cf. Chancel et al., 'Climate Inequality', 18, 30, 35–6; Chancel, 'Global Carbon', 934–5.

290 See e.g. Jarmo S. Kikstra, Alessio Mastrucci, Jihoon Min et al., 'Decent Living Gaps and Energy Needs around the World', *Environmental Research Letters* (2021) 16: 1–11; Bruckner et al., 'Impacts of Poverty', 311, 314; Chancel et al., 'Climate Inequality', 39–42.

291 Plus the lower emissions intensity of these economies, carried by said parts of the consuming public: Chancel, 'Global Carbon', 935. However, the *American* working class was nevertheless fairly saturated with carbon: the poorest half of the US population had emissions levels comparable to the middle European 40 per cent, despite being 'almost twice as poor'. Ibid., 932. If there was one proletariat caught in fossil consumption, it was this.

292 Marx, *Grundrisse*, 93 (emphasis in original).

293 Chuck Collins, Omar Ocampo and Kalena Thomhave, 'High Flyers 2023: How Ultra-Rich Private Jet Travel Costs the Rest of Us and Burns Up Our Planet', Patriotic Millionaires and Institute for Policy Studies, May 2023.

294 For a while, the topic trended on social media: see e.g. Oliver

Milman, 'A 17-Minute Flight? The Super-Rich Who Have "Absolute Disregard for the Planet"', *Guardian*, 21 July 2022; Jon Blistein, 'Kylie's 17-Minute Flight Has Nothing on the 170 Trips Taylor Swift's Private Jets Took this Year', *Rolling Stone*, 5 August 2022; Patino et al., 'Elon Musk'.

295 Damian Carrington and Pamela Duncan, 'Revealed: 5,000 Empty "Ghost Flights" in the UK since 2019, Data Shows', *Guardian*, 28 September 2022. Cf. e.g. Paul Sillers, 'Why the Sky Is Full of Empty "Ghost" Flights', *CNN*, 2 April 2022; Elias Visontay, '"Ghost Flights": Qatar Airways Flying Near-Empty Planes in Australia to Exploit Legal Loophole', *Guardian*, 8 August 2023.

296 Laura Cozzi, Apostolos Petropoulos, Leonardo Paoli et al., 'As their Sales Continue to Rise, SUVs' Global CO_2 Emissions Are Nearing 1 Billion Tonnes', IEA, 27 February 2023.

297 Oliver Wainwright, 'Pepper Spray for the School Run? The Weaponised SUV Set to Terrify America's Streets', *Guardian*, 25 January 2023; Tim Levine, 'This $250,000 Beast of an SUV Has a Cadillac Interior, Bulletproofing, and Almost 700 Horsepower – See Inside the Rezvani Vengeance', *Business Insider*, 22 August 2023. The man behind the company was Ferris Rezvani, 'whose father was an F4 Phantom fighter pilot in the Iranian air force. Unable to become a pilot himself for various health reasons, Rezvani Jr decided to start a car company to "create the same high-speed thrill of flying on the ground".' Wainwright, 'Pepper Spray'.

8. Induce the Panic

1 Meaney, 'Fortunes of the Green', 101–2.
2 Meyer, 'Texas: How'.
3 Tuttle, 'Chevron Is'.
4 McDermott Hughes, *Who Owns*, 20. For just one more example, see Tobita Chow, 'We Don't Have Time to End Capitalism – But Growth Can Still Be Green', *In These Times*, 15 April 2019.
5 Antonio Gramsci, *Prison Notebooks: Volume III* (New York: Columbia University Press, 2011 [1930–1]) 168–9.
6 See Daniel Gaido, 'The Origins of the Transitional Programme', *Historical Materialism* (2018) 26: 87–117.
7 Quoted in ibid., 93 (emphasis in original).
8 Leon Trotsky, 'The Death Agony of Capitalism and the Tasks of the Fourth International: The Transitional Program', in Leon Trotsky, Daniel Bensaïd, John Riddell et al., *Unity and Strategy: Ideas for Revolution* (Amsterdam: IIRE/London: Resistance Books, 2015) 133 (emphasis in original).

9   Although higher wages might, of course, ensue from the transition.

10 KPD programme from 1922 quoted in Gaido, 'The Origins of the Transitional', 98.

11 Trotsky, 'The Death Agony', 130.

12 Griffith, *Electrify*, 131.

13 Ibid., 134.

14 Ibid., 134–5 (emphases added).

15 Robin Blackburn, *The Overthrow of Colonial Slavery: 1776–1848* (London: Verso, 2011 [1988]), 521; see further e.g. 454–7, 490–1.

16 Kris Manjapra, 'D.C.'s Enslavers Got Reparations. Freed People Got Nothing', *Politico*, 17 June 2022; Andrew Weintraub, 'The Economics of Lincoln's Proposal for Compensated Emancipation', *American Journal of Economics and Sociology* (1973) 32: 171–77.

17 On the analogy of these forms of reparations, see Olúfẹ́mi O. Táíwò, *Reconsidering Reparations* (Oxford: Oxford University Press, 2022), e.g. 174–84. For a recent initiative on the former front, see Aamna Mohdin, 'Guyana's President Asks European Slave Traders' Descendants to Pay Reparations', *Guardian*, 25 August 2023.

18 Oliver Hailes, 'Unjust Enrichment in Investor-State Arbitration: A Principled Limit on Compensation for Future Income from Fossil Fuels', *Review of European, Comparative and International Environmental Law* (2023) 32: 358–70.

19 Cf. e.g. Kok-Chor Tan, 'Climate Reparations: Why the Polluter Pays Principle Is Neither Unfair nor Unreasonable', *WIREs Climate Change* (2023) 14: 1–6.

20 Blackburn, *The Overthrow*, 521; cf. e.g. 458, 505.

21 Griffith, *Electrify*, 134; Martí Orta-Martínez, Lorenzo Pellegrini, Murat Arsel et al., 'Unburnable Fossil Fuels and Climate Finance: Compensation for Rights Holders', *Global Environmental Politics* (2022) 22: 15–27; Hansen, 'Stranded Assets', 10.

22 'Department of Defense', usaspending.gov, 2023; cf. Táíwò, *Reconsidering Reparations*, 179.

23 Dehm, 'Legally Constituting', 273–4.

24 Orta-Martínez et al., 'Unburnable Fossil', 21–22.

25 Achim Hagen, Niko Jaakkola and Angelika Vogt, 'The Interplay between Expectations and Climate Policy: Compensation for Stranded Assets', *IAEE Energy Forum* (2019) fourth quarter: 31; Caldecott et al., 'Stranded Assets', 431; Bos and Gupta, 'Stranded Assets', 7.

26 The Editorial Board, 'Governments Should Not Foot the Bill for Stranded Assets', *Financial Times*, 21 February 2022.

27 On this as the alternative to bailouts, see Matt Huber, 'Expropriating Hydrocarbons in the 21st Century', *Environment and Planning A* (2022) 54: 1664–8.

28 Jacob Blumenfeld, 'Expropriation of the Expropriators', *Philosophy and Social Criticism* (2023) 49: 431 (emphasis in original).

29 See further ibid. on this objective-subjective dialectic of expropriation.

30 Gramsci, *Prison Notebooks*, 161.

31 Michael Lazarus and Harro van Asselt, 'Fossil Fuel Supply and Climate Policy: Exploring the Road Less Taken', *Climatic Change* (2018) 150: 1–13; Fergus Green and Richard Denniss, 'Cutting with Both Arms of the Scissors: The Economic and Political Case for Restrictive Supply-Side Climate Policies', *Climatic Change* (2018) 150: 73–87; Fergus Green, 'The Logic of Fossil Fuel Bans', *Nature Climate Change* (2018) 8: 444–53; G. B. Asheim, T. Faehn, K. Nyborg et al., 'The Case for a Supply-Side Climate Treaty', *Science* (2019) 365: 325–7; Georgia Piggot, Cleo Verkuijl, Harro van Asselt and Michael Lazarus, 'Curbing Fossil Fuel Supply to Achieve Climate Goals', *Climate Policy* (2020) 8: 881–7; Nicolas Gaulin and Philippe Le Billon, 'Climate Change and Fossil Fuel Production Cuts: Assessing Global Supply-Side Constraints and Policy Implications', *Climate Policy* (2020) 8: 888–901; Philippe Le Billon and Berit Kristoffersen, 'Just Cuts for Fossil Fuels? Supply-Side Carbon Constraints and Energy Transition', *Environment and Planning A* (2020) 52: 1072–92; Buck, *Ending Fossil*, 145–66.

32 Anthony Burke and Stefanie Fishel, 'A Coal Elimination Treaty 2030: Fast Tracking Climate Change Mitigation, Global Health and Security', *Earth System Governance* (2020) 3: 1–9; Peter Newell and Andrew Simms, 'Towards a Fossil Fuel Non-Proliferation Treaty', *Climate Policy* (2020) 20: 1043–54; Peter Newell, Harro van Asslet and Freddie Daley, 'Building a Fossil Fuel Non-Proliferation Treaty: Key Elements', *Earth System Governance* (2022) 14: 1–10; Harro van Asselt and Peter Newell, 'Pathways to an International Agreement to Leave Fossil Fuels in the Ground', *Global Environmental Politics* (2022) 22: 28–47.

33 Gaulin and Le Billon, 'Climate Change', 889; Newell et al., 'Building a Fossil', 4.

34 Green and Denniss, 'Cutting with Both'; Lazarus and Van Asselt, 'Fossil Fuel', 4; Newell and Simms, 'Towards a Fossil', 1046.

35 Cf. Angela V. Carter and Janetta McKenzie, 'Amplifying "Keep It in the Ground" First Movers: Toward a Comparative Framework', *Society and Natural Resources* (2020) 33: 1339–58; Le Billon and Kristofferson, 'Just Cuts', 1080–1; Gaulin and Le Billon, 'Climate Change', 889.

36 Cf. Oil Change International, 'The Case for Public Ownership of the Fossil Fuel Industry', April 2020; Sean Sweeney, 'There May Be No Choice but to Nationalize Oil and Gas – and Renewables, Too', *New Labor Forum* (2020) 29: 114–20; Fergus Green and Ingrid Robeyns, 'On the Merits and Limits of Nationalising the Fossil Fuel Industry', *Royal Institute of Philosophy Supplement* (2022) 91: 53–80; Buck, *Ending Fossil*, 167–75.

37 For the potential for extremely reactionary politics in such crises, see Malm and the Zetkin Collective, *White Skin*.

38 Donald V. Kingsbury, Teresa Kramarz and Kyle Jacques, 'Populism or Petrostate?: The Afterlives of Ecuador's Yasuní-ITT Initiative', *Society and Natural Resources* (2019) 32: 531.

39 Andrés López Rivera, 'Chronicle of a Schism Foretold: The State and Transnational Activism in Ecuador's Yasuní-ITT Initiative', *Environmental Sociology* (2017) 3: 228; Synneva Geitus Laastad, 'Leaving Oil in the Ground: Ecuador's Yasuní-ITT Initiative and Spatial Strategies for Supply-Side Climate Solutions', *Environment and Planning A* (2023) online first, 3–4; Benjamin K. Sovacool and Joseph Scarpaci, 'Energy Justice and the Contested Petroleum Politics of Stranded Assets: Policy Insights from the Yasuní-ITT Initiative in Ecuador', *Energy Policy* (2016) 95: 162–4; Lorenzo Pellegrini, Murat Arsel, Fander Falconi and Roldan Muradian, 'The Demise of a New Conservation and Development Policy? Exploring the Tensions of the Yasuní ITT Initiative', *Extractive Industries and Society* (2014): 284, 287.

40 Sovacool and Scarpaci, 'Energy Justice', 161, 164.

41 Ibid., 158, 162, 165, 168–9; quotation from López Rivera, 'Chronicle of a Schism', 232. The figure 0.4 refers to money collected; up to 10 per cent had been promised.

42 Manuela Andreoni and Catrin Einhorn, 'Would You Vote to Halt Drilling? In Ecuador, They're Getting the Chance', *New York Times*, 17 August 2023; Maxwell Radwin, 'Ecuador Referendum Halts Oil Extraction in Yasuní National Park', *Mongabay*, 21 August 2023.

43 Laura Peterson, 'Investing in Amazon Crude II: How the Big Three Asset Managers Actively Fund the Amazon Oil Industry', *Amazon Watch*, June 2021, 18.

44 Fitch Ratings, 'Fitch Downgrades Ecuador's Long-Term IDR to "CCC+"', 16 August 2023. See further Alejandra Padín-Dujon, 'Ecuadorians Voted to Protect Nature. A Big Three Credit Rating Agency Penalised Them For It', *LSE Blogs*, 9 October 2023.

45 Dan Collyns, 'Ecuadorians Vote to Halt Oil Drilling in Biodiverse Amazonian National Park', *Guardian*, 21 August 2023.

46 Ibid.; Radwin, 'Ecuador Referendum'; Alexandra Valencia,

'Ecuador Says It Will Honour Referendum on Yasuni Oil Project', Reuters, 24 August 2023.

47 Pablo Ospina Peralta, 'In Ecuador, Disaffected Voters Have Elected the Son of a Banana Magnate', *Jacobin*, 23 October 2023.

48 Antonia Juhasz, 'Ecuadorians Vote to "Keep the Oil in the Soil" in the Amazon', *Human Rights Watch*, 23 August 2023.

49 On sabotage, see e.g. Kelly Hearn, 'Woes Mount for Oil Firms in Ecuador', *Amazon Watch*, 9 February 2006; *The Sydney Morning Herald*, 'Ecuador Forces Recapture Oil Wells', 21 August 2005; Alexandria Valencia, 'Ecuador Private Oil Pipeline Reports Spill Following Vandalism', Reuters, 14 October 2022.

50 Carter and McKenzie, 'Amplifying "Keep"', 1343.

51 Gaulin and Le Billon, 'Climate Change', 1351.

9. Chronicle of One More Year of Madness

1 Goddard Digital Team, 'NASA Finds June 2023 Hottest on Record', 13 July 2023; Claire A. O'Shea, 'NASA Clocks July 2023 as Hottest Month on Record Ever Since 1880', NASA, 14 August 2023; National Oceanic and Atmospheric Administration, 'The World Just Sweltered through Its Hottest August on Record', 14 September 2023; Copernicus Programme, 'The 2023 Northern Hemisphere Summer Marks Record-Breaking Oceanic Events', 8 September 2023.

2 WMO, 'September Smashes Monthly Temperature Record by Record Margin', 17 October 2023.

3 Damian Carrington, '"Gobsmackingly Bananas": Scientists Stunned by Planet's Record September Heat', *Guardian*, 5 October 2023.

4 WMO, 'Storm Daniel Leads to Extreme Rain and Floods in Mediterranean, Heavy Loss of Life in Libya', 12 September 2023; Helena Smith, 'Rescue Efforts Stepped Up after Deadly Floods in Central Greece', *Guardian*, 9 September 2023; Clea Skopeliti, '"The Earth Is Sick": Storm Daniel Has Passed, but Greeks Fear Its Deathly Legacy', *Guardian*, 29 September 2023. A less vulgar, more precise nomenclature for extreme weather events was finally introduced by one American meteorologist in 2023: naming them after oil and gas companies. Oliver Milman, '"People Need to Be Riled Up": Meteorologist Names US Heatwaves after Oil and Gas Giants', *Guardian*, 20 July 2023.

5 E.g. Miriam Berger, 'Greece Struggles to Contain Europe's Largest Wildfire on Record', *Guardian*, 29 August 2023.

6 Clea Skopeliti, '"A Biblical Catastrophe": Death Toll Rises to

Six as Storm Daniel Lashes Greece', *Guardian*, 7 September 2023.

7 Independently of the Greek mayor – to all appearances – a Libyan observer also used the term 'biblical catastrophe' for what happened there: Moin Kikhia in Patrick Wintour and Luke Harding, '"Sea is Constantly Dumping Bodies": Fears Libya Flood Death Toll May Hit 20,000', *Guardian*, 13 September 2023.

8 NASA Earth Observatory, 'Torrential Rain Wreaks Havoc in Libya', 14 September 2023.

9 Ayman Werfali, Mohammed Abdusamee, Vivian Nereim and Isabella Kwai, 'More than 5,000 Dead in Libya as Collapsed Dam Worsen Flood Disaster', *New York Times*, 12 September 2023; Delger Erdenesanaa, 'Why Floods Can Turn So Deadly, So Fast', *New York Times*, 13 September 2023; WMO, 'Storm Daniel'.

10 Islam Alatrash, '"The Waters Carried My Son Away in Front of My Eyes": Anguished Libyans Mourn Lost Loved Ones', *Guardian*, 15 September 2023.

11 Wintour and Harding, '"Sea Is"'; Jahwar Ali, then in Turkey, quoted in Abdusamee et al., 'More than 5,000'.

12 Fayez Abu Amra in Nidal Al-Mughrabi, 'Palestinian Family that Fled Wars Suffers Deaths in Libya', Reuters, 14 September 2023.

13 Mariam Zachariah, Vassiliki Kotroni, Lagouvardos Kostas et al., 'Interplay of Climate Change-Exacerbated Rainfall, Exposure and Vulnerability Led to Widespread Impacts in the Mediterranean Region', Grantham Institute for Climate Change, Imperial College London, 18 September 2023.

14 Copernicus Programme, 'The 2023 Northern'.

15 Cf. Daisy Dunne and Ayesha Tandon, 'How Are Libya's "Medicane"-Fuelled Floods Linked to Climate Change?', *Carbon Brief*, 19 September 2023; AFP, 'What Are Medicanes? The "Supercharged" Mediterranean Storms that Could Become More Frequent', *Guardian*, 15 September 2023; WMO, 'Storm Daniel'.

16 Daisy Dunne, 'Africa's Extreme Weather Has Killed at Least 15,000 People in 2023', *Carbon Brief*, 25 October 2023.

17 *Al Jazeera*, 'Cyclone Freddy Death Toll in Malawi, Mozambique Passes 100', 14 March 2023; Dunne, 'Africa's Extreme'; fourth and eighth: World Data, 'The Poorest Countries in the World', June 2023.

18 Rebecca Ratcliffe and Hannah Ellis-Peterson, 'Severe Heatwave Engulfs Asia Causing Deaths and Forcing Schools to Close', *Guardian*, 19 April 2023; Rebecca Ratcliffe, '"Endless Record Heat" in Asia as Highest April Temperatures Recorded', *Guardian*, 27 April 2023.

19 Philbert Girinema, 'Heavy Rain, Floods Kill at Least 136 in Rwanda and Uganda', Reuters, 4 May 2023; *Reliefweb*, 'DR Congo: Floods and Landslides', May 2023.

20 Justin McCurry, '"Heaviest Rain Ever" Causes Deadly Floods and Landslides in Japan', 11 July 2023; Justin McCurry, 'South Korea Floods: President Urges Climate Crisis Action as Death Toll Hits 40', *Guardian*, 17 July 2023.

21 Hamida Dandoush in Ali Haj Suleiman and Husam Hezaber, 'Living in "an Oven": Heatwave Grips Displacement Camps in Syria', Al Jazeera, 16 July 2023.

22 Simon Speakman Cordall, '"We Can't Endure This": Migrants Suffer in Extreme Tunisian Heat', Al Jazeera, 24 July 2023.

23 Fedja Grulovic and Lamine Chikhi, 'Wildfires Bring Death and Destruction to Sun-scorched Mediterranean', Reuters, 26 July 2023; Helen Sullivan and Lorenzo Tondo, '"Like a Blowtorch": Mediterranean on Fire as Blazes Spread across Nine Countries', *Guardian*, 26 July 2023.

24 Kevin Hamilton, 'Hawaii's Climate Future: Dry Regions Get Drier with Global Warming, Increasing Fire Risk', *Conversation*, 16 August 2023; Stephen Culp, 'Maui Wildfires: What Are the Deadliest Wildfires in US History?', Reuters, 21 August 2023; Taylor Romine and Aya Elamroussi, 'Advanced DNA Testing Prompts Officials to Revise Maui Fires Death Toll to 97, Down from 115', *CNN*, 15 September 2023; Adeel Hassan and Anna Betts, 'Maui Wildfires Latest: Lahaina Reopens to Residents', *New York Times*, 29 September 2023.

25 Jamey Keaten, 'As Thaw Accelerates, Swiss Glaciers Have Lost 10% of their Volume in the Past 2 Years, Experts Say', *AP*, 28 September 2023.

26 Jon Henley, 'Zero-Degree Line at Record Height above Switzerland as Heat and Fire Hit Europe', *Guardian*, 21 August 2023.

27 Alession Perrone, 'Cacti Replacing Snow on Swiss Mountainsides Due to Global Heating', *Guardian*, 10 February 2023.

28 Patrick Barkham, 'Red Fire Ant Colonies Found in Italy and Could Spread across Europe, Says Study', *Guardian*, 11 September 2023.

29 Sarah Newey, 'Paris Fumigated to Stop Spread of "Break-Bone Fever"', *Telegraph*, 1 September 2023.

30 Haroon Janjua, 'A Year On, the Devastating Long-Term Effects of Pakistan's Floods Are Revealed', *Guardian*, 5 August 2023; Shehryar Fazli, 'Modern Slavery: Pakistan's Latest Climate Change Curse', *Al Jazeera*, 1 September 2023.

31 Laura Paddison, 'Catastrophic Drought That's Pushed Millions into Crisis Made 100 Times More Likely by Climate Change, Study Finds', *CNN*, 27 April 2023.

32 Manuela Andreoni, 'Uruguay Wasn't Supposed to Run Out of Water', *New York Times*, 10 August 2023.

33 Bruno Kelly and Jake Spring, 'Amazon River Falls to Lowest in Over a Century amid Brazil Drought', Reuters, 17 October 2023.

34 John Yoon, 'Flooding from Cyclone in Southern Brazil Kills at Least 37', *New York Times*, 6 September; *Euronews Green*, '"Hundreds Were Saved": Heavy Rains Cause Record Death Toll in Southern Brazil', 7 September 2023; Rachel Ramirez, 'Hurricane Otis' Explosive Intensification Is a Symptom of the Climate Crisis, Scientists Say', *CNN*, 25 October 2023; Christine Murray and Aime Williams, '"Nightmare" Hurricane Otis Leaves 27 Dead in Mexico', *Financial Times*, 26 October 2023.

35 See e.g. Ben Stockton and Amy Westervelt, 'Inside the Campaign that Put an Oil Boss in Charge of a Climate Summit', *Intercept*, 25 October 2023.

36 Sam Meredith, 'Oil Giant Led by COP28 Boss to Spend an "Eye-watering" $1 Billion a Month on Fossil Fuels this Decade, Global Witness Says', *CNBC*, 4 September 2023.

37 Oliver Klaus, 'Adnoc Closes in on 4.5 Million b/d Output Capacity', *Energy Intelligence*, 4 July 2023. Saudi Arabia and Iraq were the two largest.

38 U.S. Energy Information Administration, 'Country Analysis Brief: United Arab Emirates (UAE)', 28 August 2023; Fiona Harvey, '"I Wasn't the Obvious Choice": Meet the Oil Man Tasked with Saving the Planet', *Guardian*, 7 October 2023.

39 Stockton and Westervelt, 'Inside the Campaign'.

40 Tony Blair Institute for Global Change, 'Tony Blair Statement on COP28 Presidency', 12 January 2023.

41 Ben Stockton, 'Cop28 President's Team Accused of Wikipedia "Greenwashing"', *Guardian*, 30 May 2023.

42 Damian Carrington, 'Army of Fake Social Media Accounts Defend UAE Presidency of Climate Summit', *Guardian*, 8 June 2023.

43 Damian Carrington, '"Absolute Scandal": UAE State Oil Firm Able to Read Cop28 Climate Summit Emails', *Guardian*, 7 June 2023; Stockton and Westervelt, 'Inside the Campaign'.

44 Ben Stockton, 'UAE Oil Company Executives Working with Cop28 Team, Leak Reveals', *Guardian*, 22 September 2023.

45 Harvey, '"I Wasn't"'. '"It's the consumer who contributes to increasing CO_2 emissions, not the producer."' Ibid.

46 Quoted in Attracta Mooney and Camilla Hodgson, 'COP28 Head at Odds with Climate Leaders over Future for Fossil Fuels', *Financial Times*, 3 May 2023.

47 Quoted in Maha El Dahan, 'UAE's Jaber Says Oil, Industrial Firms

to Commit to Decarbonization at COP28', Reuters, 2 October 2023.

48 Quoted in Fiona Harvey, '"The Window Is Closing": Cop28 Must Deliver Chance of Course on Climate', *Guardian*, 2 June 2023.

49 Ajit Niranjan, 'Dutch Politician Who Worked for Shell Poised to Become EU Climate Chief', *Guardian*, 4 October 2023.

50 E.g. Sarah Marsh, Kate Abnett and Gloria Dickie, '"Greenlash" Fuels Fears for Europe's Environmental Ambitions', Reuters, 10 August 2023; Sam Meredith, 'From Washington to Warsaw, a "Greenlash" Is Picking up Steam Despite Extreme Heat', *CNBC*, 1 August 2023.

51 Frederica di Sario and Giorgio Leali, 'Macron Calls for "Regulatory Pause" in EU Environmental Laws Wink at Conservatives', *Politico*, 12 May 2023.

52 Andrew McDonald and Bethany Dawson, 'UK's Rishi Sunak Confirms Loosening of Key Green Pledges', *Politico*, 20 September 2023; Helena Horton, '"Detached from Reality": Anger as Rishi Sunak Plans to Restrict Solar Panels', *Guardian*, 7 October 2023; Jillian Ambrose, '"Biggest Clean Energy Disaster in Years": UK Auction Secures No Offshore Windfarms', *Guardian*, 8 September 2023.

53 Jillian Ambrose, 'Ukraine Built More Onshore Wind Turbines in Past Year than England', *Guardian*, 28 May 2023.

54 Maria Davidsson, 'Kraftigt ökade utsläpp med ny budget', *Svenska Dagbladet*, 20 September 2023; Miranda Bryant, 'Swedish Government Faces Backlash after Slashing Climate Budget', *Guardian*, 21 September 2023.

55 WMO, 'Global Temperatures Set to Reach New Records in Next Five Years', 17 May 2023.

56 See above, and e.g. Copernicus Programme, 'First Days of June Surpass the 1.5°C Limit', 15 June 2023.

57 IEA, *World Energy Outlook 2023*, 201; cf. e.g. 50.

58 Ibid., e.g. 19, 50, 203.

59 Ibid., 83; IEA, 'Global Coal Demand Set to Remain at Record Levels in 2023', 27 July 2023; *Economist Intelligence Unit*, 'Energy Outlook 2024: Surging Demand Defies War and High Prices', 2023.

60 E.g. Zanagee Artis, 'Oil Industry Netted Billions in Profits, Despite Global Price Dip', *National Resource Defense Council*, 10 August 2023; Jillian Ambrose and Rob Davies, 'BP's £2bn Profits Cause Anger Amid Climate Crisis', *Guardian*, 1 August 2023.

61 Mark Fulton, Paul Spedding and Mike Coffin, 'Avoiding Stranded Assets: Risk Aversion in Oil Development', *Carbon Tracker Initiative*, 15 September 2023; Nicholas Megaw and Jamie Smyth, 'US

Oil and Gas Finds Warmer Welcome in Capital Markets', *Financial Times*, 16 October 2023; Mack Wilowski, 'Oil and Gas Stocks Topped S&P 500 Returns this Quarter as Crude Prices Surged', *Investopedia*, 29 September 2023.

62 Edward Moya in Phillip Inman, 'Global Inflation Fears as Oil Price Rises towards $100 a Barrel', *Guardian*, 17 September 2023. 'Oil and gas are hot again.' Irina Slav, 'Oil and Gas Are Still Drawing in Investors Despite Transition', *Oilprice.com*, 22 October 2023.

63 Phillip Inman, 'Global Inflation Fears as Oil Price Rises towards $100 a Barrel', *Guardian*, 17 September 2023; IEA, *World Energy Outlook 2023*, 81; Jillian Ambrose, 'Why Oil Prices Are Rising amid the Israel-Hamas War', *Guardian*, 20 October 2023, Irina Slav, 'How High Could War in the Middle East Drive Oil Prices?', *Oilprice.com*, 23 October 2023. Chevron also closed the largest gas platform in the territorial waters of Israel, within range of rockets from Gaza, which contributed to a rise in gas prices: Anna Cooban and Matt Egan, 'Israel Just Shut a Gas Field Near Gaza. Here's Why that Matters', *CNN*, 10 October 2023. 'Chevron has been working on plans to expand production at the Tamar and Leviathan natural gas fields, and to add pipelines to increase gas flows from Israel to Egypt, which indirectly exports Israeli output in the form of liquefied natural gas from facilities on the Mediterranean coast. The fierce fighting could slow the pace of energy investment in the region, just as the eastern Mediterranean's prospects as an energy center have gained momentum. Israel used to be one of the few countries in the Middle East without significant discovered petroleum resources. Now, natural gas has become a mainstay of its economy', but the Palestinian resistance threatened to upend this equation. Stanley Reed, 'Chevron Shuts Down Natural Gas Platform near Gaza Strip', *New York Times*, 9 October 2023.

64 Tsvetana Paraskova, 'Oil Poised to Become U.S.' Single Largest Export Product', *Oilprice.com*, 16 October 2023.

65 The White House, 'Remarks by President Biden on the United States' Response to Hamas's Terrorist Attacks against Israel and Russia's Ongoing Brutal War against Ukraine', 20 October 2023. Obviously, these trends spelled serious difficulties for classical world-systems analysis and at least the cruder versions of theories of ecologically unequal exchange.

66 Thomas Catenacci, 'Biden Admin Approves Massive Gas [sic] Pipeline Project in Huge Blow to Climate Activists', *Fox News*, 14 April 2023; Benji Jones, 'An Oil Company Wants to Use Giant Chillers to Refreeze the Ground that Climate Change Is Thawing

in Order to Drill for More Oil – Which Will Ultimately Accelerate Global Warming', *Business Insider*, 17 August 2020.

67 April Merlaux et al., *Banking on Climate*, 22.

68 Ioulainen and Trout, *Planet Wreckers*, 17; Zahra Hirji and Ari Natter, 'Fear and Anger Follow the Path of Joe Manchin's Mountain Valley Pipeline', *Bloomberg*, 11 October 2023; Robert Rapier, 'U.S. Oil Defies Odds, Races toward Annual Production Record', *Oilprice.com*, 12 October 2023; Tim Mullaney, 'U.S. Oil Is Back, and ExxonMobil's $60 Billion Deal Isn't Even the Biggest Signal', *CNBC*, 15 October 2023.

69 Sarah Young, 'Britain Gives Go-Ahead for Biggest New North Sea Oilfield in Years', Reuters, 27 September 2023.

70 Russell Borthwick of the Aberdeen and Grampian Chamber of Commerce in Mary McCool and Nichola Rutherford, 'Rosebank Oil Field Given Go-Ahead by Regulators', *BBC*, 27 September 2023.

71 Jeff Goodell, 'Will an Oil Racket Destroy One of Africa's Most Sacred Places?', *Rolling Stone*, 26 March 2023; Surina Esterhuyse, 'Oil Drilling Threatens the Okavango River Basin, Putting Water in Namibia and Botswana at Risk', *Conversation*, 2 August 2023; Simon Watkins, 'Western Companies Look to Grow Influence in Egyptian Oil and Gas', *Oilprice.com*, 9 October 2023; Felicity Bradstock, 'New Oil and Gas Discoveries Propel South Africa's Energy Renaissance', *Oilprice.com*, 5 October 2023; Matthew Smith, 'Suriname's Oil Boom Finally Gets the Green Lights', *Oilprice.com*, 26 September 2023.

72 Christina Figueres, 'I Thought Fossil Fuel Firms Could Change. I Was Wrong', *Al Jazeera*, 6 July 2023.

73 John Otis, 'Oil-Exporting Colombia Says No to Oil Exploration', *Wall Street Journal*, 28 January 2023; Andrea Jaramillo, 'Roa Named Ecopetrol CEO to Pursue Petro's Energy Transition', *Bloomberg*, 12 April 2023.

74 On his trajectory and the context, see Forrest Hylton and Aaron Tauss, 'Colombia at the Crossroads', *New Left Review* (2022) 137: 87–125.

75 Luke Taylor, 'Colombia Announces Halt on Fossil Fuel Exploration for a Greener Economy', *Guardian*, 20 January 2023.

76 Otis, 'Oil-Exporting Colombia'.

77 After Brazil and Mexico.

78 Ibid.

79 IEA, *World Energy Outlook 2023*, 17, 36, 85, 147; Damian Carrington, '"Revolutionary" Solar Power Cell Innovations Break Key Energy Threshold', *Guardian*, 6 July 2023.

80 IEA, *World Energy Outlook 2023*, 99–100; cf. e.g. 43, 75.

81 Ibid., 61, 76; Tyler Durden, 'Lights Out: Solar Power Stocks Crash after Demand Warning across Europe', *Zero Hedge*, 20 October 2023; Hakyung Kim, 'Solar Stocks Tumble to 3-Year Low as Solaredge Drops Nearly 30% on Demand Warning', *CNBC*, 20 October 2023; Anna Hirtenstein, 'Why the Shine Has Come Off Clean-Energy Stocks', *Wall Street Journal*, 25 October 2023.

82 IEA, *World Energy Outlook 2023*, 26–9.

83 Bloomberg News, 'China's Solar Boom Is Already Accelerating Past Last Year's Record Surge', *Bloomberg*, 23 May 2023; Amy Hawkins, 'China Ramps up Coal Power Despite Carbon Neutral Pledges', *Guardian*, 24 April 2023; Helen Davidson, 'China Continues Coal Spree Despite Climate Goals', *Guardian*, 29 August 2023.

84 Friederike Otto in Damian Carrington, 'Dramatic Climate Action Needed to Curtail "Crazy" Extreme Weather', *Guardian*, 28 August 2023.

85 Claudio Descalzi in Alex Kimani, 'The Peak Oil Demand Debate is Sure to Dominate COP 28', *Oilprice.com*, 9 October 2023.

86 Fiona Harvey, 'Private Jet Sales Likely to Reach Highest Ever Level this Year, Report Says', *Guardian*, 1 May 2023; Damien Gayle, 'Private Jet Service for Rich Dog Owners Condemned by Climate Campaigners', *Guardian*, 30 September 2023.

87 Adam Golder in PR Newswire, 'It's Absolutely Paw-some: K9 JETS Announces Limited-Time Route between Los Angeles and London for the 2023 Holiday Season', *Yahoo! Finance*, 13 September 2023.

Index

denial, 147, 162, 288, 297–8,
300, 355
see also global warming
climate crisis, xi, xiii, 4, 23–4, 46,
76, 78, 88, 91–2, 110, 218–21,
237, 253, 288, 291, 311, 361.
See also crises; global warming
climate politics/policy, xiii, 31, 35,
38, 46, 52–4, 69, 71–2, 83, 88,
98, 103–4, 110–11, 126, 140,
148, 150–1, 160, 163–5, 219,
229, 233–4, 236, 243, 252–4,
300, 309, 334
as theatre, 90–1, 254, 302
'greenlash', 254
irrealism 88, 91, 196, 254
climate science, 16, 28, 36, 43–4,
46, 49–50, 56, 64, 66, 72, 86,
161, 226, 281, 334
Climáximo, 50
CNN, 286
coal, 1, 9, 11–13, 18–20, 22–3, 26,
50, 80, 99, 119, 121, 126–7,
135–6, 143, 146–8, 150–4,
156, 159, 169–70, 172, 175–7,
180–1, 191, 194–5, 202, 204–5,
209, 216, 222, 238, 242, 257,
274–5, 285–6, 366
anthracite, 19
'clean coal', 147
coke/coking coal, 19, 127, 172,
180, 238
lignite/brown coal, 12, 19, 22,
176, 178
see also fossil fuel
infrastructure/assets, coal
mines, coal plants
Colombia, 136, 256–7
commodities, 106–7, 109, 111,
118, 120, 126, 133–6, 139, 175,
179, 185, 190, 205–7, 209–13,
215, 217–19, 255, 315, 348,
363, 366, 370
commodity traders, 12, 135–6,
156, 322
renewable energy as non-
commodity, 210–13
see also capital, commercial
capital
CONAIE, 245

'Contraction and Convergence',
34, 39, 69
conservatism, 60, 65, 182–5, 188,
193
COPs, 4, 29, 32–3, 37, 39, 70, 76,
89–90, 103, 155, 162, 239, 288
COP1 (Berlin, 1995), 35–6
COP15 (Copenhagen, 2009),
33–7, 40, 66–7, 70, 92, 103,
149, 187
COP16 (Cancun, 2010), 37
COP17 (Durban 2011), 70
COP21 (Paris, 2015), 37–40,
46, 48, 54, 66, 79, 88–90, 94,
104, 137, 148–9, 167, 187,
302, 307, 333. *See also:* Paris
Agreement (2015)/Paris era
COP26 (Glasgow, 2021), 26,
29, 152
COP27 (Sharm el-Sheikh,
2022), 26, 29, 31, 90, 152,
197
COP28 (Dubai), 252–4
coral reefs, 41
Correa, Rafael, 245–6. *See also*
Ecuador
Costa Rica, 247
Covid-19 pandemic, 3, 8, 10–11,
15, 23–4, 41, 141, 151, 186–7,
189–90, 197, 208, 268
vaccines, 186, 351
crises, 24, 105, 108–9, 113, 237,
242–3, 274–5, 324, 364, 374
crisis of rationality, 220
financial crises, 136–9, 149,
151, 242, 244. *See also*
'carbon bubble'; global
financial crisis (2007–8)
scissors crisis, 204, 206, 211,
214, 218, 220, 235, 361, 365
see also climate crisis
Croatia, 251
Crusoe, Robinson, 85
Cyclone Freddy, 251

decarbonisation, 36, 234, 239,
253, 345
deforestation, xiii, 48, 92, 261,
354
afforestation, 68